Energy Consumption Analysis and Characterization of Complex Systems

Energy Consumption Analysis and Characterization of Complex Systems

Topic Editors

Md Rasheduzzaman
Yongming Han

MDPI

Basel • Beijing • Wuhan • Barcelona • Belgrade • Novi Sad • Cluj • Manchester

Topic Editors

Md Rasheduzzaman
Department of Engineering and Technology
Southeast Missouri State University
Cape Girardeau, MO
USA

Yongming Han
College of Information Science and Technology
Beijing University of Chemical Technology
Beijing
China

Editorial Office
MDPI AG
Grosspeteranlage 5
4052 Basel, Switzerland

This is a reprint of the Topic, published open access by the journals *Energies* (ISSN 1996-1073), *Processes* (ISSN 2227-9717), *Smart Cities* (ISSN 2624-6511) and *Sustainability* (ISSN 2071-1050), freely accessible at: https://www.mdpi.com/topics/80LTUYGL18.

For citation purposes, cite each article independently as indicated on the article page online and as indicated below:

Lastname, A.A.; Lastname, B.B. Article Title. *Journal Name* **Year**, *Volume Number*, Page Range.

ISBN 978-3-7258-8159-8 (Hbk)
ISBN 978-3-7258-8160-4 (PDF)
https://doi.org/10.3390/books978-3-7258-8160-4

Contents

About the Editors

Md Rasheduzzaman

Md Rasheduzzaman is currently a professor with the Department of Engineering and Technology at Southeast Missouri State University, Cape Girardeau, MO. His research interests include microgrids, power systems, and power electronics. Dr. Rashed has received a Ph.D. degree in electrical engineering with an emphasis in electrical power and energy from the Missouri University of Science and Technology, Rolla, MO, an M.S. degree in electrical and computer engineering from Purdue University, Hammond, IN, and a B.S. degree in electrical and electronic engineering from the Chittagong University of Engineering and Technology, Chittagong, Bangladesh, in 2015, 2010, and 2006, respectively.

Yongming Han

Yongming Han is a professor and doctoral supervisor at the Beijing University of Chemical Technology. He is mainly engaged in the research of energy-efficient intelligent optimization and control of chemical processes. He has published more than 120 SCI articles (13 ESI highly cited papers) in authoritative journals, such as *IEEE Transactions* and *IFAC Journal*s as the first/corresponding author, authorized 13 national invention patents, served as a member of 6 SCI journals as a guest and subject editorial board member, held more than 10 outstanding youth funds, top-level projects, and youth projects of the National Natural Science Foundation of China (NNSF), and served as the vice chairman of the special committee of Network Computing and the member of the special committee of Data-driven Control, Learning and Optimization of the Chinese Society of Automation. He also serves as the deputy director of the Network Computing Committee and a member of the Data-Driven Control, Learning and Optimization Committee of the Chinese Society of Automation, a member of the Youth Working Committee of the Chinese Society of Artificial Intelligence, a director of the Beijing Automation Society, a director of the Beijing Society of Artificial Intelligence, and more. Additionally, he has been selected as one of the top 2% scientists in the world by Stanford University, the 2020-2022 Young Talent Support Program of the Beijing Municipal Association for Science and Technology, and the Outstanding Talent Program of the Beijing Municipal Party Committee and Organization Department.

Preface

The rapid advancement of industrial infrastructure and the increasing demand for sustainable power have necessitated a deeper understanding of how modern networks utilize energy. This Reprint, titled **Energy Consumption Analysis and Characterization of Complex Systems**, compiles a series of high-quality research contributions focused on the modeling, optimization, and monitoring of energy patterns within intricate technical environments.

Scope and Purpose

The primary aim of this Reprint is to explore the intersection of control theory, data analytics, and power systems. As systems grow in complexity—incorporating distributed energy resources, automated industrial processes, and smart grid technologies—the ability to accurately characterize energy consumption becomes vital. The works collected here address critical challenges, such as load forecasting, efficiency optimization in industrial controllers, and the integration of renewable energy sources into existing frameworks.

Motivation

The motivation for this scientific work stems from the global transition toward smarter, more resilient energy systems. Traditional analysis methods often struggle to account for the dynamic behavior of non-linear loads and the stochastic nature of modern energy generation. By bringing together diverse perspectives on system characterization, this Reprint provides a comprehensive resource for understanding how energy flows can be managed more effectively to reduce waste and improve operational stability.

Audience

This Reprint is addressed to a broad audience within the scientific and engineering communities. It is intended for:

- *Researchers and Academics* focusing on power electronics, control systems, and energy management.
- *Practicing Engineers* seeking robust methodologies for industrial energy auditing and system modeling.
- *Graduate Students* looking for advanced case studies in energy characterization and complex system analysis.

It has been a privilege to serve as Topic Editor for this collection. We hope the insights provided within these pages inspire further innovation and contribute to the ongoing evolution of energy-efficient technologies.

Md Rasheduzzaman and Yongming Han
Topic Editors

processes

MDPI

Article

Toward Optimal Design of a Factory Air Conditioning System Based on Energy Consumption Prediction

Shuwei Zhu [1], Siying Lv [1], Wenping Wang [1] and Meiji Cui [2,*]

[1] School of Artificial Intelligence and Computer Science, Jiangnan University, No. 1800 Lihu Road, Wuxi 214122, China; zhushuwei@jiangnan.edu.cn (S.Z.)
[2] School of Intelligent Manufacturing, Nanjing University of Science and Technology, No. 200 Xiaolingwei Street, Nanjing 210094, China
* Correspondence: cui_mj@njust.edu.cn

Abstract: The Make-up Air Unit (MAU) is an air conditioning system which plays an important role in serving semiconductor cleanrooms. It provides constant temperature and humidity for fresh air through various sections, including fresh air filtration, preheating, precooling, humidification, recooling, reheating, air supply, and high-efficiency filtration. However, the commonly used PID control method of the MAU indicates a deficiency in energy consumption. Hence, this research introduces a proactive energy-saving optimization control method based on machine learning and intelligent optimization algorithms. Firstly, the machine learning methods are used to train historical data of the MAU, resulting in a data-driven prediction model of energy consumption for the system. Subsequently, the customized genetic algorithm (GA) is used to optimize energy in cold and hot water systems. It facilitates the dynamic adjustment of the regulating valve opening for the cold and hot water coil in the fresh air unit, responding to real-time variations in outdoor air conditions. Meanwhile, it ensures that the supply air temperature and humidification adhere to specified requirements, thereby reducing the energy consumption associated with cold and hot water usage in the MAU. The experimental results indicate that the proposed algorithm can provide significant energy conservation in the MAU.

Keywords: MAU (Make-up Air Unit); energy consumption; machine learning; intelligent optimization; evolutionary algorithm

1. Introduction

1.1. Background

Common industrial plants include the electronics, food, pharmaceutical and new energy sectors. The high-cleanness area used to manufacture and service production within these industrial facilities is called a clean room [1]. The cleanrooms share common characteristics of being structurally complex, having high technological requirements, and varying environmental requirements [2]. The characteristics of the product determine the need for cleaning in the production process. Compared to other industries, environmental impurities can damage product characteristics. As the complexity of the products increases, strict control of the production environment becomes essential to ensure product quality.

In addition, global warming, resulting from carbon emissions, has become one of the most severe and urgent climate problems. Today, the construction sector directly generates 8% of the global carbon emissions due to the use of fossil fuels, and another 19% is indirectly emitted, resulting from the electricity and heat/cooling supply. Therefore, decarbonization in the building sector is necessary and promising for the global carbon neutrality target [3]. Hence, research on energy savings techniques in the building sector, for example factory air conditioning systems, i.e., the Make-Up Air Unit (MAU) system, has gained increasing attention in this study [4].

For the MAU system, it is crucial to control various air conditioning parameters during production, such as temperature, humidity, cleanliness, flow organization, and pressure balance. Conventional cleanrooms have the following characteristics:

- High temperature and humidity requirements: $T \leq \pm 2\,^{\circ}\mathrm{C}$, $RH \leq \pm 5\%$.
- High requirements in cleanliness level.
- High air exchange rates and large fresh air supply.
- The balance in the pressure gradient.

The fresh outdoor air for the electronics plant is subjected to treatment by the MAU to meet the temperature and humidity requirements before entering the supply air manifold. Subsequently, air is transported through air supply ducts to the various production areas within the cleanroom. Figure 1 shows the operational principle of the MAU. The goal of our research in this paper is to search for the optimal cooling and heating capacities required by each coil in the air handling process under ideal conditions, while satisfying the MAU air supply requirements, thus maximizing the reduction in the MAU's energy cost.

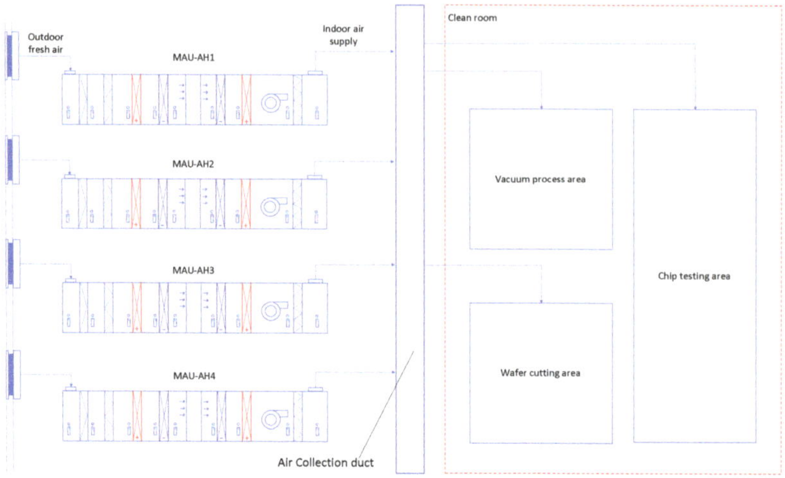

Figure 1. The operation principle of MAU.

1.2. A Brief Introduction to the MAU's Configuration and Control Strategy

As displayed in Figure 2, an MAU unit consists of five sections: precooling, preheating, humidifying, recooling, and reheating. Outdoor air is processed through these five stages to meet certain set specifications, including temperature and dew temperature, before being fed into cleanrooms. The precooling and preheating stages are not switched on at the same time, and it is usually the temperature or specific enthalpy of the outside air that determines which of the stages is switched on to process the outside air. After pre-processing, the air is humidified to raise the relative humidity to 100%. It then enters the recooling section for cooling and dehumidification to meet the set dew temperature. Finally, it enters the reheating section for heating to meet the temperature target. In summary, this is the whole process of air treatment in the MAU.

The MAU cools or heats air through hot and cold water coils, respectively, and valves are used to control the flow of water in the coils and thus the cooling or heating effect to achieve a preset target. PID control of valve openings is a common practice; however, an unreasonable setting based on experience can lead to wasted resources. For example, in winter mode, a high set value results in excessive humidification in the humidification section and and increase in the workload of the recooling section. In addition, PID control is less flexible and cannot adaptively adjust the control parameters according to the indoor and outdoor environment changes to maintain the optimal control state.

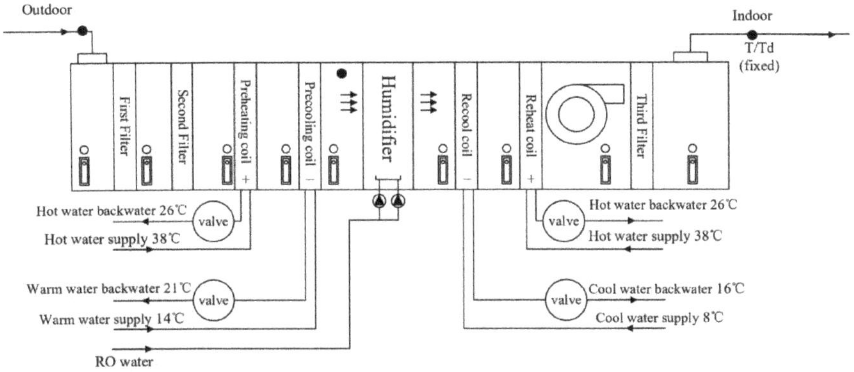

Figure 2. The configuration of MAU.

In summary, the original intention of designing this algorithm is to reduce the waste of energy without relying on expert experience to set the preset values of the pre-processing section. Hence, the MAU system can achieve adaptive adjustment according to environmental changes.

2. Related Work and the Motivation

Global warming, resulting from carbon emissions, has become one of the most severe and urgent climate problems. Nowadays, the building sector directly generates 8% of the global carbon emissions due to fossil fuel usage, and another 19% is indirectly emitted, resulting from electricity and heat/cooling supply. Therefore, decarbonization in the building sector is necessary and promising for the global carbon neutrality target [3,5–7]. Hence, the research about energy-saving techniques in the building sector, for example, the factory air conditioning systems in this study, has gained increasing attention.

Recently, an energy-saving control method for air conditioning based on support vector regression (SVR) and particle swarm optimization (PSO) was proposed [8], to optimize energy consumption while maintaining thermal comfort. The method considered indoor temperature, humidity, outdoor conditions, and thermal comfort constraints. By adjusting air conditioning settings intelligently, the method improved thermal comfort prediction accuracy and reduced energy consumption. A very new study [9] has explored energy-saving optimization for variable-flow air conditioning systems. This research aimed to enhance energy efficiency by dynamically adjusting flow rates based on demand and conditions. In [10], the energy-saving strategies for frequency-variable heat pump air conditioning systems was investigated, which focused on improving efficiency and reducing energy consumption. It is worth mentioning that the energy-saving rate of air conditioning systems varies based on urban climate. For example, drier climates tend to yield higher energy savings [11]. Also, in [12], climate change models were developed to predict meteorological parameters in the Greater Bay Area (Guangzhou, Nansha, Shenzhen, and Hong Kong), which can facilitate the climate-adaptive resilience in district buildings and cross-regional energy sharing.

The literature review and rational analysis show that the interdisciplinary between the machine learning and district energy systems shows promising prospects for the promotions of resilient and smart district energy systems [13–15]. The machine learning technique can be used in the prediction of district energy demand, robust model development, surrogate model [16,17] development for uncertainty and optimization analyses, geometrical and operating parameters' optimal design, and efficient and robust energy management and control strategy development.

In the MAU system, traditional PID control is a common strategy that does not require the energy consumption model. The effect of the PID control method depends on whether the process variables are reasonable. When the process variables are unreasonable, the MAU wastes energy in the operation process. Based on the above literature review and analysis of traditional PID control methods, research gaps can be noticed:

1. The PID control parameters are manually set based on experience; however, the system (or the parameters of PID) without professional operators may result in energy wastage. For example, a common problem is excessive humidification by the humidifier of the MAU in winter mode, followed by dehumidification through the cooling coil.
2. Following the above issue, the energy cost of the MAU has not been considered in the control criterion of traditional PID control. Moreover, the roles of the machine learning techniques in energy prediction have not been characterized well in the MAU system. Hence, there is significant room to optimize the energy consumption in the overall framework of the MAU, in addition to meeting the requirements of the supply air temperature and humidity.
3. The effect of PID control could be poor when outdoor air conditions and supply air requirements are complex. Moreover, it cannot adaptively adjust the control parameters according to changes in indoor and outdoor environments and maintain an optimal control state at all times.

To fill the aforementioned gap, we develop the data-driven optimization algorithm (DDOA) for the MAU, which is partly inspired from the recent research work of surrogate-assisted evolutionary algorithms [16,17]. The data-driven model of the MAU system can help make the operational parameters of PID more transparent. It simultaneously considers the air supply requirements and the energy cost of the MAU, through the customized optimization algorithm. To be specific, it adopts historical data and machine learning algorithms to identify patterns and make predictions without prior knowledge of the system's physical properties. The data-driven models have several advantages like automatic pattern recognition, high accuracy, and self-improvement. Hence, we can develop the optimization method of the fresh air unit system through a mechanism model and historical data [18]. This not only facilitates the energy conversion in the air treatment process of the fresh air unit, but also realizes the prediction of supply air temperature, supply air dew point, and energy cost according to the outdoor air conditions.

3. The Proposed MAU Energy-Saving Control Method

The overall process of the proposed energy conservation framework in the MAU system mainly consists of the following four steps:

1. Establish the energy consumption model of the MAU system: Based on the historical operation data of the MAU, we apply a machine learning algorithm to establish the data model. This model will correlate the coil opening with the energy supply of the MAU. Then, we can predict the temperature and the dew point of the supply air through an operation mechanism model of the MAU. Finally, the model is established to predict the MAU's energy consumption for cold or hot water.
2. Intelligent optimization method: It searches for the optimal coil opening to minimize energy consumption and achieve active energy-saving optimization.
3. Control the MAU according to the optimization results: The algorithm controls the MAU according to the optimized coil opening. It also calculates the deviation between the actual value and the set value and corrects the deviation with the PID algorithm.
4. Monitor the outdoor air temperature and humidity conditions in real time: When the temperature or humidity changes, the algorithm re-optimizes the energy saving.

The MAU energy-saving control method is shown in Figure 3. We can see that it consists of two main parts: establish the prediction model based on historical data, and apply the intelligent optimization algorithm [17,19,20]. To predict the energy cost of the

MAU more accurately, the mechanistic model and the machine learning model (data-driven model) are simultaneously used here. Then, the intelligent optimization algorithm, i.e., the customized genetic algorithm (GA) [21,22], is applied to minimize the objective function of the prediction model, so that the optimized coil opening is provided to reduce the energy costs. For convenience, the symbols used in the process of establishing the models of the MAU and their meanings are presented in Table 1. The details of the proposed method shown in Figure 3 are presented in the subsequent sections.

Table 1. The symbols used in the system modeling process and their meanings.

Symbol	Meaning
h	Air-specific enthalpy
Q	The actual amount of cooling or heating provided by the coil to the air per unit time
m	Air mass flow rate
T	Air temperature
d	Air moisture content, kg water/kg dry air
C_i	$C_8 = -5800.2206$, $C_9 = 1.3914993$, $C_{10} = -0.04860239$, $C_{11} = -5800.2206$, $C_{12} = 1.3914993$, $C_{13} = -0.04860239$
α	Heat transfer efficiency
c	The specific heat capacity of water
ρ	The density of water
f(TCV)	The flow characteristic of the coil control valve

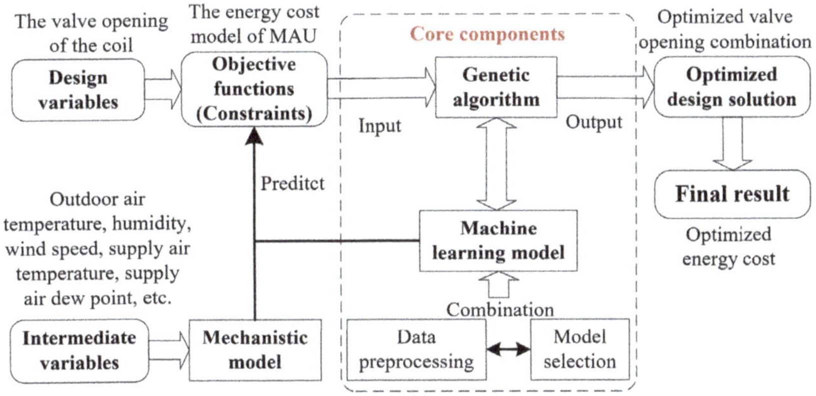

Figure 3. The MAU energy-saving control method.

Figure 3 shows a close relationship between machine learning models, mechanism models, and objective functions. The machine learning model obtains the energy values corresponding to the outdoor air conditions and valve openings by training historical data. The mechanism model uses these energy values to determine whether the solution searched by the intelligent optimization algorithm meets the final air supply temperature and dew point temperature requirements. The objective function calculates the energy consumption by optimizing the water output corresponding to the valve opening, such that the intelligent optimization algorithm searches for the lowest energy consumption solution by minimizing the objective function.

3.1. Mathematical Model of MAU

MAU system modeling aims to describe the energy conversion process during MAU operation in mathematical models and provide a mathematical basis for subsequent MAU energy optimization. Therefore, briefly introducing the mathematical model of the MAU is necessary before describing our algorithm. We will take the phase of precooling as an example.

Based on the different cooling capacities provided by the precooling coil, outdoor air treatment in the precooling stage can be divided into two types: cooling only and dehumidification. To determine whether the air in the precooling stage is being dehumidified, it is first necessary to calculate the cooling capacity Q_{dh} required to cool the outdoor air from its initial temperature to its dew point temperature (Td) within a unit of time. This cooling capacity is designated as Q_{dh}. By comparing Q_{dh} with the cooling capacity provided by the precooling coil, $Q_{precooling}$, we can determine whether the coil's capacity is sufficient for dehumidifying the air. The temperature at which air reaches saturation while maintaining moisture content is known as the dew point temperature. Given the outdoor air temperature (T_{oa}) and the outdoor air humidity (RH_{oa}), the energy conversion process involved in cooling the outdoor air to its dew point can be expressed as follows [23]:

$$m \times h_{oa} - Q_{dh} = m \times h_{Td} \tag{1}$$

$$h_{Td} = 1.01 \times Td + (2501 + 1.84 \times Td) \times d_{oa}/1000 \tag{2}$$

$$h_{oa} = 1.01 \times T_{oa} + (2501 + 1.84 \times T_{oa}) \times d_{oa}/1000 \tag{3}$$

$$d_{oa} = 622 \times P_q/(101325 - P_q) \tag{4}$$

$$\begin{aligned} ln(P_q) =& C_8/(Td + 273.15) + C_9 + C_{10} \times (Td + 273.15) + C_{11} \times (Td + 273.15)^2 \\ &+ C_{12} \times (Td + 273.15)^3 + C_{13} \times ln(Td + 273.15) \end{aligned} \tag{5}$$

The upper formulas are simultaneously solved to obtain the cooling capacity Q_{dh}. The outdoor air's moisture content d_{oa} is calculated based on the outdoor air temperature T_{oa} and outdoor air humidity RH_{oa}.

$$RHoa = P_q/(P_q, b) \tag{6}$$

$$\begin{aligned} ln(P_q, b) =& C_8/(T_{oa} + 273.15) + C_9 + C_{10} \times (T_{oa} + 273.15) + C_{11} \times (T_{oa} + 273.15)^2 \\ &+ C_{12} \times (T_{oa} + 273.15)^3 + C_{13} \times ln(T_{oa} + 273.15) \end{aligned} \tag{7}$$

$$d_{oa} = 622 \times P_q/(101325 - P_q) \tag{8}$$

Moreover, the dew point is calculated based on the moisture content.

$$\begin{aligned} ln(101325 \times d/(622 + d)) = & C_8/(Td + 273.15) + C_9 + C_{10} \times (Td + 273.15) \\ &+ C_{11} \times (Td + 273.15)^2 + C_{12} \times (Td + 273.15)^3 \\ &+ C_{13} \times ln(Td + 273.15) \end{aligned} \tag{9}$$

When the cooling capacity is less than the dehumidification capacity, i.e., $Q_{precooling} < Q_{dh}$, the precooling coil is only used for cooling treatment. Otherwise, the precooling coil performs both cooling and dehumidification treatment. More details are provided as follows.

- Cooling only: The cooling treatment provided by the precooling coil only reduces the air temperature. Given the cooling capacity of the precooling coil denoted as $Q_{precooling}$, the energy conversion process in the precooling section can be expressed as follows:

$$m \times h_{oa} - Q_{precooling} = m \times h_{precooled} \tag{10}$$

The air-specific enthalpy $h_{precooled}$ can be solved based on the above equation.

- Cooling with dehumidifying: The precooling coil performs both cooling and dehumidification treatment, which simultaneously reduces the air temperature and decreases the moisture content of the air. During this process, condensate water is produced, and given the cooling capacity of the precooling coil, denoted as $Q_{precooling}$, the energy conversion process in the precooling section can be expressed as follows:

$$m \times h_{oa} - Q_{precooling} = m \times h_{precooled} + m \times c \times (d_{oa} - d_{precooled}) \times T_{precooled} \tag{11}$$

$$h_{precooled} = 1.01 \times T_{precooled} + (2501 + 1.84 \times T_{precooled}) \times d_{precooled}/1000 \quad (12)$$

$$d_{precooled} = 622 \times P_q/(101325 - P_q) \quad (13)$$

$$\begin{aligned} ln(P_q) =& C_8/(T_{precooled} + 273.15) + C_9 + C_{10} \times (T_{precooled} + 273.15) \\ &+ C_{11} \times (T_{precooled} + 273.15)^2 + C_{12} \times (T_{precooled} + 273.15)^3 \\ &+ C_{13} \times ln(T_{precooled} + 273.15) \end{aligned} \quad (14)$$

Therefore, the simultaneous upper formula is solved to obtain the moisture content and the air-specific enthalpy after precooling, i.e., $d_{precooled}$ and $h_{precooled}$.

3.2. *The Data-Driven Model Based on Machine Learning Methods*

According to the heat transfer equation, the amount of heat provided by the coil per unit of time is related to the specific heat capacity of water, the mass of the water circulating through the coil in a unit of time, and the temperature change in the water (the temperature difference between the inlet and outlet water of the coil). Among them, the mass of the water circulating through the coil in a unit of time is controlled by the coil opening and is related to the flow characteristics of the coil-regulating valve. The ice machine controls the temperature difference between the coil's inlet and outlet water, which can be considered constant. Since there will be losses in the process of heat transfer from the coil to the air, the actual amount of heat provided to the air by the coil also needs to take into account the heat exchange efficiency of the coil. In short, each coil's actual energy supplied to the air can be expressed as follows:

$$Q = \alpha \times c \times \rho \times f(TCV) \times \Delta T \quad (15)$$

In practical projects, due to the difficulty of accurately obtaining the flow rate of hot and cold water in the coil ($f(TCV)$) and the heat exchange efficiency of the coil (α), it is often infeasible to directly apply Equation (2.1) to the calculation of coil energy supply based on the coil opening. Therefore, the relationship between coil openings and coil cooling and heating outputs was modeled using the Random Forest algorithm based on historical data. This approach allows for the calculation of coil cooling and heating outputs based on coil control valve openings, as shown in Figure 4. The next discussion will focus on three aspects: data pre-processing, comparison of different machine learning algorithms, and analysis of the best predictive model.

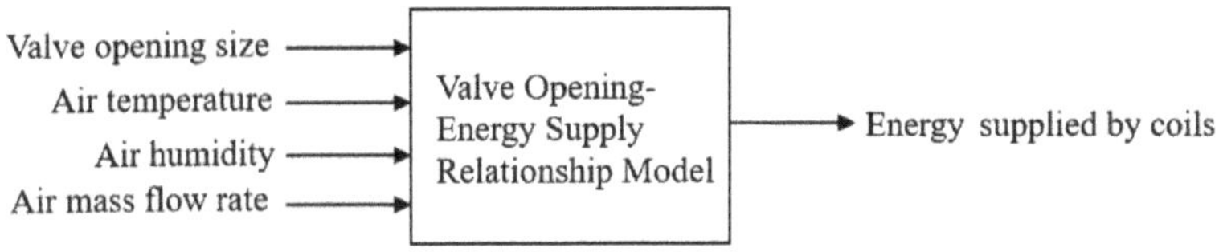

Figure 4. Schematic diagram of coil model.

3.2.1. Data Analysis

The dataset comprises 51,000 records, encompassing temperature, humidity, airflow, and valve openness. Recognizing the inherent limitations and logical boundaries of these parameters—such as the valve position being constrained within the range of 0–100—enabled a thorough cleansing of the raw data. This involved the elimination of erroneous or implausible entries that did not adhere to these established norms. Post-cleaning, the refined dataset was significantly condensed yet markedly improved in quality, yielding a final tally of 25,000 valid and meaningful records. This curated dataset forms the robust foundation upon which subsequent analysis and modeling efforts are based, ensuring that the insights derived are reliable and representative of real-world conditions.

3.2.2. Data Pre-Processing

To achieve a relatively accurate prediction for full opening, it is essential to enhance the training data in advance [24]. As mentioned in Section 3.2.1, the original dataset comprises 51,000 records, while 25,000 valid and meaningful records are left after the post-cleaning stage, i.e., the refined dataset. It forms 12,500 pairs of date points, where each pair serves as the basis for linear interpolation within the valve opening range of 0–100. This process generates additional data points within the full opening range, as illustrated in Figure 5.

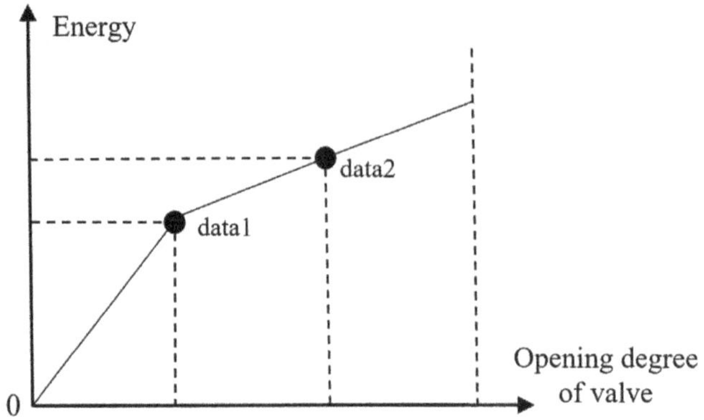

Figure 5. Linear interpolation of data.

3.3. *Machine Learning Algorithms*

This study employed eight regression algorithms as the core of machine learning [25], including linear regression [26], ridge regression [27], Lasso regression [28], K-nearest neighbors regression [29], decision tree regression [30], random forest regression [31], gradient boosting regression [32,33], and AdaBoost regression [34].

3.3.1. Linear Regression

Linear regression is a fundamental machine learning algorithm used for predictive modeling. It models the relationship between a dependent variable and one or more independent variables by fitting a linear equation to observed data. The equation can be written as follows:

$$y = \beta_0 + \beta_1 x_1 + \beta_2 x_2 + \ldots + \beta_n x_n + \epsilon$$

where y is the dependent variable, $x_1, x_2, \ldots, x_n$ are the independent variables, $\beta_0, \beta_1, \beta_2, \ldots, \beta_n$ are the coefficients, and ϵ is the error term.

3.3.2. Ridge Regression

Ridge regression, also known as Tikhonov regularization, is a variant of linear regression that includes a regularization term to prevent overfitting. It adds a penalty equal to the square of the magnitude of coefficients to the loss function:

$$\text{Loss} = \sum_{i=1}^{n} (y_i - \hat{y}_i)^2 + \lambda \sum_{j=1}^{p} \beta_j^2$$

where λ is the regularization parameter controlling the strength of the penalty.

3.3.3. Lasso Regression

Lasso (Least Absolute Shrinkage and Selection Operator) regression is another regularization technique that performs both variable selection and regularization to enhance the prediction accuracy. It adds a penalty equal to the absolute value of the magnitude of coefficients:

$$\text{Loss} = \sum_{i=1}^{n}(y_i - \hat{y}_i)^2 + \lambda \sum_{j=1}^{p} |\beta_j|$$

The λ parameter controls the sparsity of the coefficients.

3.3.4. K-Nearest Neighbors Regression

K-nearest neighbors (KNN) regression is a non-parametric method used for regression and classification. In KNN regression, the output is the average of the values of its K-nearest neighbors:

$$\hat{y} = \frac{1}{K} \sum_{i=1}^{K} y_i$$

where y_i are the target values of the K-nearest neighbors to the input point.

3.3.5. Decision Tree Regression

Decision tree regression models the target variable by learning decision rules inferred from the features. It creates a tree structure where internal nodes represent decision rules based on features, branches represent outcomes, and each leaf node represents a predicted value:

$$\hat{y} = \sum_{m=1}^{M} c_m I(x \in R_m)$$

where c_m is the predicted value in region R_m, and I is the indicator function.

3.3.6. Random Forest Regression

Random forest regression is an ensemble learning method that constructs multiple decision trees during training and outputs the average prediction of the individual trees. It reduces overfitting and improves generalization:

$$\hat{y} = \frac{1}{B} \sum_{b=1}^{B} \hat{y}^{(b)}$$

where B is the number of trees, and $\hat{y}^{(b)}$ is the prediction of the b-th tree.

3.3.7. Gradient Boosting Regression

Gradient boosting regression builds an ensemble of trees sequentially, where each tree corrects the errors of its predecessor. It minimizes a loss function by adding weak learners using gradient descent:

$$\hat{y}_m(x) = \hat{y}_{m-1}(x) + \nu h_m(x)$$

where $h_m(x)$ is the new tree trained on the residuals of the previous model, $\hat{y}_{m-1}(x)$, and ν is the learning rate.

3.3.8. AdaBoost Regression

AdaBoost (Adaptive Boosting) regression improves the performance of weak learners by focusing on the errors of previous learners. It combines the predictions of multiple learners by assigning weights to each learner based on their performance:

$$\hat{y} = \sum_{m=1}^{M} \alpha_m h_m(x)$$

where α_m is the weight assigned to the m-th weak learner $h_m(x)$.

3.4. Model Configuration and Performance Evaluation

In the proposed algorithm, the random forest algorithm was chosen due to its superior performance and robustness across various predictive tasks. One of its key advantages is the ability to mitigate overfitting through the aggregation of multiple decision trees, each trained on random subsets of the data and features. This ensemble method ensures that the final model generalizes well to unseen data, providing more reliable predictions compared to single models. Additionally, the random forest algorithm excels in handling high-dimensional datasets, which is crucial given the numerous attributes in our dataset that collectively influence the predictive outcomes. It does not require extensive feature selection or pre-processing, making it both efficient and effective without compromising accuracy. Empirical results in Table 2 demonstrate that the random forest algorithm achieved the lowest root mean square error (RMSE) and the highest R Squared values among all tested models, reinforcing its selection as the optimal algorithm for our regression tasks.

Table 2. Comparison of regression algorithm models on RMSE and R Squared.

Model	RMSE	R Squared
Linear Regression	15.892662	0.995577
Ridge Regression	15.892681	0.995577
Lasso Regression	15.928640	0.995557
K Neighbors Regressor	12.711077	0.997171
Decision Tree Regressor	13.526912	0.996800
Random Forest Regressor	11.162434	0.997822
Gradient Boosting Regressor	13.578922	0.996771
Adaboost Regressor	25.567316	0.988897

Usually, some key hyperparameters of the random forest algorithm should be carefully chosen to optimize the model's predictive capabilities. Specifically, the number of trees was set to 100, and the maximum depth of the trees was left unspecified to allow trees to grow fully. The minimum number of samples required to split an internal node was set to 2, while the minimum number of samples required at each leaf node was set to 1. To balance the trade-off between bias and variance, the maximum number of features considered for splitting a node was set to the square root of the total number of features. The model was trained using the mean squared error as the criterion for measuring the quality of splits. The dataset was split into 80% for training and 20% for testing, with a fixed random state of 42 for reproducibility. No explicit feature selection was conducted, and the model was trained on a standard desktop computer equipped with an Intel Core i7 processor and 32 GB of RAM.

The regression algorithm models utilized RMSE (root mean square error) and R Squared as the evaluation metrics for model accuracy [35]. RMSE measures the square root of the ratio of the squared deviations between predicted and actual values to the number of observations (n). It gauges the deviation between predicted and actual values and is sensitive to outliers in the data. On the other hand, R Squared, the coefficient of determination, is used in statistics to measure the proportion of the variability in the dependent variable that is explained by the independent variable. It serves to assess the explanatory power of the regression model. A lower RMSE indicates a better model, while a higher R Squared suggests a better model. The experimental results, as depicted in Table 2, show that the random forest algorithm yielded the optimal performance.

3.5. Definition of the Optimization Problem

Before applying intelligent optimization algorithms [36], it is crucial to design the proactive energy-saving optimization control of the fresh air handling unit as an optimization problem [13]. Assuming the costs of chilled water, low-temperature chilled water, and hot water per cubic unit are represented as a, b, and c, respectively, $f(x)$ represents the flow through the coil control valve at an opening of x, measured in cubic meters per hour. TCV represents the opening of the coil control valve, indicating the degree to which the valve is open. pc, rc, ph, and rh represent the four sections in the model. The energy cost (E) of the MAU for chilled and hot water at steady state can be expressed as follows:

$$E = a \times f(TCV_{pc}) + b \times f(TCV_{rc}) + c \times \left[f(TCV_{ph}) + f(TCV_{rh}) \right] \tag{16}$$

Through the above model for calculating the energy cost of chilled and hot water usage in the MAU, it is possible to predict the energy cost under different combinations of control valve openings for the MAU's chilled and hot water systems.

Given a set supply air temperature and supply air dew point, there may be multiple combinations of coil openings that satisfy the requirements. To optimize the energy cost of chilled and hot water usage in the MAU, it is necessary to find a combination of control valve openings $(TCV_{ph}, TCV_{pc}, TCV_{rc}, TCV_{rh})$ that, while meeting the required supply air temperature and supply air dew point, minimizes the energy cost(E). Let the function for calculating the supply air temperature be denoted as $f_T(TCV_{ph}, TCV_{pc}, TCV_{rc}, TCV_{rh})$ and the function for calculating the supply air dew point be denoted as $f_d(TCV_{ph}, TCV_{pc}, TCV_{rc}, TCV_{rh})$. Then, the optimization problem [37] for the MAU's energy cost of chilled and hot water usage can be expressed as follows:

$$\begin{aligned} & P : minE \\ & s.t. \\ & C1 : f_T(TCV_{ph}, TCV_{pc}, TCV_{rc}, TCV_{rh}) = T \\ & C2 : f_d(TCV_{ph}, TCV_{pc}, TCV_{rc}, TCV_{rh}) = Td \end{aligned} \tag{17}$$

By applying intelligent optimization algorithms to solve the aforementioned optimization problem of the MAU's energy cost for chilled and hot water usage, it is possible to obtain the coil opening that minimizes the energy cost while satisfying the set supply air temperature and dew point.

This paper customizes genetic operators [21,38] to address the specific characteristics of the real-world problem. The following sections provide detailed descriptions of population initialization [39] and environmental selection [20].

3.6. Direction-Based Optimization Algorithm

Algorithm 1 gives the specific implementation details of the direction-based optimization algorithm. The problem to be optimized E is the minimum energy consumption that satisfies the constraints and has been defined above.

For the initialization of population (line 1), it is possible to sample based on the distribution of control valve openings in the dataset. The sampled results can be used as the initial population. The distribution illustrates the frequency of different control valve openings, where a higher frequency indicates more frequent usage and a greater likelihood of being a feasible solution. This approach allows for the quicker identification of feasible solutions, but caution should be taken to avoid the population becoming trapped in local optima. The population size is set to 50; therefore, 50 individuals need to be sampled as the initial population. Figure 6 depicts the sampling results at various sections.

Algorithm 1 Direction-based optimization algorithm

Input: Problem E, Population size pop_size, Max generations $max\_generations$
Output: Best solution $best\_individual$

1: Initialize population P with pop_size individuals;
2: **for** generation $g = 1$ to $max\_generations$ **do**
3:     **for** each individual ind in population P **do**
4:         Evaluate fitness $f(ind)$;
5:     **end for**
6:     **if** convergence criteria is not met **then**
7:         Selected individuals $Q \leftarrow$ Select the top half of population P based on fitness;
8:         %Generate new population P_{new} below;
9:         **while** $|P_{new}| < pop\_size$ **do**
10:             Randomly select two parents s_1, s_2 from Q;
11:             Perform directional crossover on s_1 and s_2 to produce children q_1, q_2;
12:             Apply directional mutation on q_m;
13:             Add q_1, q_2 and q_m to P_{new};
14:         **end while**
15:         Set $P = P_{new}$;
16:         Print the best fitness in current generation;
17:     **else**
18:         **break**;
19:     **end if**
20: **end for**

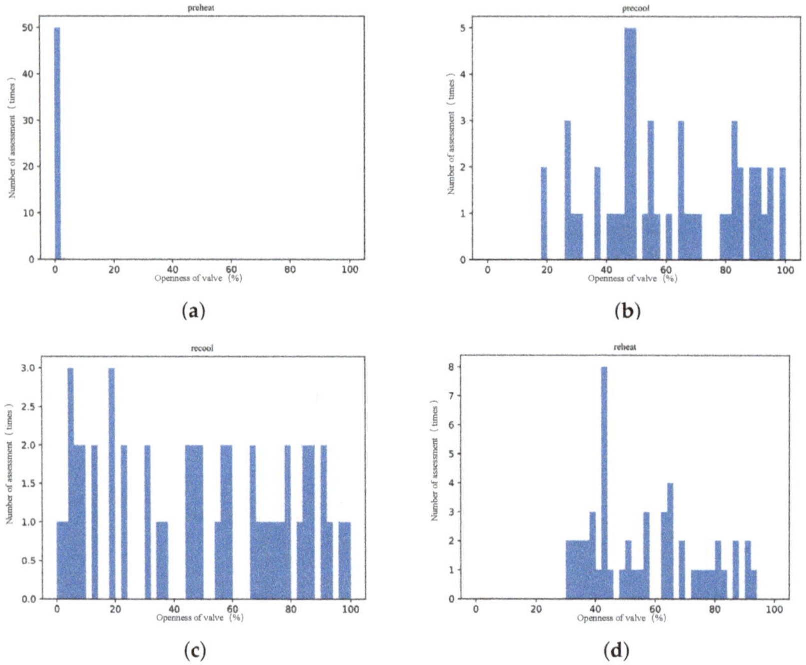

Figure 6. Sampling results at various sections: (**a**) preheat, (**b**) precool, (**c**) recool and (**d**) reheat.

The number of iterations of the algorithm is predefined in the main loop of the algorithm (lines 2–19). And in each iteration, the fitness of the population individuals is calculated and new individuals are generated based on the better individuals using a directed learning strategy until the termination condition is reached. For constrained single-objective optimization problems, the fitness evaluation of individuals considers both the objective function value and the degree of constraint violation (lines 2–3). We use the penalty function method to transform the constraint violations into penalty values and combine them with the objective function value to compute the fitness of each individual. The formula for calculating fitness is as follows:

$$P(ind) = \sum_{i=1}^{m} w_i \cdot v_i(ind) \tag{17}$$

$$f(ind) = f_{obj}(ind) + P(ind) \tag{18}$$

where $P(ind)$ is the sum of weighted violations of all constraints, forming the total penalty function value. Specifically, each constraint is applied to the individual, with the resulting violation multiplied by its corresponding weight. Weights range from 0 to 1, reflecting constraint importance. For instance, if two constraints are deemed equally significant, a weight of 0.5 may be assigned to both.

In addition, we improve on the traditional genetic algorithm by designing directed crossover and directed mutation operators, which can realize fast solution finding in the solution space and improve the convergence speed of the population (lines 11–16). The directional crossover process generates offspring guided by the fitness values of the parents.

3.6.1. Directional Crossover Operator

Assuming a population size of N and n decision variables, the optimization process allows both crossover and mutation to be performed in variable directions. The crossover operator primarily involves four parameters: the crossover rate p_c, the probability of variable direction p_{cv}, the direction probability p_d, and the multiplication factor α. Assuming two unequal parent solutions s_1^j and s_2^j (where j ranges from 1 to n and is allowed to crossover), and given s_{best}^j and s_{mean}^j as the best and mean solutions in the current population, respectively, if $s_{\text{best}}^j \geq s_{\text{mean}}^j$, then two offspring solutions q_1 and q_2 are constructed using the following formulas:

$$\text{val} = 1 - (0.5)^{e^{\left[\frac{s_1^j - s_2^j}{s_u^j - s_l^j}\right]}} \tag{19}$$

$$\beta = \frac{r_1}{\alpha^2} \tag{20}$$

$$\begin{cases} q_1 = \text{val} \times (s_1^j + s_2^j) + \alpha^{r_1} \times e^{(1-\beta)} \times (1 - \text{val}) \times |s_1^j - s_2^j| \\ q_2 = (1 - \text{val}) \times (s_1^j + s_2^j) - \alpha^{(1-r_1)} \times e^{(-\beta)} \times \text{val} \times |s_1^j - s_2^j| \end{cases} \text{, if } r_2 \leq p_d \tag{21}$$

$$\begin{cases} q_1 = \text{val} \times (s_1^j + s_2^j) - \alpha^{r_1} \times e^{(1-\beta)} \times (1 - \text{val}) \times |s_1^j - s_2^j| \\ q_2 = (1 - \text{val}) \times (s_1^j + s_2^j) + \alpha^{(1-r_1)} \times e^{(-\beta)} \times \text{val} \times |s_1^j - s_2^j| \end{cases} \text{, if } r_2 > p_d \tag{22}$$

where r_1 and r_2 be two distinct random numbers generated within the range [0, 1], val and β are two intermediate parameters, s_u^j and s_l^j denote the upper and lower bounds of the j-th variable, respectively. In one case, if $s_{\text{best}}^j < s_{\text{mean}}^j$, the offspring solutions are constructed using the following formulas:

$$\begin{cases} q_1 = \text{val} \times (s_1^j + s_2^j) - \alpha^{r_1} \times e^{(1-\beta)} \times (1 - \text{val}) \times |s_1^j - s_2^j| \\ q_2 = (1 - \text{val}) \times (s_1^j + s_2^j) + \alpha^{(1-r_1)} \times e^{(-\beta)} \times \text{val} \times |s_1^j - s_2^j| \end{cases} \text{, if } r_2 \leq p_d \tag{23}$$

$$\begin{cases} q_1 = \text{val} \times (s_1^j + s_2^j) + \alpha^{r_1} \times e^{(1-\beta)} \times (1 - \text{val}) \times |s_1^j - s_2^j| \\ q_2 = (1 - \text{val}) \times (s_1^j + s_2^j) - \alpha^{(1-r_1)} \times e^{(-\beta)} \times \text{val} \times |s_1^j - s_2^j| \end{cases} \text{, if } r_2 > p_d \tag{24}$$

During the directed crossover process, newly created offspring solutions are constrained by variable boundaries to avoid generating infeasible solutions, while at the same time learning from better solutions in the population to generate new ones. Additionally, the probability that q_1 is considered as the first offspring solution or the second offspring solution is 50%. If q_1 is chosen as the first offspring solution, then q_2 is regarded as the second offspring solution, and vice versa. This step is referred to as the offspring identification condition.

3.6.2. Directed Mutation Operator

Similar to the directed crossover operator, the directed mutation operator can adjust the directional information for the optimization problem. For the parent solution s_i^j and its mutated solution s_m, where i represents an individual in the population from 1 to N, and j represents a decision variable from 1 to n. A random number r within the range [0, 1] is generated. If this random number is less than or equal to the mutation probability p_m, the parent solution s_i^j is allowed to participate in the mutation operation.

The directed mutation operator requires directional information, which is collected before using the operator. The way to collect this information is by comparing the current parent solution s_i^j with the best solution in the population s_{best}^j. When it is found that the fitness of s_{best}^j is greater than or equal to the fitness of s_i^j, the mutated solution q_m is created using the following formula:

$$\beta_1 = e^{\left(2r - \frac{2}{r}\right)}, \tag{25}$$

$$\beta_2 = e^{\left(r - \frac{2}{r}\right)}, \tag{26}$$

$$q_m = \begin{cases} s_i^j + \beta_1 \times \left(s_u^j - s_i^j\right), & \text{if } r_3 \le p_d \\ s_i^j - \beta_2 \times \left(s_i^j - s_l^j\right), & \text{otherwise} \end{cases} \tag{27}$$

where β_1 and β_2 are two intermediate parameters used to determine q_m. r_3 is a random number generated within the range [0, 1]. s_u^j and s_l^j represent the upper and lower bounds of the j-th variable, respectively. p_d denotes the direction probability, typically set between 0.5 and 1. In the case wherein the fitness of s_{best}^j is worse compared to s_i^j, the mutated solution q_m can be updated using the following formulas:

$$q_m = \begin{cases} s_i^j - \beta_1 \times \left(s_i^j - s_l^j\right), & \text{if } r_3 \le p_d \\ s_j^i + \beta_2 \times \left(s_u^j - s_i^j\right), & \text{otherwise.} \end{cases} \tag{28}$$

It is evident from the above formulas that the mutated solution is always created within the variable boundary constraints. Therefore, in this case, no boundary handling techniques are needed.

By using directed crossover and directed mutation strategies, the population can converge quickly. When the convergence degree meets the predefined requirements, it jumps out of the main loop and outputs the best individual in the population. In this article, the predetermined convergence termination condition is that, if all individuals in the population meet the constraint conditions, it can be terminated in advance. Otherwise, it will continue to run until the set number of iterations is used up.

4. Simulink Simulation Design

4.1. Simulation Model

In this section, we use Simulink to perform simulation of the designed system. Figure 7 presents the overview of our simulation model. Firstly, specific environmental conditions are given as system inputs, including temperature (T_env) and dew point temperature (Td_env). We then set our desired control results, including air temperature (T_send), air dew point temperature (Td_send), and the pre-processing stage setpoint (T_SP). Finally, we can observe how the intermediate variables change by running the system. Note that, for the setting of the setpoint in the pre-processing stage, the conventional PID control relies entirely on experience, while we use the results obtained by the intelligent optimization algorithm, which is the most significant advantage of our algorithm.

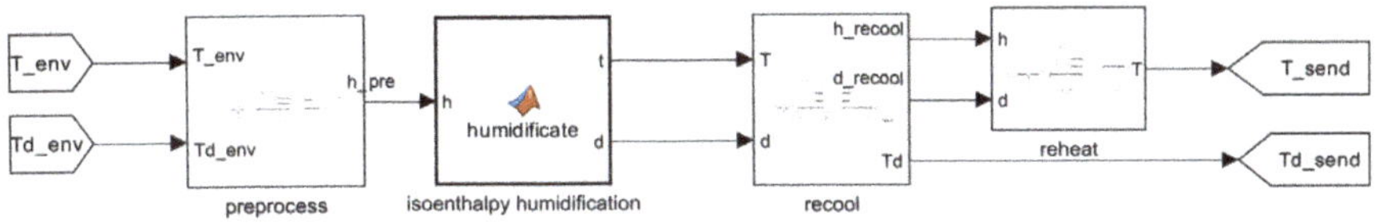

Figure 7. Simulation model overview.

4.2. *Verification of the Model's Reliability*

In Section 3.1, we described the working principle of the precooling stage of the MAU and the changes of air in this stage from a mathematical point of view. The pre-processing stage in the simulation model is built based on the mathematical descriptions. The precooling stage is chosen for detailed description because it is the most complex mechanism, and in this subsection we will use the precooling stage as a representative to verify the reliability of the overall model.

Table 3 shows data recorded in a real production environment describing the change in properties after the energy exchange between the outdoor air and the precooling coil, which can be checked for correctness by the mathematical model in Section 3.1. The operation of the precooling stage in the simulation model is shown in Figure 8, where it can be observed that our model can correctly calculate the properties of the treated air based on the input outside air temperature, dew point temperature, and the work accomplished by the precooling coil.

Table 3. Actual data describing changes in air properties during the precooling phase.

Air Temperature (°C)	Air Dew Point (°C)	Q (kJ/s)	Temperature After Cooling (°C)	Specific Enthalpy After Cooling (kJ/kg)
33.39	24.0880	1408.389	49.7023	17.48

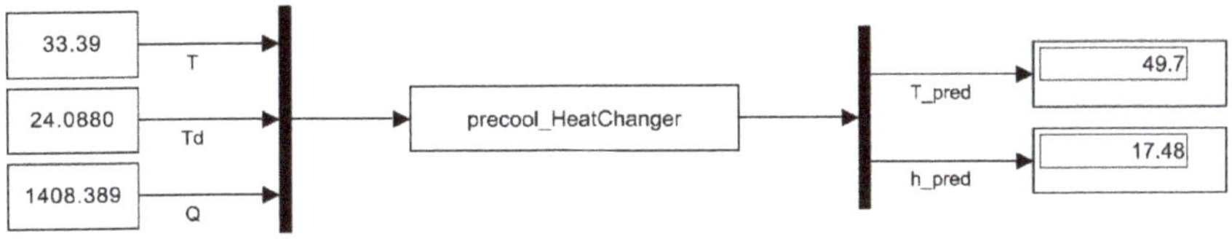

Figure 8. Verification of precool phase.

5. Experiment

In order to investigate the performance of the intelligent control method proposed in this paper from multiple perspectives, two experiments are designed in this section, including theoretical and simulation experiments. In the rest of this section, we first introduce the theoretical experimental content and then the simulation experimental content.

5.1. *Data Visualization*

The pre-trained random forest regression model was applied to predict the preprocessed data, and prediction was used to assess the accuracy and effectiveness of the model. The following sections will present separate analyses for the preheating, precooling, recooling, and reheating sections in the prediction model.

1. Preheating: In the preheating section, out of 25,000 historical datasets, 563 did not achieve an accuracy of 90% or higher in predictions, constituting approximately 2.1% of the total data sample. The prediction accuracy in the preheating section is illustrated in Figure 9a, while the specific comparison between the predicted values and true values for both the overall prediction accuracy and the instances wherein the standard was not met is illustrated in Figure 9b.
2. Precooling: In the precooling section, out of 25,000 historical datasets, 965 did not achieve an accuracy of 90% or higher in predictions, constituting approximately 3.6% of the total data sample. The specific comparison between the predicted values and true values for both the overall prediction accuracy and the instances wherein the standard was not met is illustrated in Figure 10.
3. Recooling: In the recooling section, out of 25,000 historical datasets, 352 did not achieve an accuracy of 90% or higher in predictions, constituting approximately 1.3% of the total data sample. The specific comparison between the predicted values and true values for the overall prediction accuracy and the instances wherein the standard was not met is illustrated in Figure 11.
4. Reheating: In the reheating stage, all 25,000 historical datasets achieved an accuracy of 90% or higher in predictions. The specific comparison between the predicted values and true values for the overall prediction accuracy and the instances wherein the standard was not met is illustrated in Figure 12.

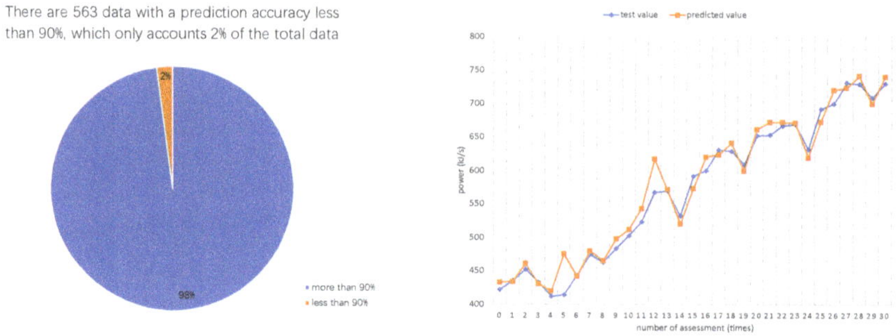

(a) (b)

Figure 9. Prediction accuracy in the preheating section. (**a**) Overall prediction accuracy. (**b**) Comparison of the predicted value and the actual value of the non-compliant data.

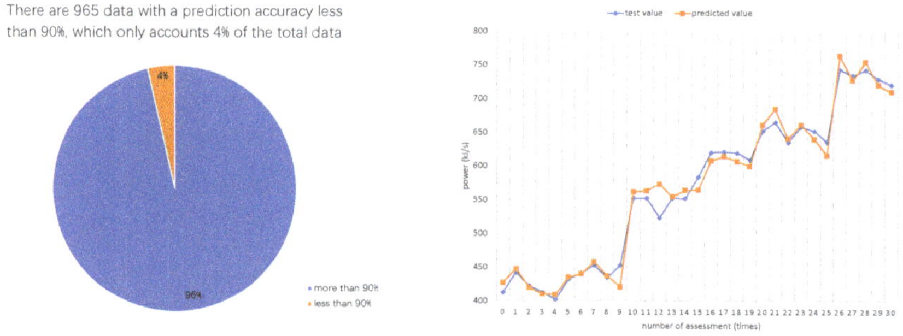

(a) (b)

Figure 10. Prediction accuracy in the precooling section. (**a**) Overall prediction accuracy. (**b**) Comparison of the predicted value and the actual value of the non-compliant data.

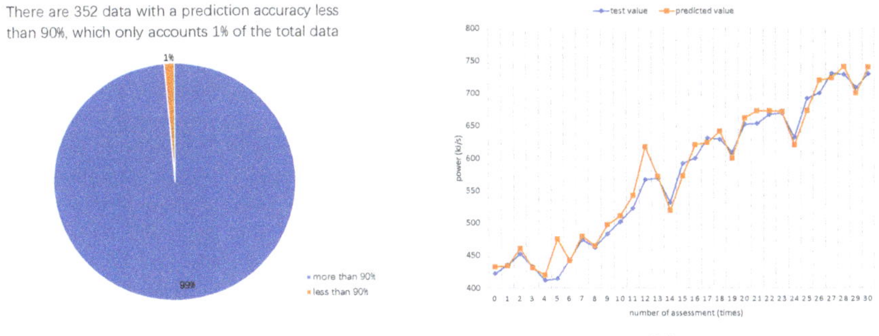

(a) (b)

Figure 11. Prediction accuracy in the recooling section. (**a**) Overall prediction accuracy. (**b**) Comparison of the predicted value and the actual value of the non-compliant data.

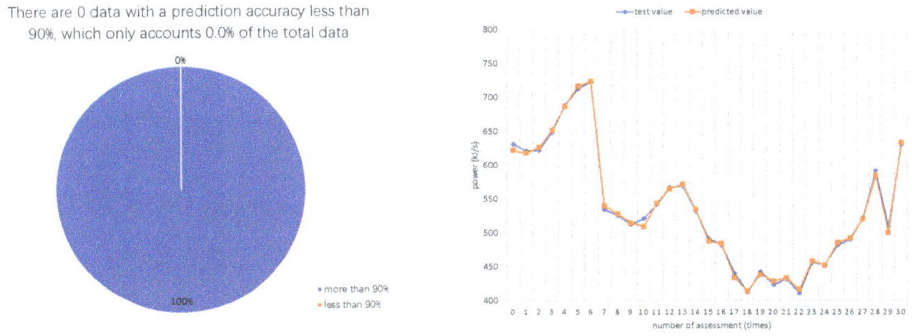

(a) (b)

Figure 12. Prediction accuracy in the reheating section. (**a**) Overall prediction accuracy. (**b**) Comparison of the predicted value and the actual value of the non-compliant data.

In addition, we can see that the test and estimated values are separated at points 5 and 12 in Figures 9 and 11. This separation also occurs at points 9 and 12 in Figure 10. This is mainly due to two factors: (1) There are in total 25,000 pieces of data from the actual field tests, which are subject to environmental impacts; hence, there are inevitably some unpredictable errors. (2) It is inevitable that there are some errors in machine learning-based prediction models, since it is usually hard or impossible to train the machine learning models that can fit all test datasets. Moreover, it is worth pointing out that the machine learning models in precooling mode is really not as effective as in other modes (preheating, recooling, and reheating) in terms of the prediction accuracy, as partly shown in Figure 10. This may be owing to the fact that the data collected in the precooling mode are not good enough and the precooling valve is more active.

5.2. Real Installation Design

In the real installation and deployment, we conducted tests in the central control room of a factory air conditioning system in Deqing, Zhejiang Province, China. The test time was in August. The local temperature range for a day was of 30–39 degrees Celsius, which belonged to the summer mode. The valves that should be activated were precooling, recooling, and reheating. Since the ambient temperature fluctuates during the day, in order to test the actual effect of the proposed method, we set the single test period to 24 h. The factory air conditioning system must first meet stable output temperature and dew point to ensure the normal operation of the factory. Therefore, in the actual test, we incorporated traditional PID control into the execution process. In PID control, each adjustment is made according to the current environmental conditions. In theory, even if the control effect of the intelligent optimization algorithm does not meet the requirements, it will provide a good initial result for PID control, which can reduce the PID control time and achieve energy saving.

Figure 13 shows the execution flowchart in real deployment. First, start the PLC connection and read the outside air environment data, including the outside air temperature and dew point temperature, etc. At the same time, the set temperature and dew point temperature are also obtained. Then, the intelligent optimization algorithm module is executed. The algorithm uses the Pymoo framework [40] and machine learning models to perform optimization; thereafter, the optimized valve opening can be output. The valve openings obtained by the intelligent algorithm will be written into the corresponding PLC points, and the air device system will start the corresponding valve settings to control the temperature and humidity. If the output result of the intelligent optimization algorithm cannot meet the preset temperature and dew point temperature requirements, PID control will be started to prioritize the stable temperature and dew point temperature requirements. The energy consumption at this time is calculated and written into the corresponding PLC point as a historical record. Due to the fact that it only takes about 15 s for the valve to move and achieve temperature control, in order to ensure that the current round of control is completed thoroughly, we wait for 30 s for the control to be completed. After the regulation is completed, a new round of intelligent optimization algorithm optimization control is started until it stops after 24 h of operation and all energy consumption records are output. In order to calculate the energy saving effect, we repeated the above test several times. Finally, the energy consumption generated by the proposed method was compared with the energy consumption required by the previous PID control alone.

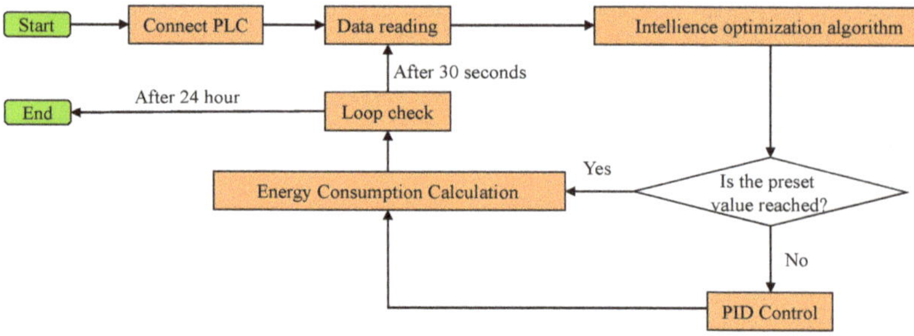

Figure 13. Real deployment execution flowchart.

5.3. The Experiment of Optimization Process

The experiment of optimization process is conducted with evolutionary algorithms based on the Pymoo platform [40]. The experiment starts by randomly selecting 30 original data items from the dataset. The original data include environmental temperature (T_env), environmental humidity (d_env), air supply temperature (T_send), air supply dew point (Td_send), and consumption under PID control (yuan/hour). Based on the above data, our proposed algorithm AESOC-GA can be conducted to obtain the corresponding consumption. The proposed algorithm is independently run 20 times on each data item. Instead of embedded GA in the proposed framework, we also consider Differential Evolution (DE) [41,42] and PSO [43,44] to perform the optimization process for comparison.

Table 4 shows the statistical results of the traditional PID control method and our proposed intelligent control method on the 20 original data items. It can be observed that, compared to the traditional PID control method, our proposed intelligent control method significantly reduces resource consumption, regardless of whether the embedded algorithm is GA, DE, or PSO. Moreover, it can obviously show the advantage of our GA-based method using directed generic operators, i.e., AESOC-GA, compared to the other other two competitive algorithms AESOC-DE and AESOC-PSO.

Table 4. Statistics of energy consumption results (mean and standard deviation) obtained by PID and DDOA with different embedded algorithms on thirty chosen original data items. The best result in each row is highlighted in boldface.

T_env	d_env	T_SP	Td_SP	kWh			
				AESOC-GA	AESOC-DE	AESOC-PSO	PID
32.25	16.01	19.97	9.18	152.45 (25.19)	244.20 (6.713)	156.63 (19.26)	226.30
27.55	19.41	19.96	10.16	143.51 (2.42)	168.03 (24.10)	198.80 (13.80)	162.53
31.65	19.12	20.12	9.36	223.20 (0.82)	222.02 (0.02)	213.88 (49.91)	228.23
23.06	10.80	21.06	11.11	17.07(7.06)	15.66 (7.27)	20.94 (18.79)	76.15
22.85	15.45	18.88	10.19	51.42 (54.68)	73.87 (25.89)	72.128 (6.95)	108.13
24.41	17.82	19.15	12.02	43.24 (36.31)	61.04 (32.70)	127.51 (57.28)	54.50
23.66	10.52	21.20	10.86	13.15(3.23)	9.06 (1.68)	5.00 (9.88)	77.28
26.87	14.55	18.81	9.65	107.55 (2.42)	106.49 (12.12)	110.2 (12.89)	96.01
34.68	21.36	20.19	8.97	137.35 (169.08)	310.75 (103.84)	268.75 (96.47)	224.86
29.63	13.10	20.84	10.27	72.40 (3.06)	66.64 (4.86)	88.42 (26.67)	237.67
25.91	15.98	21.13	10.57	85.53 (20.72)	73.65 (31.73)	55.76 (51.66)	241.46
26.10	17.13	20.13	9.93	125.51 (1.74)	124.68 (4.31)	128.64 (31.70)	192.16
24.28	17.43	19.11	11.98	39.49 (40.32)	60.20 (16.66)	145.86 (52.97)	54.49
26.23	19.94	19.99	9.90	150.19 (2.59)	175.76 (21.39)	201.05 (17.57)	152.64
24.05	10.02	20.74	10.60	28.60 (14.53)	20.03 (12.63)	2.79 (8.55)	81.50
27.64	21.14	20.02	9.59	226.25 (1.07)	214.30 (49.35)	229.63 (9.67)	230.27
31.49	12.24	21.09	10.16	127.99 (1.41)	128.14 (8.49)	123.05 (29.56)	231.95
30.04	17.07	21.14	10.83	131.6 (4.22)	136.37 (11.97)	135.86 (81.54)	222.94
31.92	10.08	19.85	10.31	56.01 (17.39)	57.23 (10.03)	57.63 (2.92)	208.78
30.03	16.80	21.13	10.77	116.41 (51.03)	142.02 (14.12)	131.84 (6.18)	222.94
	+/−			17/3	13/7	15/5	

5.4. Simulation Experiment

The simulation experiment plan is based on Simulink to compare the energy consumption of traditional PID control methods and intelligent control methods. In this experiment, the main difference between the two control methods lies in the determination of the setpoint during the pre-processing stage: in the traditional PID control method, the setpoint is determined by expert experience, while in the intelligent control method, the setpoint is determined by an intelligent optimization algorithm. Subsequently, the experimental method will be introduced in detail, and the results are presented.

The experimental data used in Table 5 were measured by the sensors attached to the MAUs in the actual production environment, where the dew point temperature and temperature of the outdoor ambient air were measured by the dew point/temperature sensors of the model EE210-HT6xPCxDx-Td048-Tx003M + HA010502, and the data of the air in the other MAUs were measured by the air sensors of the corresponding positions of the model EE210- HT6xPCxDx-Txd004-Tx004M + EE215-HTxD model dew point/temperature sensors in the corresponding locations. The PID controller coefficients were tuned using the Z-N method, based on time response and experience [45,46]. The PID controller coefficients for the pre-process section, recooling section, and reheat section are set as shown in Table 6, with the same settings used for both the traditional PID control method and DDOA.

Table 5. The data used are collected from real-world scenarios. For the traditional PID control, the setpoint for the pre-processing stage is set to 21 °C, as used on-site. For the intelligent optimization control, the algorithm determines the setpoint for the pre-processing stage.

Air Temperature (°C)	Air Dew Point (°C)	Post-Processing Temperature (°C)	After-Cooling Temperature (°C)	Delivery Air Temperature (°C)	Delivery Air Dew Point (°C)
29.93	16.40	17.70	11.06	20.99	11.06

Table 6. The setting of the PID controller coefficients.

	Proportional (P)	Integral (I)	Derivative (D)
preprocess	1	1	0
recool	1	1.2	0
reheat	1	1.2	0

There are three control variables which are the setpoints of the PIDs for the different stages, i.e., the temperature setpoint for the pre-processing stage, the supply air dew point temperature setpoint for the re-cooling stage, and the supply air temperature setpoint for the reheating stage.

1. Experiment method:
   (a) Traditional PID control:
      i. Select several representative data from the dataset for the experiment.
      ii. Run the simulation system based on the selected data using the traditional PID control method. This allows us to observe the energy consumption of the PID control method.
   (b) Intelligent optimization control:
      i. Run the proposed algorithm based on the selected data to obtain the setpoint for the pre-processing stage.
      ii. Run the simulation system based on the selected data and the setpoint obtained from the pre-processing stage. Then, we can observe the energy consumption of the intelligent control method.
2. Experiment result:
   (a) Experiment data:
   (b) Experimental configuration for traditional PID control:
      i. Preprocessing stage setpoint: 21.00 °C;
      ii. Supply air dew point: 11.06 °C;
      iii. Supply air temperature: 21.00 °C.
   (c) Experimental configuration for intelligent optimization control:
      i. Preprocessing stage setpoint: 15.07 °C;
      ii. Supply air dew point: 11.06 °C;
      iii. Supply air temperature: 21.00 °C.
   (d) Comparison of experimental results:

Figure 14 shows the control comparison of the air temperature variables under conventional PID control (precool, recool, and reheat) and under DDOA-based intelligent control (precool_DDOA, recool_DDOA, and reheat_DDOA), respectively. It can be observed that, under DDOA control, the air temperature in the recooling section converges to the set value more quickly, which is likely due to the lower setpoint of the precooling section PID. This also explains why the energy consumption is lower under the proposed DDOA control. Moreover, Figure 15 shows the comparison of the energy consumption curve between PID and DDOA control methods. It is obvious that the proposed DDOA method can reduce the energy consumption significantly, as compared to the PID control. To be specifically, the DDOA method can achieve up to 27.3% of energy savings as well as the related amount of overall saved money.

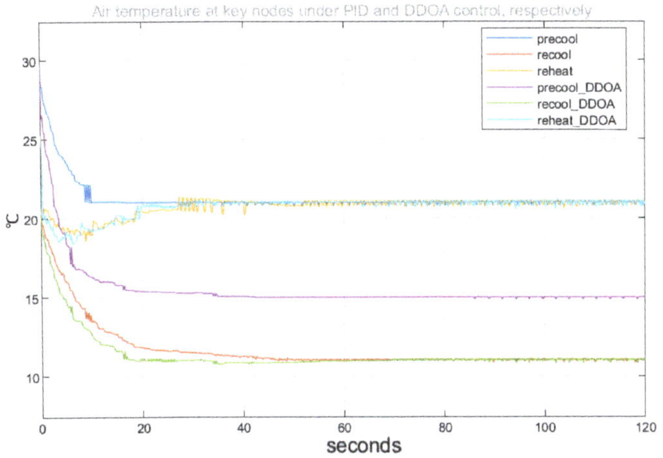

Figure 14. Comparison of the control result of air temperature variables (precool, recool, and reheat temperatures) between PID and the DDOA control methods.

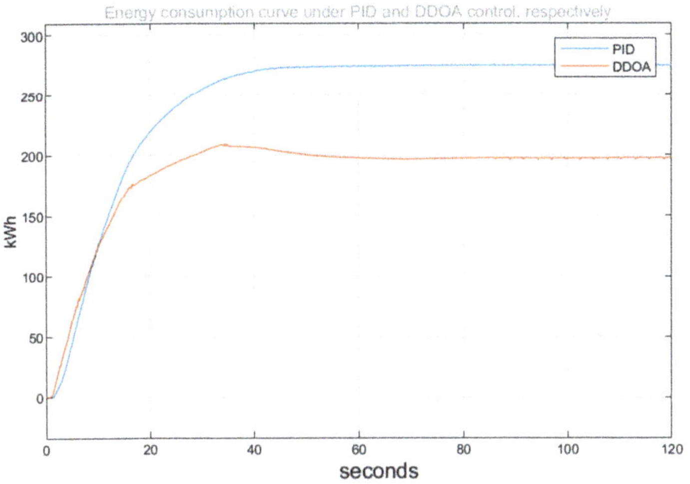

Figure 15. Comparison of the energy consumption curve between PID and DDOA control methods.

5.5. Detailed Analysis and Discussion of Experimental Results

We have conducted two kinds of experiments in this study: (1) the theoretical experiment based on Pymoo, and (2) the simulation experiment based on Simulink. These two experiments complement each other. The theoretical experiment based on Pymoo directly uses the optimal solution obtained by the intelligent optimization algorithm as the final result. However, due to the limitation in the accuracy of the prediction model, there is a slight deviation between this result and the actual outcome. Therefore, it is necessary to conduct a simulation experiment to further analyze the performance of the proposed algorithm. Using the result from the intelligent optimization algorithm as the foundation, we fine-tune it with a PID controller in the simulation environment to better match the actual production conditions, and then compare it with the traditional pure PID control method. The results are shown in Figures 14 and 15. In Figure 15, it is visually evident that the proposed DDOA control method requires lower energy consumption compared to the traditional PID control algorithm, while Figure 14 explains the reason. For the traditional PID control method, the setpoints are typically determined based on expert experience, with the setpoints for the preprocessing stage set to be around 20 °C, as shown in recool, recool, and reheat of Figure 14. In contrast, the result obtained by the DDOA method adjusts the PID setpoints for the pre-processing stage to be around 15°C, as shown in recool_DDOA, recool_DDOA, and reheat_DDOA of Figure 14. Since the saturation vapor pressure of air increases exponentially with temperature, higher temperatures allow air to hold more water vapor. Therefore, setting the pre-processing stage setpoint to 20 °C leads to excessive humidification during the humidification stage, which results in the need for more energy for dehumidification during the cooling stage. The proposed DDOA method replaces expert experience with the intelligent optimization algorithm, and hence make the more reasonable PID setpoints, which helps avoid the aforementioned issue. Therefore, it is more energy-efficient compared to the traditional PID control algorithm.

6. Conclusions

This algorithm includes a method for modeling the relationship between the cold and hot coil opening and the energy based on machine learning and an optimization method for minimizing the energy cost of cold and hot water usage based on intelligent optimization algorithms. It successfully achieves real-time adjustment of the coil opening of the MAU in response to changes in outdoor air conditions. This optimization ensures that the supply air temperature and dew point meet the requirements, ultimately reducing the energy cost of the cold and hot water usage in the MAU.

The experimental results indicate that the intelligent optimization algorithm can output the coil opening, which minimizes the energy cost of the cold and hot water usage in the MAU. Additionally, the designed energy-saving optimization control method is compared with the commonly used PID control method through simulation experiments. The results demonstrate improvements in adjustment speed and energy cost of cold and hot water usage for the energy-saving control method compared to the PID control method. To be specific, the proposed strategy can achieve up to 27.3% of saved energy and the related amount of money.

Furthermore, it is very interesting to explore the integration of other energy-saving techniques into the MAU system. For example, natural cooling energy [47] and high-efficient radiant cooling techniques [48] are essential to mitigate energy resources shortage crises, a breakthrough for traditional convective air conditioning systems in isolation between indoor and outdoor environment with high indoor air quality.

Author Contributions: Conceptualization, S.Z.; Methodology, S.Z. and M.C.; Software, S.L. and W.W.; Validation, S.L. and W.W.; Formal analysis, W.W. and M.C.; Writing—original draft, S.Z.; Writing—review and editing, S.Z., W.W. and M.C.; Supervision, M.C.; Funding acquisition, S.Z. All authors have read and agreed to the published version of the manuscript.

Funding: This research was supported by the Natural Science Foundation of China (No. 62206113) and the Natural Science Foundation of Jiangsu Province, China (No. BK20230923 and No. BK20221067).

Data Availability Statement: Data are contained with the article.

Conflicts of Interest: The authors declare no conflicts of interest. The funders had no role in the design of the study; in the collection, analyses, or interpretation of the data; in the writing of the manuscript; nor in the decision to publish the results.

References

1. Zhuang, C.; Shan, K.; Wang, S. Coordinated demand-controlled ventilation strategy for energy-efficient operation in multi-zone cleanroom air-conditioning systems. *Build. Environ.* **2021**, *191*, 107588. [CrossRef]
2. Zhao, W.; Li, H.; Wang, S. Energy performance and energy conservation technologies for high-tech cleanrooms: State of the art and future perspectives. *Renew. Sustain. Energy Rev.* **2023**, *183*, 113532. [CrossRef]

3. Zhang, X.; Zhou, Y. Waste-to-energy (W2E) for renewable-battery-FCEV-building multi-energy systems with combined thermal/power, absorption chiller and demand-side flexibility in subtropical climates. *Energy Build.* **2024**, *307*, 113949. [CrossRef]
4. Grassi, B.; Piana, E.A.; Lezzi, A.M.; Pilotelli, M. A review of recent literature on systems and methods for the control of thermal comfort in buildings. *Appl. Sci.* **2022**, *12*, 5473. [CrossRef]
5. Li, L.; Gao, Y.; Ning, D.Z.; Yuan, Z.M. Development of a constraint non-causal wave energy control algorithm based on artificial intelligence. *Renew. Sustain. Energy Rev.* **2021**, *138*, 110519. [CrossRef]
6. Prince; Hati, A.S. A comprehensive review of energy-efficiency of ventilation system using Artificial Intelligence. *Renew. Sustain. Energy Rev.* **2021**, *146*, 111153. [CrossRef]
7. Sayed, E.T.; Olabi, A.G.; Elsaid, K.; Al Radi, M.; Semeraro, C.; Doranehgard, M.H.; Eltayeb, M.E.; Abdelkareem, M.A. Application of artificial intelligence techniques for modeling, optimizing, and controlling desalination systems powered by renewable energy resources. *J. Clean. Prod.* **2023**, *413*, 137486. [CrossRef]
8. Wang, X.; Liu, S.; Xiong, L.; Wu, D.; Zhang, Y. Research on intelligent regulation of air conditioning energy saving based on human thermal comfort. *J. Ambient. Intell. Humaniz. Comput.* **2021**, 1–14. [CrossRef]
9. Bao, L.; Li, N.; Qin, H. Research on Energy-Saving Optimization of Variable Flow Air Conditioning Chilled Water System. In *Future Energy: Challenge, Opportunity, and, Sustainability*; Springer: Berlin/Heidelberg, Germany, 2023; pp. 187–198.
10. Peng, Q.; Yang, G.; Du, Q. Energy-saving Optimization of Frequency-variable Heat Pump Air Conditioning System for Electric Vehicles Based on a Genetic Algorithm. *Int. J. Automot. Technol.* **2023**, *24*, 1543–1551. [CrossRef]
11. Sun, T.; Huang, X.; Liang, C.; Liu, R.; Yan, Y. Energy Consumption and Energy Saving Analysis of Air-Conditioning Systems of Data Centers in Typical Cities in China. *Sustainability* **2023**, *15*, 7826. [CrossRef]
12. Zhou, Y.; Dan, Z.; Yu, X. Climate-adaptive resilience in district buildings and cross-regional energy sharing in Guangzhou-Shenzhen-Hong Kong Greater Bay Area. *Energy Build.* **2024**, *308*, 114004. [CrossRef]
13. Zhu, S.; Sun, N.; Lv, S.; Chen, K.; Fang, W.; Cao, L. Research progress on intelligent optimization techniques for energy-efficient design of ship hull forms. *J. Membr. Comput.* **2024**, 1–17. [CrossRef]
14. Fattahi, M.; Sharbatdar, M. Machine-learning-based personal thermal comfort modeling for heat recovery using environmental parameters. *Sustain. Energy Technol. Assessments* **2023**, *57*, 103294. [CrossRef]
15. Zhou, Y.; Zheng, S.; Hensen, J.L. Machine learning-based digital district heating/cooling with renewable integrations and advanced low-carbon transition. *Renew. Sustain. Energy Rev.* **2024**, *199*, 114466. [CrossRef]
16. Cui, M.; Li, L.; Zhou, M.; Abusorrah, A. Surrogate-assisted autoencoder-embedded evolutionary optimization algorithm to solve high-dimensional expensive problems. *IEEE Trans. Evol. Comput.* **2022**, *26*, 676–689. [CrossRef]
17. Zhou, M.; Cui, M.; Xu, D.; Zhu, S.; Zhao, Z.; Abusorrah, A. Evolutionary optimization methods for high-dimensional expensive problems: A survey. *IEEE/CAA J. Autom. Sin.* **2024**, *11*, 1092–1105. [CrossRef]
18. Sun, N.; Zhu, S. Modeling and Optimization of Energy Consumption in Data-driven Factory Air Conditioning System. In Proceedings of the Genetic and Evolutionary Computation Conference Companion, Melbourne, VIC, Australia, 14–18 July 2024; pp. 107–108.
19. Zhu, S.; Xu, L.; Goodman, E.D. Hierarchical Topology-Based Cluster Representation for Scalable Evolutionary Multiobjective Clustering. *IEEE Trans. Cybern.* **2022**, *52*, 9846–9860. [CrossRef]
20. Zhu, S.; Xu, L.; Goodman, E.D.; Lu, Z. A New Many-Objective Evolutionary Algorithm Based on Generalized Pareto Dominance. *IEEE Trans. Cybern.* **2022**, *52*, 7776–7790. [CrossRef]
21. Maskooki, A.; Deb, K.; Kallio, M. A customized genetic algorithm for bi-objective routing in a dynamic network. *Eur. J. Oper. Res.* **2022**, *297*, 615–629. [CrossRef]
22. Zhao, Z.; Zhou, M.; Liu, S. Iterated greedy algorithms for flow-shop scheduling problems: A tutorial. *IEEE Trans. Autom. Sci. Eng.* **2022**, *19*, 1941–1959. [CrossRef]
23. Tan, Y.; Lian, L. *Engineering Thermodynamics*; Chemical Industry Press: Beijing, China, 2010.
24. Kamiran, F.; Calders, T. Data preprocessing techniques for classification without discrimination. *Knowl. Inf. Syst.* **2011**, *33*, 1–33. [CrossRef]
25. Noi, P.T.; Kappas, M. Comparison of Random Forest, k-Nearest Neighbor, and Support Vector Machine Classifiers for Land Cover Classification Using Sentinel-2 Imagery. *Sensors* **2017**, *18*, 18. [CrossRef] [PubMed]
26. Gray, J.B. Introduction to Linear Regression Analysis. *Technometrics* **2002**, *44*, 191–192. [CrossRef]
27. Hoerl, A.E.; Kennard, R.W. Ridge Regression: Biased Estimation for Nonorthogonal Problems. *Technometrics* **2000**, *42*, 80–86. [CrossRef]
28. Lee, J.H.; Shi, Z.; Gao, Z. On LASSO for predictive regression. *J. Econom.* **2022**, *229*, 322–349. [CrossRef]
29. Beskopylny, A.N.; Stel'makh, S.A.; Shcherban', E.M.; Mailyan, L.R.; Meskhi, B.; Razveeva, I.; Chernil'nik, A.; Beskopylny, N. Concrete strength prediction using machine learning methods CatBoost, k-nearest neighbors, support vector regression. *Appl. Sci.* **2022**, *12*, 10864. [CrossRef]
30. Charbuty, B.; Abdulazeez, A.M. Classification Based on Decision Tree Algorithm for Machine Learning. *J. Appl. Sci. Technol. Trends* **2021**, *2*, 20–28. [CrossRef]
31. Belgiu, M.; Drăguţ, L.D. Random forest in remote sensing: A review of applications and future directions. *Isprs J. Photogramm. Remote. Sens.* **2016**, *114*, 24–31. [CrossRef]

32. Chen, T.; Guestrin, C. XGBoost: A Scalable Tree Boosting System. In Proceedings of the 22nd ACM SIGKDD International Conference on Knowledge Discovery and Data Mining, San Francisco, CA, USA, 13–17 August 2016; pp. 785–794. [CrossRef]
33. Ke, G.; Meng, Q.; Finley, T.; Wang, T.; Chen, W.; Ma, W.; Ye, Q.; Liu, T.Y. Lightgbm: A highly efficient gradient boosting decision tree. In Proceedings of the 31st International Conference on Neural Information Processing Systems (NIPS'17), Long Beach, CA, USA, 4–9 December 2017; Curran Associates Inc., Red Hook, NY, USA, 2017; pp. 3149–3157.
34. Hastie, T.J.; Rosset, S.; Zhu, J.; Zou, H. Multi-class AdaBoost. *Stat. Its Interface* **2009**, *2*, 349–360. [CrossRef]
35. Smola, A.; Scholkopf, B. A tutorial on support vector regression. *Stat. Comput.* **2004**, *14*, 199–222. [CrossRef]
36. Deb, K.; Pratap, A.; Agarwal, S.; Meyarivan, T. A fast and elitist multiobjective genetic algorithm: NSGA-II. *IEEE Trans. Evol. Comput.* **2002**, *6*, 182–197. [CrossRef]
37. Cheng, R.; Jin, Y.; Olhofer, M.; sendhoff, B. Test Problems for Large-Scale Multiobjective and Many-Objective Optimization. *IEEE Trans. Cybern.* **2017**, *47*, 4108–4121. [CrossRef] [PubMed]
38. Ghosh, A.; Deb, K.; Goodman, E.; Averill, R. A user-guided innovization-based evolutionary algorithm framework for practical multi-objective optimization problems. *Eng. Optim.* **2023**, *55*, 2084–2096. [CrossRef]
39. Zhang, Q.; Li, H. MOEA/D: A Multiobjective Evolutionary Algorithm Based on Decomposition. *IEEE Trans. Evol. Comput.* **2007**, *11*, 712–731. [CrossRef]
40. Blank, J.; Deb, K. Pymoo: Multi-objective optimization in python. *IEEE Access* **2020**, *8*, 89497–89509. [CrossRef]
41. Price, K.V. Differential Evolution. In *Handbook of Optimization: From Classical to Modern Approach*; Springer: Berlin/Heidelberg, Germany, 2013; pp. 187–214. [CrossRef]
42. Cai, Z.; Gao, S.; Yang, X.; Zhou, M. Multiselection-Based Differential Evolution. *IEEE Trans. Syst. Man, Cybern. Syst.* **2024**. [CrossRef]
43. Sun, J.; Fang, W.; Wu, X.; Palade, V.; Xu, W. Quantum-behaved particle swarm optimization: Analysis of individual particle behavior and parameter selection. *Evol. Comput.* **2012**, *20*, 349–393. [CrossRef]
44. Li, X.; Fang, W.; Zhu, S. An improved binary quantum-behaved particle swarm optimization algorithm for knapsack problems. *Inf. Sci.* **2023**, *648*, 119529. [CrossRef]
45. Meshram, P.M.; Kanojiya, R.G. Tuning of PID controller using Ziegler-Nichols method for speed control of DC motor. In Proceedings of the IEEE-International Conference On Advances In Engineering, Science And Management (ICAESM-2012), Nagapattinam, India, 30–31 March 2012; pp. 117–122.
46. Borase, R.P.; Maghade, D.K.; Sondkar, S.Y.; Pawar, S.N. A review of PID control, tuning methods and applications. *Int. J. Dyn. Control.* **2021**, *9*, 818–827. [CrossRef]
47. Lyu, W.; Li, X.; Shi, W.; Wang, B.; Huang, X. A general method to evaluate the applicability of natural energy for building cooling and heating: Revised degree hours. *Energy Build.* **2021**, *250*, 111277. [CrossRef]
48. Hassan, M.A.; Abdelaziz, O. Best practices and recent advances in hydronic radiant cooling systems–Part II: Simulation, control, and integration. *Energy Build.* **2020**, *224*, 110263. [CrossRef]

sustainability

MDPI

Article

Research on the Optimized Design of Medium and Deep Ground-Source Heat Pump Systems Considering End-Load Variation

Jianlin Li [1], Xupeng Qi [1,2,*], Xiaoli Li [2], Huijie Huang [3,*] and Jian Gao [4]

1 Beijing Future Electrochemical Energy Storage System Integration Technology Innovation Center, North China University of Technology, Beijing 100144, China; dkyljl@163.com
2 Department of Information, Beijing University of Technology, Beijing 100124, China; lixiaolibjut@bjut.edu.cn
3 School of Civil Engineering, Liaoning Engineering Technology University, Fuxin 123000, China
4 School of Resources and Civil Engineering, Liaoning Institute of Science and Technology, Benxi 117004, China; lntugaojian@outlook.com
* Correspondence: qixupeng007@163.com (X.Q.); huanghuijie1022@163.com (H.H.)

Abstract: Ground-source heat pump (GSHP) systems with medium-depth and deeply buried pipes in cold regions are highly important for addressing global climate change and the energy crisis because of their efficient, clean, and sustainable energy characteristics. However, unique geological conditions in cold climates pose serious challenges to the heat transfer efficiency, long-term stability, and adaptability of systems. This study comprehensively analyses the effects of various factors, including well depth, inner-to-outer tube diameter ratios, cementing material, the thermal conductivity of the inner tube, the flow rate, and the start–stop ratio, on the performance of a medium-depth coaxial borehole heat exchanger. Field tests, numerical simulations, and sensitivity analyses are combined to determine the full-cycle thermal performance and heat-transfer properties of medium-depth geological formations and their relationships with system performance. The results show that the source water temperature increases by approximately 4 °C and that the heat transfer increases by 50 kW for every 500 m increase in well depth. The optimization of the inner and outer pipe diameter ratios effectively improves the heat-exchange efficiency, and a larger pipe diameter ratio design can significantly reduce the flow resistance and improve system stability. When the thermal conductivity of the cementing cement increases from 1 W/(m·K) to 2 W/(m·K), the outlet water temperature at the source side increases by approximately 1 °C, and the heat transfer increases by 13 kW. However, the improvement effect of further increasing the thermal conductivity on the heat-exchange efficiency gradually decreases. When the flow rate is 0.7 m/s, the heat transfer is stable at approximately 250 kW, and the system economy and heat-transfer efficiency reach a balance. These findings provide a robust scientific basis for promoting medium-deep geothermal energy heating systems in cold regions and offer valuable references for the green and low-carbon transition in building heating systems.

Keywords: medium deep; rock and soil mass heat transfer; cold region; ground-source heat pump

1. Introduction

Energy shortages and climate change have become global challenges, posing serious threats to sustainable socioeconomic development and necessitating an urgent transition in energy systems [1,2]. According to statistics, the thermal energy used for cooling and heating accounts for more than 50% of the global end-use energy consumption [3]. In the context of the global zero-carbon emission target, reducing energy consumption for building heating is widely recognized as a critical strategy for achieving low-carbon development [4]. Building heating systems, as an important part of energy consumption, constitute not only one of the main areas of energy consumption [5] but also one of the main sources of carbon emissions. However, conventional building heating continues to depend heavily on fossil fuels, resulting in substantial greenhouse gas emissions and worsening energy security concerns [6]. Therefore, it is imperative to introduce clean and renewable energy sources to replace traditional fossil fuel-based heating systems to mitigate carbon emissions and ease environmental pressures effectively.

Geothermal energy, as an abundant and sustainable clean energy source, has garnered increasing attention because of its widespread availability and minimal environmental footprint [6]. In recent years, ground-source heat pump (GSHP) technology, an important approach for developing and utilizing geothermal energy, has undergone rapid expansion in its applications across China and globally, particularly in building heating for cold regions, demonstrating significant advantages [7]. However, the diffusion of traditional groundwater source heat pump systems is geographically constrained by their dependence on groundwater resources and is severely limited in areas with scarce water resources or extreme climates [1]. Moreover, although shallow (depths ranging from 20–200 m) ground-source heat pump systems have shown better economic viability and feasibility in certain regions, the imbalance of the shallow heat source load in northern cold regions has significantly compromised their long-term performance and stability because of the significantly higher heating demand in winter than in summer [8]. Consequently, the development of medium and deep ground-source heat pump systems, which are better suited to severely cold climates, has emerged as a major focus and key trend in current research.

As an energy development method between shallow geothermal energy and deep high-temperature geothermal energy, the development depth of medium-depth geothermal energy heating is generally in the range of 200–2000 m, which can effectively balance the cost of resource development and the efficiency of energy output [9], and it has become a potentially preferable option for heating in severely cold areas. In recent years, domestic and international research on medium and deep ground-source heat pump systems has gradually increased, with a focus on the optimization of heat exchanger structures [10], energy efficiency improvement [11], and engineering economic analysis [12]. For example, Ma et al. [10] evaluated the thermal performance of deep coaxial borehole heat exchangers in the Songliao Basin via OpenGeoSys numerical simulations, providing a scientific basis for converting abandoned water wells into geothermal heat exchanger wells in Northeast China. Norouzi et al. [13] assessed the effectiveness of ground-source heat pumps in mitigating ground settlement in the permafrost zone by establishing a two-dimensional thermo-water mechanical coupled finite element model. Jathunge et al. [14] evaluated the long-term thermal performance of biased tube thermally active foundation piles in a residential GSHP system through numerical simulations, revealing the effects of different installation configurations and building loads on ground temperature variations and system efficiency. However, most current studies rely on theoretical models and laboratory simulations, and research on heat transfer characteristics under long-term operating conditions in severely cold regions is insufficient. Furthermore, the unique geological conditions and climatic environments of cold regions pose significant challenges to medium–deep ground-source heat pump systems, and improving the adaptability of these systems in these regions remains a critical scientific challenge that requires urgent attention.

In a medium-depth ground-source heat pump system, the borehole heat exchanger serves as the core heat exchanger device, facilitating geothermal energy extraction and transfer through heat exchange between the circulating fluid and the surrounding geological formation [6,15]. The structural design and operating conditions of a borehole heat exchanger directly impact heat-exchange efficiency [16], with well depth, the inner-to-outer pipe diameter ratio, the thermal conductivity of the inner pipe material, the flow rate, and the start–stop operation ratio being key factors influencing performance [8,17]. Especially in cold regions, the heat transfer characteristics of source-side geological formations are influenced not only by the borehole heat exchanger structure and operating conditions but also by the coupling action of complex factors such as freeze–thaw cycles and variations in soil thermophysical properties [18]. Feng et al. [8] evaluated the heating performance of a medium–deep ground-source heat pump system in a cold and arid region through experimental tests and numerical simulations, validating a new 3D numerical model that illustrates the effects of the insulation layer, well diameter, flow rate, and well depth on system performance. Shen et al. [17] investigated the effects of thermal conductivity, tube diameter, and flow rate on the heat extraction capacity of a U-shaped deep well heat exchanger via a segmented finite line source method. Li et al. [19] reported that increasing the thickness of the inner tube insulation and the outer diameter of the outer tube is an effective strategy for enhancing the thermal performance of a coaxial ground heat exchanger as demonstrated through simulations. Luo et al. [20] reported that by selecting the appropriate system design parameters, energy savings in a ground-source heat pump system based on deep-well heat exchangers can reach 21.8%. However, most of the current studies are based on theoretical models and laboratory simulations [21] and address the heat transfer characteristics under long-term operating conditions in cold areas. In addition, the special geological conditions and climate environment in cold regions pose unique challenges to medium and deep ground-source heat pump systems, and how to improve the adaptability of the system in these regions is still a key scientific problem that urgently needs to be solved.

The unique contribution of this paper is that we meet the special requirements of the deep ground-source heat pump system in the middle of cold areas and optimize the operation mode and design parameters of the system on the basis of annual dynamic load simulation. Unlike existing studies that focus on optimizing the heat exchanger design and improving energy efficiency, the innovation of this study is as follows: on the one hand, on the basis of traditional design parameters (such as well depth and pipe diameter ratio), several new operating parameters, such as cement thermal conductivity, flow velocity, and the start–stop ratio, are introduced; on the other hand, through sensitivity analysis, we quantify the influence of these key design parameters on the thermal efficiency and long-term stability of the system and propose specific optimization suggestions for cold areas. In particular, with respect to the application of a dynamic control strategy, this study, through the simulation of buildings by load and accurate adjustments of heat source heat-intake characteristics, significantly improve the stability and energy efficiency of the system, and for ground-source heat pumps in cold areas, it provides scientific guidance, scientific support for green low-carbon building heating, and the popularization and application of this technology in cold areas.

2. Numerical Modelling and Computational Methods

2.1. Calculation Model Fluency

As shown in Figure 1, this study is based on the structure and heat-transfer mechanism of the deep coaxial casing system of a ground-source heat pump. Moreover, to reduce the calculation cost, the three-dimensional model is simplified to establish a two-dimensional axisymmetric physical model. The simulation system is composed of a coaxial casing heat exchanger, a heat pump unit, and terminal equipment on the ground-source side. The heat exchanger consists of inner and outer tubes forming a coaxial casing, which facilitates heat exchange with the surrounding rock mass through thermal convection in the annular cavity between the outer and inner tubes.

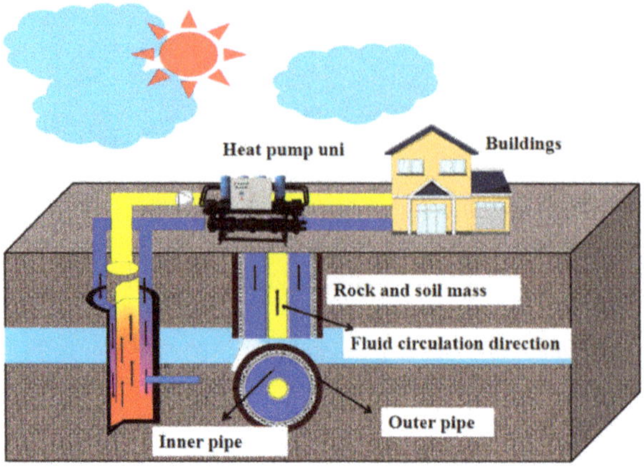

Figure 1. Schematic diagram of heat transfer medium circulation and heat transfer in a medium–deep casing ground-source heat pump system.

The heat exchange area of the coaxial casing heat exchanger (with a length of 2000 m, model height of 2200 m, outer tube radius of 177.8 mm, and inner tube radius of 110 mm) is axisymmetric. Combining the actual geological conditions and design parameters of the heat exchanger, this study uses CFD (Fluent) software (https://www.ansys.com/) to construct a two-dimensional physical model to simulate the heat-transfer process between the heat exchanger and the surrounding soil [22]. The model is discretized on a tetrahedral grid and implements local encryption processing for key areas. To reduce the calculation cost and ensure calculation accuracy, the deviation of the outlet within 1.5% is 100%, and the total grid reaches 2.73 million (as shown in Figure 2, where (a) is the overall grid model, (b) is the ground local grid model, and (c) is the ground structure model). This meshing strategy not only ensures high accuracy in heat transfer simulations but also significantly enhances computational efficiency. In terms of the grid division method, this study adopts a structured grid, which can effectively reduce the storage requirement and increase the computational speed because of its advantages in the processing of internode relationships. Compared with unstructured grids, structured grids exhibit a more regular topology and superior computational stability, making them especially suitable for accurate simulations of heat-transfer processes in local heat pump systems [23]. On the basis of the fixed grid, 150 s, 300 s, 600 s, and 1200 s are the time steps to calculate the deviation of the coaxial casing heat exchanger within 1.5% of the standard, and the calculation time step is selected to be 300 s.

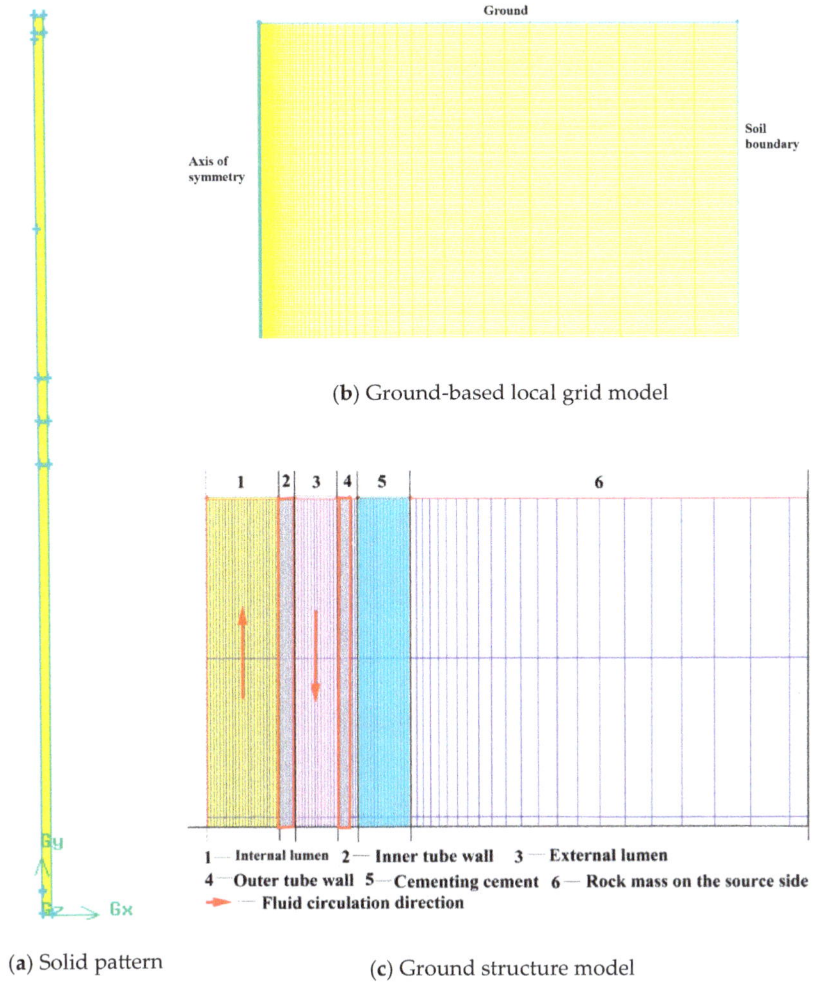

Figure 2. The erical model of buried pipe.

2.2. Model Assumptions

To increase the computational efficiency and versatility of the model while ensuring its applicability to actual engineering, the following pragmatic assumptions and simplifications are made in the modelling in this study [24]:

(1) The surface temperature is assumed to remain constant, and the effects of subsurface seepage are disregarded. Consequently, geotechnical heat transfer is modelled as a purely thermal conduction process. This assumption holds for regions with minimal seepage effects, significantly simplifying the coupled convection–heat conduction model and ensuring computational efficiency without compromising the simulation accuracy of heat-transfer behaviour.

(2) Stratigraphy with similar lithologies is combined and modelled as a homogeneous horizontally layered structure, which serves as the basis of the model. This simplification reduces the complexity of stratigraphic zoning while retaining the essential effects of major thermophysical differences and is commonly used in studies under similar geological conditions.

(3) The initial temperatures of the fluid in the casing, the backfill material, and the buried pipe are assumed to match the same horizontal geotechnical temperature, and the measured soil temperature is assumed to be uniform. This setting is based on the thermal equilibrium state, which accurately represents the initial thermal coupling relationship between the system and the geotechnical soil and simplifies the specification of the initial conditions.

2.3. Calculation Method

An analysis of the heat-transfer process between the heat exchanger and the geotechnical body revealed that the fluid flows turbulently inside the heat exchanger. To simulate the fluid heat-transfer characteristics within the heat exchanger accurately, the standard κ-ε turbulence model is employed in this study. Compared with large eddy simulation (LES), the standard κ-ε model significantly reduces the computational volume while ensuring accuracy, making it suitable for efficiently solving complex engineering problems. To further increase computational efficiency and stability, the model is solved via a simple method for pressure–velocity coupling [25–27]. The governing equations of the standard κ-ε model are as follows:

The continuity equation for the flow of water in the casing is given by

$$\frac{\partial \rho}{\partial t} + \frac{\partial}{\partial x_i}(\rho v_i) + \frac{\partial}{\partial x_i}(\rho v_j) = 0. \tag{1}$$

The momentum equation is as follows:

$$\frac{\partial(\rho v_i)}{\partial t} + \frac{\partial}{\partial x_j}(\rho v_i v_j) = -\frac{\partial P}{\partial x_i} + \frac{\partial}{\partial x_j}\left(\mu \frac{\partial v_i}{\partial x_j}\right) - \frac{\partial \tau_{ij}}{\partial x_j}. \tag{2}$$

Included among these is $\tau_{ij} = -\rho\overline{v'_i v'_j}$.

The energy equation is as follows:

$$\frac{\partial(\rho T)}{\partial t} + \frac{\partial}{\partial x_j}(\rho v_i T) = \lambda / C_P \frac{\partial}{\partial x_j}(\mu \frac{\partial T}{\partial x_j}), \tag{3}$$

where ρ is the fluid density, kg/m^3; t is the time, s; ij is the velocity vector (i, j = 1, 2); x_i, x_j is the coordinate direction (i, j = 1, 2), m/s; CP is the constant-pressure specific heat capacity; P is the pressure acting on the fluid micrometabolite, Pa; μ is the dynamic viscosity; τij is the Reynolds stress, N; and $\overline{v'_i}$ and $\overline{v'_j}$ are the mean velocities in the i, j directions, m/s.

Since the above system of equations is incomplete, certain assumptions must be made about the Reynolds stress, meaning that a new turbulence model is introduced to relate the pulsation and time-averaged turbulence values, which are used to close the system of equations [23]. Depending on the assumption and treatment of the Reynolds stress, the introduced turbulence models include the Reynolds stress model, zero-equation model, one-equation model, and two-equation model. In this study, the standard κ-ε model, one of the two-equation models, is used for the analysis. The standard κ-ε model introduces an equation for turbulent dissipation ε on the basis of the one-equation model, which has become the primary closure method in engineering turbulence owing to its inclusion of part of the history effect. The model is derived from simple turbulent flows and assumes isotropic turbulent viscosity, making it more suitable for simpler turbulent flows such as jets, pipe flows, and weak cyclones. The transport equations for the turbulent kinetic energy κ and turbulent dissipation ε in the standard κ–ε model are as follows:

$$\frac{\partial(\rho\kappa)}{\partial t}+\frac{\partial(\rho\kappa v_i)}{\partial x_i}=\frac{\partial}{\partial x_j}\left[\left(\mu+\frac{\mu_t}{\sigma_\kappa}\right)\frac{\partial\kappa}{\partial x_j}\right]+G_\kappa-\rho\varepsilon \tag{4}$$

$$\frac{\partial(\rho\varepsilon)}{\partial t}+\frac{\partial(\rho\varepsilon v_i)}{\partial x_i}=\frac{\partial}{\partial x_j}\left[\left(\mu+\frac{\mu_t}{\sigma_\varepsilon}\right)\frac{\partial\varepsilon}{\partial x_j}\right]+C_{1\varepsilon}\frac{\varepsilon}{\kappa}G_\kappa-C_{2\varepsilon}\rho\varepsilon\frac{\varepsilon^2}{\kappa} \tag{5}$$

Included among these are

$$\mu_t=\mu+\rho C_\mu\kappa^2/\varepsilon;\ G_\kappa=\mu_t\left(\partial v_i/\partial x_j+\partial v_j/\partial x_i\right)/\left(\partial v_i/\partial x_j\right)$$

where μ_t is the turbulent viscosity; C_μ is the viscosity coefficient, where $C_\mu = 0.09$; G_κ is the production term of κ caused by the mean velocity gradient; σ_κ and σ_ε are the turbulent kinetic energy κ and turbulence dissipation ε corresponding to the Prandtl number, where $\sigma_\kappa = 1.0$ and $\sigma_\varepsilon = 1.3$; and $C_{1\varepsilon}$ and $C_{2\varepsilon}$ are model constants, where $C_{1\varepsilon} = 1.44$ and $C_{2\varepsilon} = 1.92$.

The standard κ-ε turbulence model is used to simulate the flow and heat transfer in a middle-deep coaxial bushing ground-source heat pump system. The model was chosen on the basis of its widely validated and reliable performance in dealing with high Reynolds number turbulent flows and heat exchange problems [28]. By solving the transport equation of turbulent kinetic energy (κ) and its dissipation rate (ε), the standard κ-ε model can effectively characterize the turbulent properties of fluids. Compared with more complex turbulence models such as LES, this model significantly reduces computational costs while ensuring computational accuracy, and it is particularly suitable for large-scale engineering simulations. In addition, the successful application of the model to pipeline flow and heat exchanger turbulence problems [29] further confirms its applicability in this study.

Under the flow conditions involved in this study, the fluid presents typical turbulent characteristics in a coaxial tube heat exchanger. Considering the high flow rate characteristics of the system, the standard κ-ε model can accurately capture the dynamic characteristics of turbulence in the heat exchanger, especially the coupling effect of thermal convection and thermal conductivity [30]. The model effectively describes the local heat exchange effect caused by turbulence, which is highly important for optimizing the heat-transfer efficiency and evaluating the stability of the system. Although the prediction accuracy of the standard κ-ε model in the near-wall flow and low-Reynolds-number regions has several limitations, the flow in this study mainly occurs in the high-Reynolds-number region ($Re > 10^4$), and the flow in the heat exchanger is relatively uniform, so the model can provide sufficient calculation accuracy.

Compared with direct numerical simulation (DNS) and large eddy simulation (LES), the standard κ-ε model has significant advantages in computational efficiency and is particularly suitable for large-scale engineering simulations. Considering that the main goals of this study are system design optimization and long-term stability analysis, the standard κ-ε model can significantly reduce computational resource consumption while ensuring sufficient accuracy. The results show that the model has good applicability in describing the turbulent flow characteristics and heat-transfer behaviour in heat exchangers and achieves an optimal balance between computational efficiency and accuracy [31]. Therefore, the standard κ-ε turbulence model used for numerical simulation in this study can effectively meet the research requirements and provide reliable results, laying a theoretical foundation for subsequent system optimization and performance evaluation.

2.4. Fixed Solution Conditions

In this study, a medium-depth coaxial casing ground-source heat pump system in the Changchun region (43°05′–45°15′ N, 124°18′–127°05′ E) of China was employed as a case study, with field engineering and drilling construction logging data used to set the model parameters. Changchun is located in the southwest of the Songliao Basin and is characterized by cold winters, an average annual temperature of 4.6 °C, and warm summers, which are typical of cold regions. The climate, geophysical parameters, and boundary conditions in the model were established on the basis of the actual geological and climatic characteristics of Changchun, as detailed in Table 1.

Table 1. Climatic parameters in the Changchun area.

Thermal Parameters	Summertime	Winters
Calculated outdoor temperature for air conditioning, °C	30.5	−24.3
Average outdoor wind speed, m/s	3.2	3.7
Atmospheric pressure, Hpa	978.4	994.4
Air-conditioning indoor design temperature, °C	26	20
Operating cycle, day	--	169

In setting the thermophysical parameters, particular attention was given to the variations in the thermal conductivity and specific heat capacity of the surrounding rock at different depths, and the stratigraphic temperature gradient (0.0272 K/m) was precisely defined, as shown in Table 2. The high accuracy of the heat-transfer model was ensured by the detailed parameterization of components such as geotechnical layers, fluids, and pipes.

Table 2. Mid-depth casing heat exchanger constituent materials and thermal–physical parameters of the surrounding rock.

Serial Number	Part Name	Typology	Value of Each Parameter		
			Densities (kg/m³)	Specific Heat (J/(kg·k))	Thermal Conductivity (W/(m·k))
1	Water	fluids	998.2	4182	0.6
2	Perimeter rock 0–40 m	solid	1850	1840	1.9
3	Perimeter rock 0–850 m	solid	2600	1580	2.42
4	Perimeter rock 850–95 m	solid	2600	1580	2.42
5	Perimeter rock 950–1050 m	solid	2600	1580	2.75
6	Perimeter rock 1050–2050 m	solid	2600	1580	2.93
7	Thermal insulation (rubber and plastic)	solid	100	850	0.033
8	Inner tube (PE)	solid	950	2300	0.42
9	Outer pipe (J55 steel pipe)	solid	950	3800	58

In setting the numerical simulation conditions (see Table 3), the study included key factors such as the inlet and outlet flow rates, boundary temperature, and heat-transfer coefficient to ensure the reliability and representativeness of the simulation results. The inlet temperature was set to 278.15 K, the flow rate was set to 0.3 m/s, the ground heat-transfer coefficient was set to 15 W/(m^2·K), and the bottom heat flow was set to 0.0717 W/m^2, among other values, to fully account for the influence of actual operating conditions on system performance.

Table 3. Numerical simulation condition settings.

Serial Number	Part Name	Mold	Number
1	Inlet	Velocity inlet	T = 278.15 K; v = 0.3 m/s
2	Exits	Outflow	Flow rate weighting = 1
3	Ground level	Wall	h = 15 W/(m^2·K); T = 258.15 K
4	Bottoms	Wall	Q = 0.0717 W/m^2
5	Central axis	Axis	--
6	Soil boundaries	Wall	Q = 0 W/m^2
7	Initial stratigraphic temperature	--	T = 281.2 + 0.02737 × H

2.5. Model Validation

To verify the accuracy and reliability of the numerical model, the actual operation data of medium and deep coaxial-tube heat exchangers in Shandong were used for comparative analysis. Specifically, we chose a dataset that actually operates in the region, which contains measured values for the source-side outlet temperature. By comparing the simulation results with the measured data, the adaptability and accuracy of the numerical model under different operating conditions were verified.

First, the CFD software Fluent was used for numerical simulation, and the boundary conditions and operating parameters were set to be the same as those used in the actual project, including the inlet temperature, flow rate, thermal conductivity of rock and soil, pipe material and size, etc. To ensure the representativeness of the simulation, all model parameters were set according to the actual geological conditions and climate characteristics of Shandong.

In the verification process, we focused on the trend of the source-side outlet temperature, especially the temperature change over different time periods. The simulation results are highly consistent with the measured data, and the maximum temperature difference is only 273.57 K. Figure 3 shows the comparison between the simulated results of the source-side outlet temperature and the measured data. The verification shows that the model can accurately capture the dynamic changes in the heat-transfer process, and the simulated temperature change curve is consistent with the change trend of the measured data [8], which proves the reliability of the proposed model.

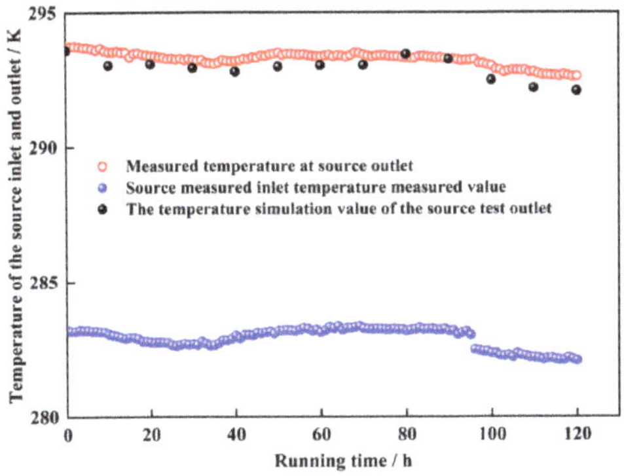

Figure 3. Simulated and measured fits of the outlet temperature at the source side.

3. Results and Analyses

In this study, 30 and 160 days were selected as the time scales for both the initial operation phase of the simulation system and the long-term operation throughout the heating season. The choice of both timescales aimed to capture the heat-transfer properties and performance changes of the system at different stages. The 30-day time scale was mainly used to simulate the system during the adaptation period after the start, evaluating the effects of thermal inertia and load fluctuations in the initial stage on the system response, especially as the system moves from the thermal equilibrium of the cold start. The 160-day time scale covered the whole heating season, reflecting the stability and efficiency of the system in long-term operation, especially in how the system maintains thermal balance and efficient operation in the case of large load fluctuations. Through simulations on both timescales, we were able to comprehensively assess the short-term dynamic response and long-term stability of the system, providing strong support for design optimization.

3.1. *Effects of Well Depth and the Internal-to-External Pipe Diameter Ratio*

Figure 4a illustrates the impact of varying well depths on the source-side effluent temperature of the medium-depth cased geothermal heat pump system. The source-side effluent temperature displays a decreasing trend that is consistent with a power function as the operation time progresses. In the first 7 days, the effluent temperature of the source side decreased rapidly, accounting for approximately 80% of the total temperature drop, indicating that the system was affected by a large thermal inertia when it entered the thermal equilibrium state at the initial stage. As the operation time continues, the temperature change gradually diminishes, and the source-side effluent temperature becomes nearly stable after 30 days, indicating that the system has reached thermal equilibrium. With increasing well depth from 1000 m to 2500 m, the formation temperature at the bottom increased from 308.54 K to 349.59 K, and the increase in well depth significantly increased the temperature of the underground thermal reservoir, thus improving the heat-exchange efficiency. With increasing well depth, the heat exchange amount between the casing heat exchanger and the rock and soil mass gradually increases, leading to a linear increase in the outlet water temperature on the source side. According to the experimental results, for every 500 m increase in well depth, the source-side outlet temperature increased by approximately 4 °C, indicating that the influence relationship of well depth on the effluent temperature on the source side is linear. At the beginning of operation, the difference in the source-side outlet temperatures among the different well depths was approximately 10.2 °C, which was 10.5 °C after 7 days and increased to 10.8 °C after 30 days. We show that the effect of well depth changes on the source-side outlet temperature stabilizes [9] with running time.

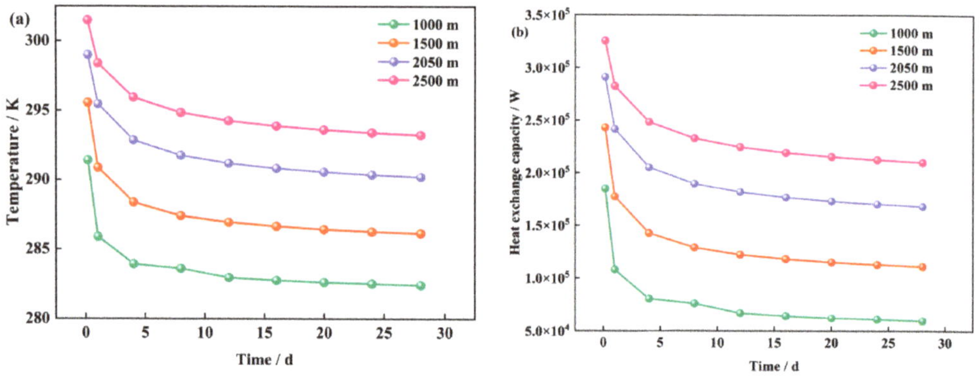

Figure 4. Effects of different well depth variations on the heat-transfer performance of the casing: (**a**) source-side effluent temperature and (**b**) heat-exchange capacity.

Figure 4b shows the influence of well depth on the heat exchange of the casing heat exchanger, and the change trend of heat exchange is consistent with the outlet temperature of the source side. In the initial stage of system operation, the decrease in heat exchange is large, accounting for approximately 80% of the total decrease, which is related to the thermal inertia and initial adaptation process of the system. As time progresses, the heat-exchange capacity stabilizes, indicating that the heat-exchange efficiency of the system gradually improves. Under different well depths, the change trends of heat change are similar and increase with increasing well depth. In the initial stage, when the well depth increases from 1000 m to 2500 m, the heat change decreases from 70 kW to 40 kW, with a decrease of 30 W/m. After 7 days, the heat-exchange capacity begins to stabilize with an increase of approximately 50 kW, and after 30 days, the heat-exchange capacity stabilizes at approximately 50 kW. This shows that the temperature difference in the system is significant, but after entering the stability period, the temperature difference is stable, and the heat exchange is maintained at a relatively stable level. After entering the stability period, the temperature difference tends to stabilize, and the heat exchange also remains relatively stable. Under steady-state operation, the heat-transfer capacity increases by approximately 50 kW for every 500 m increase in well depth, and the final heat-transfer capacity increases from 59.6 kW to 210.3 kW, with a change of approximately 10 W/m. Research has shown that with increasing well depth [24], the heat-transfer efficiency significantly improves, and heat exchange increases linearly after stable operation.

Figure 5a illustrates the impact of different internal and external pipe diameter ratios on the source-side outlet water temperature. With increasing heat transfer time, the effluent temperature on the source side tends to decrease, and the initial temperature rapidly decreases. This phenomenon is related to the thermal inertia and initial regulation of the system. Under all other identical conditions, the outer diameters of 168.3 mm and 75 mm presented high source-side outlet temperatures compared with those of the other two diameter configurations. Specifically, the outlet temperatures of the three pipe diameter configurations stabilized with the system after 30 days of operation, and the temperature changed slightly and basically remained unchanged during the subsequent heating period (after 30 days of operation). This indicates that the impact of the pipe diameter configuration on the source-side outlet temperature gradually decreases during the long-term operation of the ground-source heat pump system. However, configurations with larger pipe diameters are more effective at maintaining a higher outlet temperature during the initial stage, which has certain advantages.

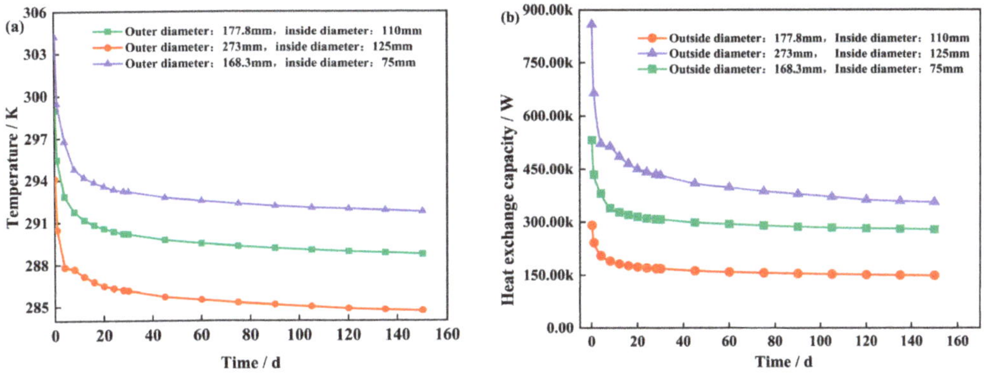

Figure 5. Influences of different inner and outer pipe diameter ratios on the heat-transfer performance of the casing: (**a**) source-side outlet water temperature and (**b**) heat-exchange capacity.

Figure 5b shows the influence of different inner and outer pipe diameter ratios on heat transfer. The heat transfer showed a power function decline with time, and the change was relatively rapid in the initial stage, which was related mainly to the change rate of the temperature difference. The results show that under the same conditions, the heat transfer of the combination of the outer diameter of 273 mm and the inner diameter of 125 mm is significantly greater than that of the other two configurations. The reason for this phenomenon is that a larger diameter ratio can provide a greater flow rate, thereby improving the heat-exchange efficiency. The large inner-to-outer pipe diameter ratio design increases the fluid flow area, reduces the flow rate, and enhances the heat-exchange capacity between the heat exchanger and the underground rock mass, thereby increasing heat transfer. However, the relationship between velocity and heat transfer is not purely linear. In this study, a lower flow rate helps reduce the thermal short-circuit effect caused by excessive turbulence and optimizes the flow path of the fluid, thereby improving the heat-exchange efficiency. The large inner-to-outer pipe diameter ratio provides a large flow rate and flow area, and in the initial stage, the temperature difference is large, resulting in a rapid increase in heat exchange. However, a smaller pipe diameter (such as an outer pipe diameter of 168.3 mm and an inner pipe diameter of 75 mm) also results in a better heat exchange effect under certain conditions. A smaller pipe diameter increases the flow rate, improves the degree of turbulence, and improves the heat-exchange efficiency, especially at moderate flow rates, which can maintain high heat exchange during the initial and stable stages. In the initial stage of the first 30 days, owing to the large temperature difference, the heat exchange between the fluid and the rock or soil body was relatively rapid, resulting in a rapid increase in heat exchange. After 30 days, the heat transfer changes of the various pipe diameter configurations tended to be stable, and the system entered the stable operation stage. During the subsequent heating period (from 30 days to 169 days), the heat transfer hardly changed and remained at a constant state. This shows that the influence of the pipe diameter on heat transfer is reflected mainly in the initial stage of the system, and a larger pipe diameter ratio can significantly improve the initial heat transfer and maintain the thermal efficiency of the system during long-term operation.

3.2. Effects of the Thermal Conductivity of the Cementing Cement and the Thermal Conductivity of the Inner Tubes

Figure 6a illustrates the impact of different solid-well cement thermal conductivities on the source-side effluent temperature. With the extension of running time, the effluent temperature on the source side decreases by a power function, and the temperature decreases rapidly in the initial stage, which indicates that it takes a long time for the system to reach thermal equilibrium in the starting stage. After 30 days of operation, the source-side effluent temperature stabilized, indicating that the system had reached a state of long-term stability. As the thermal conductivity of the cementing cement increased from 1 W/(mK) to 3 W/(mK), the outlet temperature on the source side increased accordingly. This may be because the increase in the thermal conductivity of the cementing cement improves the heat conduction efficiency, thus reducing the loss of heat in the system. However, when the thermal conductivity increases from 1 W/(mK) to 2 W/(mK), the source-side water temperature increases by approximately 1 °C after 30 days, but when the thermal conductivity increases to 3 W/(mK), the temperature increases by only approximately 0.3 °C. The thermal conductivity of the cementing cement has a certain nonlinear effect on the effluent temperature of the source side, especially after the thermal conductivity reaches a certain value, and the increase gradually decreases. The analysis indicates that since the integrated thermal conductivity of the geotechnical body at the source side is 1.93 W/(m-K), and the cementing cement layer is relatively thin, the thermal conductivity of the cementing cement has a limited impact on the heat-exchange efficiency when its value is not lower than that of the geotechnical body. The main function of cementing cement is to serve as a heat transmission channel rather than directly as a heat source, so the increase in its thermal conductivity has a gradual limit on the improvement in the overall heat-exchange efficiency.

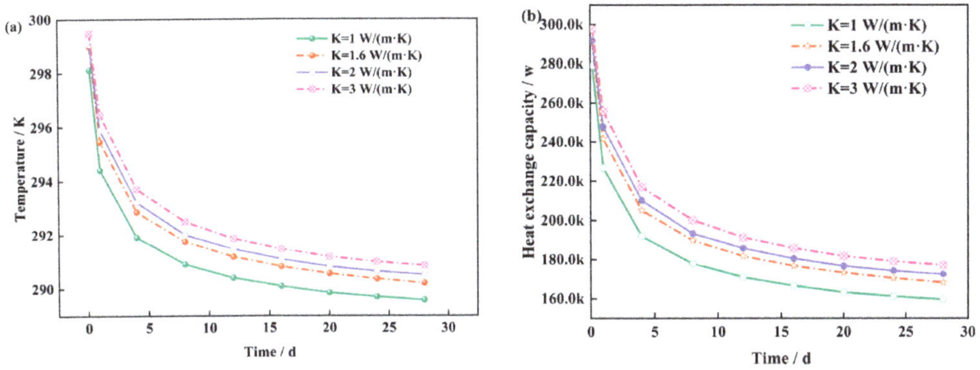

Figure 6. Influence of different cementing cement thermal conductivities on the casing heat-transfer performance: (**a**) source-side outlet water temperature and (**b**) heat-exchange volume.

Figure 6b shows the influence of different cementing cement thermal conductivities on the heat exchange of the casing heat exchanger. The trend in heat-exchange capacity mirrors that of the source-side effluent temperature, following a power function decrease, with the rate of change being more rapid during the initial phase. The change decreases after approximately 7 days and tends to stabilize after 30 days, indicating that the heat-exchange volume is closely related to the stability of the system. With increasing thermal conductivity of the cementing cement, heat exchange between the casing heat exchanger and the subsurface geotechnical body progressively increases. However, when the thermal conductivity of the cementing cement approaches or exceeds that of the formation, the rate of increase in heat-exchange capacity decreases significantly. When the thermal conductivity of the cementing cement increased from 1 W/(m-K) to 2 W/(m-K), the heat-exchange capacity increased by approximately 13 kW, whereas a further increase to 3 W/(m-K) resulted in only a 4 kW increase. This suggests that at higher thermal conductivities, the effectiveness of cementing cement tends to be saturated, and further increases in the thermal conductivity offer limited further enhancement of the heat-exchange capacity.

Figure 7a illustrates the impact of different inner pipe thermal conductivities on the source-side outlet water temperature. From the figure, it is evident that the source-side effluent temperature follows a power function decreasing trend with increasing operation time, experiencing a more rapid decrease during the initial phase, which indicates that the system undergoes a significant heat-transfer process during the start-up phase and gradually stabilizes, a phenomenon in agreement with [32]. After 30 days of operation, the source-side effluent temperature had levelled off, indicating that the system reached a state of long-term stability. The source-side effluent temperature decreased significantly with increasing thermal conductivity of the inner pipe from 0.21 W/(m-K) to 0.45 W/(m-K). Specifically, the source-side outlet temperature decreased by approximately 1.15 °C with every 0.12 W/(m-K) increase in thermal conductivity. This phenomenon can be attributed to the increase in the thermal conductivity of the inner pipe, which enhances heat exchange between the inner and outer pipes, amplifying the thermal short-circuit effect of the fluid inside the pipe [33], and thus, accelerating heat transfer and decreasing the source-side outlet temperature.

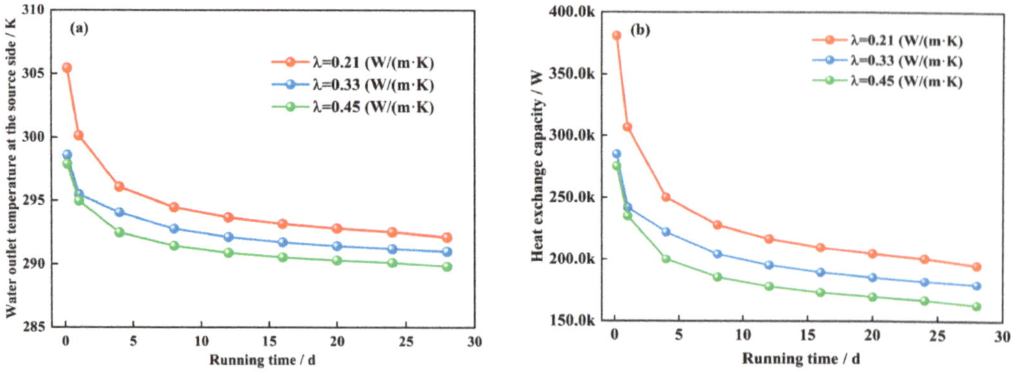

Figure 7. Effects of different inner tube thermal conductivities on the heat-transfer performance of the casing: (**a**) source-side outlet water temperature and (**b**) heat-exchange capacity.

Figure 7b further illustrates the impact of different inner pipe thermal conductivities on the heat-exchange volume. The trend of the heat-exchange volume is consistent with that of the source-side effluent temperature, which also decreases as a power function, with a more rapid change during the initial phase (approximately 7 days), after which it stabilizes around the 30th day. This trend indicates that the variation in heat-exchange volume is closely related to the thermal equilibrium state of the system. As the thermal conductivity of the inner pipe increases, the heat exchange between the inner and outer pipes increases, thereby increasing the heat-exchange efficiency of the system. However, although the increase in the thermal conductivity of the inner pipe increased the heat-exchange capacity, after 30 days of operation, the heat-exchange capacity of the buried pipe decreased by approximately 16 kW, as the thermal conductivity ranged from 0.21 to 0.45 W/(m-K). This phenomenon can be explained by the thermal short-circuit effect. As the thermal conductivity increases, heat transfer within the pipe accelerates, leading to a portion of the heat energy being lost in the transfer process, which affects the final heat-transfer effect. In conclusion, the increase in the thermal conductivity of the inner pipe has a dual effect on the ground-source heat pump system. On the one hand, increasing the thermal conductivity of the inner pipe can enhance heat exchange between the inner and outer pipes, thus improving the heat-transfer capability; on the other hand, it may intensify the thermal short-circuit effect, reducing the source-side outlet temperature and diminishing the heat-exchange capacity.

3.3. Effects of Variations in the Flow Rate and Start–Stop Ratio

Figure 8a shows the impact of different flow rates on the source-side effluent temperature. The source-side effluent temperature follows a distinct power function decreasing trend as the operation time increases, with the temperature decreasing more rapidly during the initial phase (day 1). This indicates that the source-side effluent temperature undergoes rapid heat exchange during the system's start-up period. With the change in flow rate, the relationship between the source-side effluent temperature and the flow rate also shows a dynamic change. Specifically, in the initial stage (approximately 1 to 7 days), the flow rate is directly correlated with the source-side effluent temperature; i.e., an increase in the flow rate leads to a decrease in the source-side effluent temperature [22]. In particular, when the flow rate increased from 0.3 m/s to 0.7 m/s, the source-side outlet temperature decreased by approximately 2 °C with each 0.2 m/s increase in flow rate. However, the relationship between the flow rate and outlet temperature gradually shifted over time, and after entering the stabilization stage, the effect of the flow rate on the source-side outlet temperature decreased, with the temperature difference tending to level off. After 30 days, the temperature difference decreased from 2.39 °C to 1.38 °C when the flow rate increased from 0.3 m/s to 0.7 m/s. This phenomenon shows that the change in flow velocity has a significant effect on the temperature, but during long-term operation, the influence of the flow velocity on the water temperature gradually weakens [22] and becomes stable.

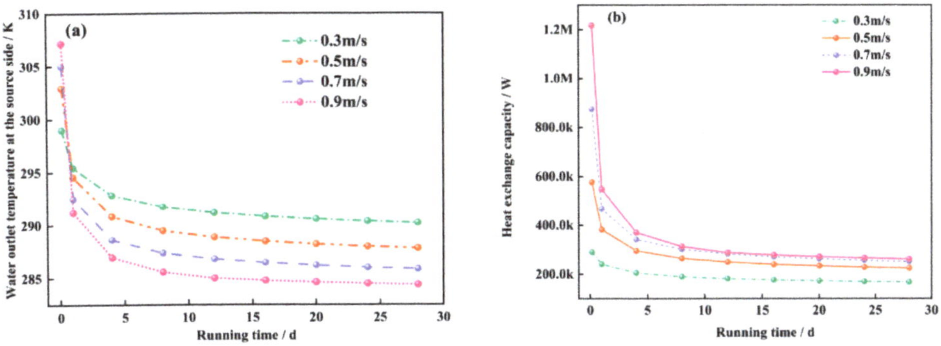

Figure 8. Effects of different water flow rates on the heat-transfer performance of the casing: (**a**) source-side outlet water temperature (**b**) heat-exchange capacity.

Figure 8b illustrates the impact of different flow rates on the heat-exchange capacity of the buried pipe. With the extension of the system running time, the heat change decreases in the power function and changes rapidly in the early stage (day 1). This shows that when the heat pump is started, the heat change is more drastic, but with increasing flow rate, the heat-exchange efficiency between the casing heat exchanger and the underground rock and soil mass gradually increases. After 30 days of operation, the heat-exchange capacity levelled off between 168 kW and 261 kW, with an increase in the flow rate leading to an overall increase in the heat-exchange capacity. However, the increase gradually decreased, especially when the flow rate increased from 0.7 m/s to 0.9 m/s, and the difference in heat change was only 10 kW. This indicates that although increasing the flow rate helps improve the heat exchange quantity, the increase in heat exchange quantity tends to stabilize once a certain flow rate is reached, and the effect of increasing the flow rate on the heat exchange effect begins to weaken [11]. While ensuring the efficient operation of the heat pump system, the choice of flow rate also needs to consider the balance of economy and energy efficiency. Since the energy consumption of the circulating pump is directly correlated with the flow rate, an excessive flow rate increases energy consumption, thus affecting the overall system economy. Therefore, setting the flow rate to 0.7 m/s is recommended. At this time, the heat exchange of the casing heat exchanger is stable at approximately 250 kW, which not only can ensure high heat-exchange efficiency but also optimize energy consumption and improve the economy and operation efficiency of the system.

Figure 9a illustrates the impact of varying start–stop ratios on the source-side outlet temperature of the buried pipe. The figure shows that the source-side outlet temperature of the ground-source heat pump system decreases following a power-function trend under continuous 24 h operation with a faster decrease in the initial period (i.e., the first 7 days). In contrast, the source-side outlet temperature under continuous operation with varying start–stop ratios (16:8 (2), 12:12 (1), and 8:16 (0.5)) exhibits an overall decreasing trend, although the change curve shows irregular fluctuations. This fluctuation is attributed primarily to the nonlinear coupling between water flow heat transfer in the casing, soil heat conduction, and start–stop variations. When the start–stop ratio is 2, the source-side outlet temperature tends to decrease continuously during the first 7 days of operation, but the temperature gradually begins to recover after 1 week. When the start–stop ratio is 1, the source-side temperature tends to stabilize overall, although a sudden increase is observed on the fifth day of operation. When the start–stop ratio is 0.5, the source-side temperature gradually decreases during the first four days, then increases steadily, returns to the initial temperature after one week, and eventually exceeds the initial value, where it continues to increase. The analysis indicated that when the heat exchanger operated continuously for 24 h, continuous heat exchange occurred between the casing heat exchanger and the source-side geotechnical body. Owing to the low thermal conductivity of the geotechnical body, heat replenishment at the far end of the stratum occurs more slowly, and heat exchanger extraction exceeds backfill replenishment. As the heat exchanger's ground temperature decreases to a certain point, the heat from the stratum's backfill reaches equilibrium with the extracted heat, causing the outlet temperature to stabilize. When the heat exchanger starts and stops periodically, it extracts heat from the ground during operation, while fluid circulation stops during the stopping period, with heat exchange gradually tending toward stabilization. However, heat transfer from the far end of the ground to the vicinity of the casing heat exchanger tube wall continues, resulting in a higher source-side outlet temperature when the heat pump resumes operation after a hiatus than during continuous operation. Additionally, during the shutdown process, high-temperature fluid at the bottom flows upwards, contributing to heat flow circulation within the heat exchanger, as the temperature of the underlying geotechnical body is higher than that of the upper layer. The fluid at the top of the heat exchanger, which is at a higher temperature than the source-side geotechnical body, begins to reverse the heat transfer to the lower-temperature geotechnical body, resulting in localized heat buildup in the stratum. Therefore, when the start–stop ratio is 0.5, the stopping time exceeds the running time, and the source-side outlet temperature may rise above the initial temperature after a period of operation.

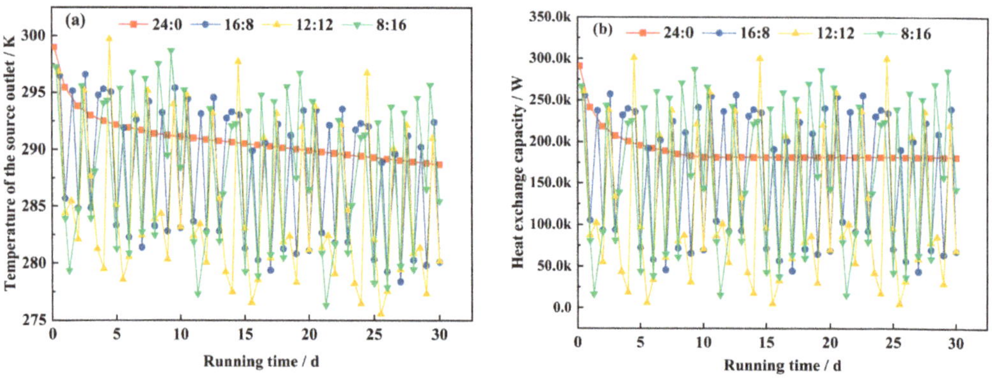

Figure 9. Effects of different start–stop ratios on the heat-transfer performance of the casing: (**a**) source-side effluent temperature and (**b**) heat-exchange capacity.

Figure 9b illustrates the effect of varying the start–stop ratio on the heat-exchange capacity of the buried pipe. The figure shows that under continuous 24 h operation, the heat-exchange capacity at the source side of the geothermal heat pump system follows a decreasing power-function trend with a more rapid decrease during the initial operation period (i.e., the first 7 days). The heat-exchange capacity of the buried pipe irregularly fluctuates and is influenced by the start–stop cycle of the heat pump. As the start–stop ratio gradually decreases, the heat extraction time on the source side decreases, and the heat-exchange volume gradually increases upon activation, stabilizing over time. Therefore, it is crucial to optimize the operation duration of the heat pump to increase the overall heat-exchange capacity of the system.

4. Discussion

4.1. Analysis of the Heat-Transfer Characteristics of the Radial Geotechnical Body on the Source Side During Full Heating Cycle Operation

The heat-exchange capacity of a heat exchanger is a key indicator for assessing its heat-transfer efficiency. The numerical simulation results for the source-side outlet temperature of the ground-source heat pump system throughout the heating season are presented in Figure 10a. As illustrated in the figure, the source-side outlet temperature initially decreases rapidly at the beginning of the heating season and then tends to stabilize. At the onset of operation, the source-side outlet temperature was highest at 299.01 K. However, as the system operated, heat was continuously extracted from the surrounding soil, resulting in a rapid decrease in the outlet temperature. In particular, the most significant temperature drop occurred within the first 30 days of operation, accounting for more than 80% of the total decrease, with the change being particularly dramatic during the first week. This shows that the system extracts heat at a high rate during the initial operation period, which has a significant effect on the thermal field of the surrounding soil. As the operation time increased, the temperature gradient of the surrounding soil gradually decreased, the heat-transfer rate decreased, and the outlet temperature stabilized within the range of 288.77–289.78 K, with an average decrease rate of only 0.008 K/d. This trend illustrates the difference between the thermal conductivity of the surrounding soil and the heat replenishment mechanism during the initial and long-term operation phases. The rapid cooling in the initial stage is due primarily to the quick extraction of the initial high-temperature heat storage in the soil, whereas the stabilization in the later stage reflects the buffering effect of the soil's thermal conductivity on long-term heat transfer. Figure 10b shows the trend of heat-exchange volume over time, which is consistent with the change in the source-side outlet temperature. At the beginning of operation, heat exchange was as high as 291 kW, but it decreased rapidly during the first 30 days of the heating season, with a decrease of approximately 90% of the total decrease, which corresponds to the rapid phase of the outlet temperature change. After one month of operation, the heat-exchange capacity stabilized between 148 kW and 162 kW, with an average decrease rate of approximately 0.117 kW/d, and it finally remained at approximately 155 kW during the stable operation phase. The rapid decline in heat-exchange capacity primarily results from the initial release of heat from the soil and the high heat-transfer demand during the heat pump's high-load operation. With the gradual establishment of the heat balance of the soil, the heat-exchange capacity of the heat exchanger subsequently tends to stabilize, indicating that the system has entered long-term sustainable operation. The heat-exchange performance and temperature change results in Figure 10 are based on the coaxial heat exchanger configuration with outer tube diameters of 177.8 mm and 110 mm, as mentioned in Section 2.1. Moreover, the thermal conductivity, velocity, and fluid properties are consistent with the assumptions in Section 2.1. Thus, the results of Figure 10 reflect the heat-transfer performance of this coaxial heat exchanger under specific operating conditions.

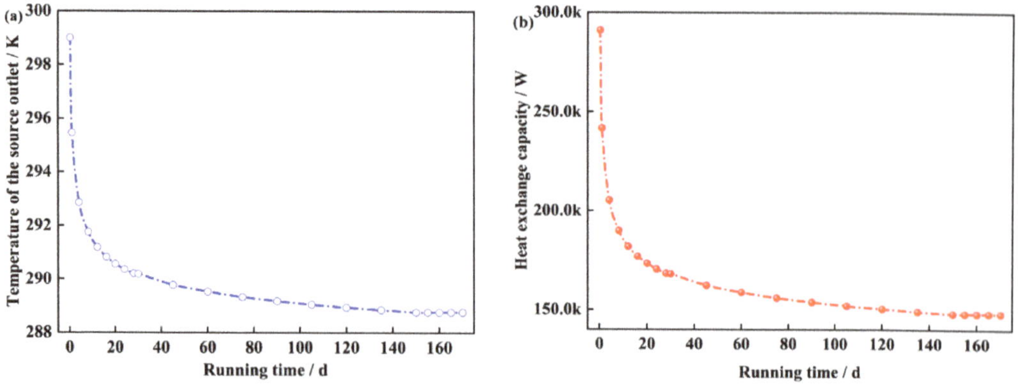

Figure 10. Heat-transfer performance during the full heating season. (**a**) Value of source-side outlet temperatures. (**b**) Amount of heat transferred.

4.2. Joint Operational Control Analysis Based on Year-Round Dynamic Simulation

The cooling and heating load of a building is constantly changing due to internal and external disturbances, especially in cold areas. Whether the control strategy of the ground-source heat pump system can effectively adapt to fluctuations in the building load and maintain the stability, energy savings, and economy of the system while meeting the needs of users is an important criterion for evaluating its effect.

The joint operation control strategy adopted in this study aims to adapt dynamically to the heating and cooling needs of buildings during different seasons and load fluctuations by automatically adjusting the type and quantity of heat sources. To verify the operation effect of this control strategy, we used Transy software (Version 18.0) to complete the building (the project is located in Jilin, China and covers an area of 95,721 m^2, with a construction area of 300,000 m^2, including hotels, offices, corporate culture halls, etc.). The annual hourly load was simulated. By establishing a mathematical model based on the actual building and inputting the climate parameters of the area where the building is located, the thermal parameters of the building envelope (see Table 1), and actual load data such as personnel and equipment lighting, the simulation results can accurately reflect the load characteristics of the building in different time periods. In this process, we took the thermal parameters and outdoor climate conditions of the building envelope as the external disturbance input, and at the same time, according to the hourly load requirements in the "Energy-saving Design Standard for Public Buildings" (GB50189–2015) [34], the personnel, equipment, and lighting were used as the internal disturbance inputs. In the end, 8760 h of energy consumption were calculated for the building throughout the year (Figure 11). Figure 12 shows the distribution of the building load throughout the year. The simulation results reveal that the annual cumulative heat load of the office building and exhibition hall is 17.3 million kWh, the cumulative cooling load is 4.25 million kWh, and the remaining load is generated by the hotel. These results provide basic data support for the control strategy proposed in this study and provide a theoretical basis for load regulation and energy savings optimization in practical engineering applications.

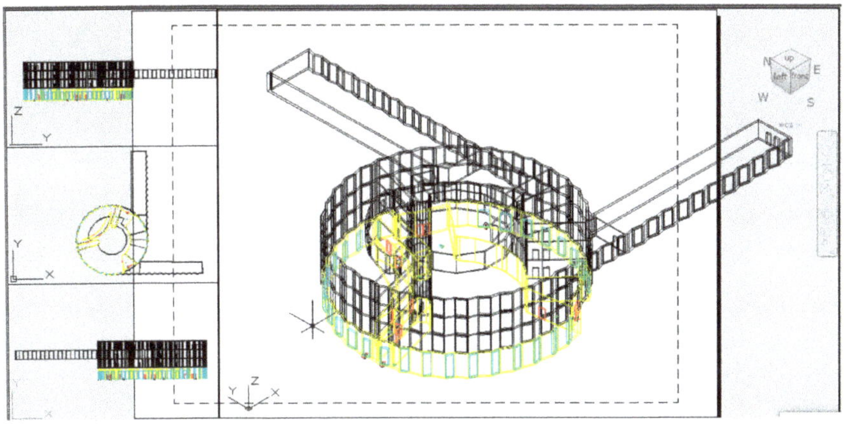

Figure 11. Architectural model of the project exhibition hall.

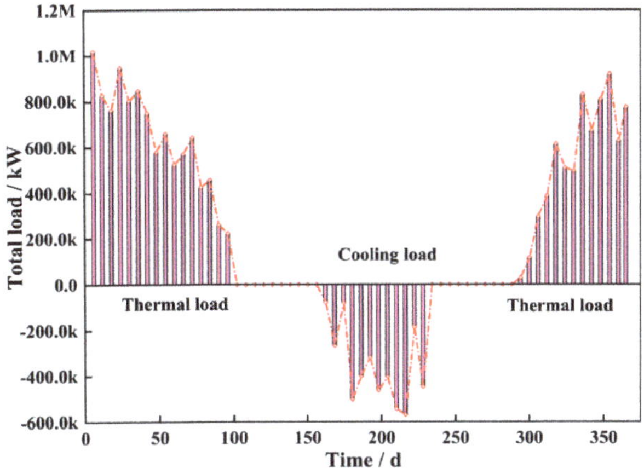

Figure 12. Year-round load distributions for office and showroom buildings.

4.3. Load-Side and Source-Side Heat Balance Analyses That Are Based on the Control Strategy

In the middle-depth and shallow ground-source heat pump and boiler combined heating system, the heating load not only is closely related to the load demand of the building but also has a direct effect on the cold and heat balance of the underground temperature field, thus affecting the energy efficiency of the ground-source heat pump unit and the operating efficiency of the whole system. In this study, the combined heat source system includes multiple ground-source heat pumps and boilers. The power of the ground-source heat pump is 7806.6 kW, which is mainly responsible for the heating load, whereas the boiler power is 2342 kW, which is used to supplement the heat in cold weather, plays the role of a peak load, and ensures the safety and stability of the system. The system's geothermal exchangers include shallow and medium-depth geothermal exchangers, with shallow geothermal exchangers operating at depths of 20 m to 200 m for low load requirements, and medium-depth geothermal exchangers operating at depths of 200 m to 2000 m for higher load requirements. The design and operation characteristics of all the geothermal exchangers are consistent with those of the coaxial heat exchangers analysed in this paper, especially in terms of heat-exchange efficiency and fluid flow mode.

To ensure the efficient operation of the system, the joint operation control strategy in this study adopts a three-layer control framework: an operation strategy, a control strategy, and an operation mode. The control policy is responsible for selecting and implementing a specific mode of operation. It is based on a combination of a series of operating modes under specific season, weather, and load conditions to achieve the goal of meeting user needs, saving operating costs, and reducing energy consumption. The control strategy of the project is based on the principle of ensuring the thermal balance of the underground temperature field and the operating efficiency of the complex energy system. In particular, a deep ground-source heat pump is preferred under heating conditions. Under air-conditioning conditions, a plate heat exchanger is preferred for cooling. During the summer and transition seasons, deep ground-source heat pumps are used to recharge heat (as shown in Figure 13). The implementation of the control strategy has a significant effect on the behaviour of the heat exchanger, especially the annual evolution of the source-side outlet temperature. In terms of heat source selection, the control strategy ensures the stability of the system under different seasons and load conditions by dynamically adjusting the operation of the deep ground-source heat pump and the shallow ground-source heat pump. Specifically, in the winter heating season, deep ground-source heat pumps are used more frequently to provide a stable heat source and ensure that the source-side outlet temperature is maintained within a reasonable range. In contrast, the use of shallow ground-source heat pumps and other heat sources during the transition season or summer improves the flexibility of the system and avoids unnecessary heat waste. Through this dynamic adjustment, the fluctuation range of the source-side outlet temperature is effectively controlled, especially during periods of large load fluctuations. The control strategy can quickly respond to the change in load demand, optimize the heat source configuration, stabilize the source-side outlet temperature, and reduce the impact of temperature fluctuations on system efficiency.

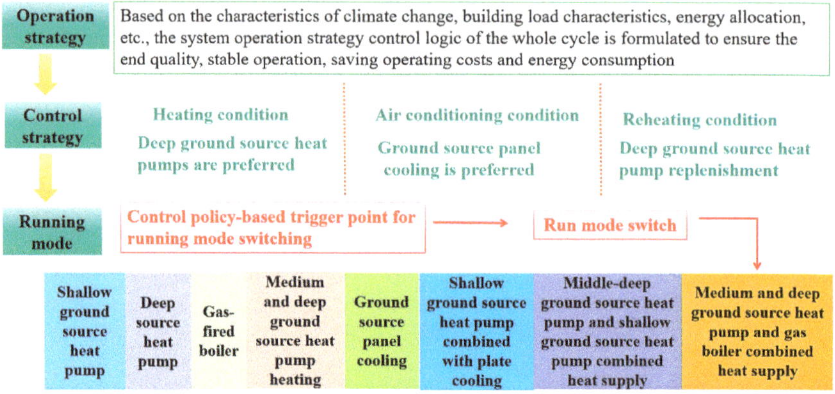

Figure 13. Control strategy for the combined operation of meddle-depth and shallow heat pumps and boilers.

The seasonal distributions of heat extraction and compensation in the multisource joint operation strategy of this project are shown in Figure 14. During winter operation, the medium-depth and shallow ground-source heat pumps, combined with boilers, jointly supply heat, with the shallow ground-source heat pump withdrawing a total of 6,024,000 kWh of heat. In contrast, during the summer, the heat recharge from the plate heat exchanger and shallow ground-source heat pump totals 4,763,000 kWh, resulting in a heat deficit of 1,261,300 kWh. This deficit could cause an imbalance in the subsurface temperature field, particularly in the shallow ground-source heat pump area. To address this, this study proposes a deep ground-source heat exchange supplementation scheme that adds

982,500 kWh of heat, particularly in autumn, significantly alleviating the imbalance in heat extraction between winter and summer. This strategy demonstrates the flexibility and efficiency of multisource joint operations. During peak load periods, deep ground-source heat pumps are prioritized to ensure stable performance, whereas a deep heat exchanger is used for supplementary heating during the transition seasons. This approach not only improves the overall efficiency of the composite energy system but also reduces the reliance on boiler heating during winter peaks, offering innovative solutions for sustainable energy management in cold regions.

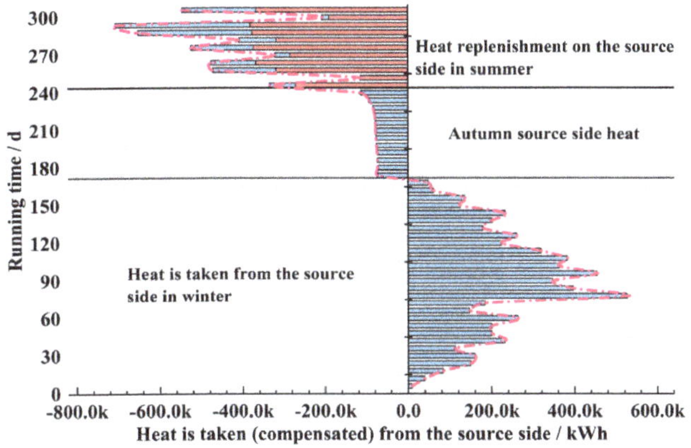

Figure 14. Distribution of source-side heat withdrawal and compensation for the combined operation of medium-depth and shallow heat pumps and boilers.

The study revealed that the heat (982,500 kWh) supplied by the deep ground-source heat exchanger in autumn is crucial for maintaining the long-term thermal balance of the subsurface temperature field. This can be explained by the thermal decay law of the deep heat exchanger, where dynamic changes in the thermal conductivity and thermal resistance of the rock formation significantly affect the heat-exchange efficiency as the operating time increases. Therefore, the heat deficit in the underground temperature field can be mitigated by dynamically adjusting the heat input from the deep heat exchanger during the transition season. In addition, insufficient summer heat recharge from shallow ground-source heat pumps may exacerbate the cooling effect on the shallow subsurface temperature field, which in turn impacts the long-term stability of the system. For this reason, a strategy of appropriately increasing the deep heat exchange make-up heat in summer and spring is proposed to compensate for the remaining heat difference of 278,800 kWh. This dynamic optimization strategy maintains the underground temperature field balance while avoiding the potential negative impacts of low shallow-ground temperature on the operational efficiency of the system.

On the basis of the experimental results, the multi-heat source joint operation strategy of this project has considerable advantages in composite energy systems in cold regions. First, the dynamic priority operation of deep and shallow ground-source heat pumps precisely aligns with seasonal load variations, ensuring efficient system operation. Second, the use of a deep heat transfer heat supplementation strategy effectively mitigates the difference in heat withdrawal and supplementation between winter and summer and ensures the long-term thermal equilibrium of the underground temperature field. In contrast to prior studies on shallow ground-source heat pumps [35,36], this study verifies the effectiveness of the make-up heat strategy for medium and deep ground-source heat pumps in long-term operation in cold regions and quantifies the specific contribution of autumn make-up heat to the annual heat balance. This result not only provides a scientific basis for the optimal design of the composite energy system but also opens up a new path for the sustainable operation of the energy system in cold regions.

5. Conclusions

In this work, the heat-transfer characteristics and operational performance of a ground–source heat pump system with deep underground pipes in cold regions are systematically analysed, with a focus on the influences of key design parameters and operational conditions on heat-transfer efficiency and heat balance adaptability. The following conclusions are drawn:

(1) Increasing the well depth significantly enhances the utilization efficiency of the underground heat storage resources, with a 50 kW increase in heat-transfer capacity for every 500 m increase in well depth, demonstrating a linear growth trend. Additionally, optimizing the internal-to-external pipe diameter ratio effectively enhances the heat-exchange capacity between the fluid and the geotechnical body. The appropriate selection of a larger pipe diameter ratio can significantly reduce the flow resistance and improve the system heat-exchange efficiency.

(2) When the thermal conductivity of cement is close to that of rock and soil, the heat-transfer performance is significantly improved. Specifically, when the thermal conductivity of the cementing cement increases from 1 W/(mK) to 2 W/(mK), the water outlet temperature on the source side increases by approximately 1 °C, and the heat change increases by 13 kW. However, the gain in system efficiency gradually increases. Although an increase in the thermal conductivity of the inner tube can increase the heat transfer rate, it may cause a thermal short-circuit effect and then have a negative impact on the outlet temperature. Therefore, optimizing the thermal conductivity should consider the balance between thermal performance and heat loss to maximize the overall efficiency of the system.

(3) The flow rate has a significant effect on the initial increase in the heat exchange of the system, but with time, the heat transfer effect tends to be saturated. When the operating flow rate is controlled at 0.7 m/s, the heat change is stable at approximately 250 kW, and the system economy can be optimized under the premise of ensuring efficient heat exchange. The adjustment of the start–stop ratio not only improves the energy consumption distribution of the system but also effectively optimizes the long-term thermal balance of the system by strengthening the thermal compensation capacity, which is especially suitable for heating scenarios with large load fluctuations.

(4) The heat release rate of the geotechnical body is high during initial system operation. However, as the heating cycle progresses, the geotechnical body's heat balance capability gradually emerges, and the outlet temperature and heat-exchange volume stabilize, demonstrating the long-term applicability of medium and deep geothermal energy resources in cold regions.

From the perspective of heat-transfer mechanisms and engineering practice, this study proposes optimized design recommendations for medium and deep ground-source heat pump systems in cold regions. The findings not only address a research gap in ground-source heat pumps in extremely cold regions but also provide scientific support for promoting green and low-carbon development in building heating. Future work should focus on integrating multifactor coupling models to further optimize system performance under complex geological and climatic conditions in cold regions. Moreover, more intelligent operation and control strategies should be developed for different building types and load fluctuations to achieve lifecycle optimization design and facilitate the widespread adoption of medium and deep ground-source heat pump systems in cold regions worldwide.

Author Contributions: J.L.: Resources, supervision, project administration, and formal analysis. X.Q.: Methodology, investigation, validation, formal analysis, writing—original draft. X.L.: Visualization, software. H.H.: Conceptualization, visualization, writing—review and editing. J.G.: Resources, funding acquisition, project administration. All authors have read and agreed to the published version of the manuscript.

Funding: This work was funded by the State Grid Corporation's Science and Technology Program (5400–202318247A–1–1-ZN).

Institutional Review Board Statement: Not applicable.

Informed Consent Statement: Informed consent was obtained from all subjects involved in the study.

Data Availability Statement: Data are contained within the article.

Conflicts of Interest: The authors declare that they have no known competing financial interests or personal relationships that could have appeared to influence the work reported in this paper.

References

1. Wang, X.; Zhou, C.; Ni, L. Experimental investigation on heat extraction performance of deep borehole heat exchanger for ground source heat pump systems in severe cold region. *Geothermics* **2022**, *105*, 102539. [CrossRef]
2. Seo, Y.; Seo, U.J. Ground source heat pump (GSHP) systems for horticulture greenhouses adjacent to highway interchanges: A case study in South Korea. *Renew. Sustain. Energy Rev.* **2021**, *135*, 110194. [CrossRef]
3. Sovacool, B.K.; Cabeza, L.F.; Pisello, A.L.; Colladon, A.F.; Larijani, H.M.; Dawoud, B.; Martiskainen, M. Decarbonizing household heating: Reviewing demographics, geography and low-carbon practices and preferences in five European countries. *Renew. Sustain. Energy Rev.* **2021**, *139*, 110703. [CrossRef]
4. Liu, S.; Guo, Y.; Wagner, F.; Liu, H.; Cui, R.Y.; Mauzerall, D.L. Diversifying heat sources in China's urban district heating systems will reduce risk of carbon lock-in. *Nat. Energy* **2024**, *9*, 1021–1031. [CrossRef]
5. Waite, M.; Modi, V. Electricity load implications of space heating decarbonization pathways. *Joule* **2020**, *4*, 376–394. [CrossRef]
6. Du, D.; Li, Y.; Wang, K.; Zhao, Y.; Hu, Z.; Zhang, W.; Wang, Q. Experimental and numerical simulation research on heat transfer performance of coaxial casing heat exchanger in 3500 m-deep geothermal well in Weihe Basin. *Geothermics* **2023**, *109*, 102658. [CrossRef]
7. Rosenow, J.; Gibb, D.; Nowak, T.; Lowes, R. Heating up the global heat pump market. *Nat. Energy* **2022**, *7*, 901–904. [CrossRef]
8. Feng, H.; Xu, H.; Feng, H.; Gao, Y. Numerical simulation and experimental study of medium and deep ground source heat pump system in a cold and arid region. *Case Stud. Therm. Eng.* **2024**, *64*, 105473. [CrossRef]
9. Li, J.; Lu, F.; Xu, W.; Li, J.; Sun, Z.; Qiao, B.; Sun, Z.; Zheng, F.; Xiang, Z.; Zhang, G.; et al. Configuration method for medium-deep ground source heat pump system considering renewable energy consumption and smart grid interaction. *Energy Build.* **2024**, *318*, 114432. [CrossRef]
10. Ma, Y.; Yang, F.; Zhu, R.; Zhou, X.; Liu, G.; Yuan, L.; Wang, X.; Dong, J.; Lü, H.; Li, C.; et al. A numerical study on the sustainability and efficiency of deep coaxial borehole heat exchanger systems in the cold region of northeast China. *Renew. Energy* **2024**, *237*, 121562. [CrossRef]
11. Yang, Q.; Zhang, C.; Jiang, W.; Lu, W.; You, S. Heat transfer of mid-deep ground source heat pump for crude oil gathering and transportation. *Int. Commun. Heat Mass Transf.* **2024**, *159*, 108237. [CrossRef]
12. Alshehri, F.; Beck, S.; Ingham, D.; Ma, L.; Pourkashanian, M. Techno-economic analysis of ground and air source heat pumps in hot dry climates. *J. Build. Eng.* **2019**, *26*, 100825. [CrossRef]
13. Norouzi, E.; Li, B.; Wang, L.L.; Raymond, J.; Gaur, A.; Zou, J. Numerical evaluation of ground source heat pumps in a thawing permafrost region. *J. Build. Eng.* **2024**, *98*, 111035. [CrossRef]
14. Jathunge, C.B.; Darbandi, A.; Dworkin, S.B.; Mwesigye, A. Numerical investigation of the long-term thermal performance of a novel thermoactive foundation pile coupled with a ground source heat pump in a cold-climate. *Energy* **2024**, *292*, 130497. [CrossRef]
15. Huang, S.; Lin, D.; Dong, J.; Li, J. Effects of building load characteristics on heating performance of the medium-deep U-type borehole heat exchanger coupled heat pumps: A coupled dynamic simulation. *Appl. Energy* **2025**, *377*, 124405. [CrossRef]
16. Lund, A.E. Effect of heat and mass transfer related parameters on the performance of deep borehole heat exchangers. *Appl. Therm. Eng.* **2024**, *253*, 123764. [CrossRef]
17. Shen, J.; Zhou, C.; Luo, Y.; Tian, Z.; Zhang, S.; Fan, J.; Ling, Z. Comprehensive thermal performance analysis and optimization study on U-type deep borehole ground source heat pump systems based on a new analytical model. *Energy* **2023**, *274*, 127367. [CrossRef]

18. Liu, J.; Zhang, Y.; Zhang, S.; Zhang, X.; Ma, D. Heat transfer and rock–soil temperature distribution characteristics in whole life cycle for different structure sizes of medium-deep U-shaped butted well. *Case Stud. Therm. Eng.* **2024**, *60*, 104705. [CrossRef]
19. Li, W.; Xu, J.; Chen, Y.; Chen, Z. Heat transfer performance and optimal design of shallow coaxial ground heat exchangers. *Appl. Therm. Eng.* **2024**, *250*, 123571. [CrossRef]
20. Luo, Y.; Xu, G.; Yan, T. Performance evaluation and optimization design of deep ground source heat pump with nonuniform internal insulation based on analytical solutions. *Energy Build.* **2020**, *229*, 110495. [CrossRef]
21. Dong, J.; Wang, H.; Huang, S.; Jiang, Y.; Liu, J. Advances in the effects of multiple factors on the thermal performance of deep borehole heat exchangers: A systematic review. *J. Build. Eng.* **2024**, *96*, 110488. [CrossRef]
22. Li, C.; Jiang, C.; Guan, Y.; Chen, H.; Yang, R.; Wan, R.; Shen, L. Comparison of the experimental and numerical results of coaxial-type and U-type deeply buried pipes' heat transfer performances. *Renew. Energy* **2023**, *210*, 95–106. [CrossRef]
23. Qi, X.; Yi, F.; Jin Hg Meng, F.; Yu, H.; Bai, W.; Bao, H. Analysis of influencing factors of casing heat transfer in deep geothermal well in cold region. *J. Shenzhen Univ. Sci. Technol. Ed.* **2022**, *1*, 42–50.
24. Li, M.; Shi, Y.; Chen, H.; Liu, C.; Li, H. Study on the heat transfer performance of coaxial casing heat exchanger for medium and deep geothermal energy in cold regions. *Renew. Energy* **2024**, *237*, 121666. [CrossRef]
25. Qin, L.; Lu, D.; Zheng, H.; Wang, C.; Dong, W. A new stacking model method to solve an inverse flow and heat coupling problem for aero-engine turbine blades. *Case Stud. Therm. Eng.* **2024**, *56*, 104209. [CrossRef]
26. Wang, T.; Xuan, Y.; Han, X. Investigation on hybrid thermal features of aero-engines from combustor to turbine. *Int. J. Heat Mass Transf.* **2023**, *200*, 123559. [CrossRef]
27. Luo, Z.; Liang, K.; Chen, X.; Li, Z.; Wang, X.; Zhou, H.; Wang, Y. Numerical investigation of the effect of vibration on spray cooling heat transfer in the nonboiling region. *Case Stud. Therm. Eng.* **2023**, *51*, 103518. [CrossRef]
28. Motaghian, S.; Rayegan, S.; Pasdarshahri, H.; Ahmadi, P.; Rosen, M.A. Comprehensive performance assessment of a solid desiccant wheel using an artificial neural network approach. *Int. J. Heat Mass Transf.* **2021**, *165*, 120657. [CrossRef]
29. Alsaleh, M.; Abdul-Rahim, A.S. The pathway toward pollution mitigation in EU28 region: Does hydropower growth make a difference? *Renew. Energy* **2022**, *185*, 291–301. [CrossRef]
30. Aydin, M.; Pata, U.K. Are shocks to disaggregated renewable energy consumption permanent or temporary for the USA? Wavelet based unit root test with smooth structural shifts. *Energy* **2020**, *207*, 118245. [CrossRef]
31. Chen, C.; Wang, C.; Zhang, Z. Numerical investigation of the water transport and performance of proton exchange membrane fuel cell with an imitating river flow field. *Energy Convers. Manag.* **2023**, *276*, 116532. [CrossRef]
32. Chen, H.; Liu, H.; Yang, F.; Tan, H.; Wang, B. Field measurements and numerical investigation on heat transfer characteristics and long-term performance of deep borehole heat exchangers. *Renew. Energy* **2023**, *205*, 1125–1136. [CrossRef]
33. Han, Y.; Wang, K.; Wang, Y.; Gou, L.; Yang, Y.; Lei, Y.; Shen, H. Study on multifactor influence of heat transfer characteristics of middle-deep geothermal Wells. *Coal Geol. Explor.* **2024**, *1*, 104–116.
34. *GB 50189-2015*; Design Standard for Energy Efficiency of Public Buildings. China Architecture & Building Industry Press: Beijing, China, 2015.
35. Emmi, G.; Bottarelli, M. Enhancement of shallow ground heat exchanger with phase change material. *Renew. Energy* **2023**, *206*, 828–837. [CrossRef]
36. Figueira, J.S.; Gil, A.G.; Vieira, A.; Michopoulos, A.K.; Boon, D.P.; Loveridge, F.; Cecinato, F.; Götzl, G.; Epting, J.; Zosseder, K.; et al. Shallow geothermal energy systems for district heating and cooling networks: Review and technological progression through case studies. *Renew. Energy* **2024**, *236*, 121436. [CrossRef]

sustainability

MDPI

Article

Fuel Consumption Prediction for Full Flight Phases Toward Sustainable Aviation: A DMPSO-LSTM Model Using Quick Access Recorder (QAR) Data

Jing Xiong [1,*], Chunling Zou [1,2,*], Yongbing Wan [3], Youchao Sun [4] and Gang Yu [5]

1 College of Air Transportation, Shanghai University of Engineering Science, Shanghai 201620, China
2 Scania China Manufacturing Co., Ltd., Nantong 226500, China
3 Shanghai Shentong Metro Group Co., Ltd., Technical Center, Shanghai 201103, China; wybingsh@163.com
4 College of Civil Aviation, Nanjing University of Aeronautics and Astronautics, Nanjing 210016, China; sunyc@nuaa.edu.cn
5 SILC Business School, Shanghai University, Shanghai 201800, China; gyu@shu.edu.cn
* Correspondence: xjsh@sues.edu.cn (J.X.); zouchunling1314@163.com (C.Z.); Tel.: +86-021-67791109 (J.X.)

Abstract: Reducing emissions in the aviation industry remains a critical challenge for global low-carbon transition. Accurate fuel consumption prediction is essential to achieving emission reduction targets and advancing sustainable development in aviation. Aircraft fuel consumption is influenced by numerous complex factors during flight, resulting in significant nonlinear relationships between segment-specific variables and fuel usage. Traditional statistical and econometric models struggle to capture these relationships effectively. This article first focuses on the different characteristics of QAR data and uses the Adaptive Noise Ensemble Empirical Mode Decomposition (CEEMDAN) method to obtain more significant potential features of QAR data, solving the problems of mode aliasing and uneven mode gaps that may occur in traditional decomposition methods when processing non-stationary signals. Secondly, a dynamic multidimensional particle swarm optimization algorithm (DMPSO) was constructed using an adaptive adjustment dynamic change method of inertia weight and learning factor, which solved the problem of local extremum and low search accuracy in the solution space that PSO algorithm is prone to during the optimization process. Then, a DMPSO-LSTM aircraft fuel consumption model was established to achieve fuel consumption prediction for three flight segments: climb, cruise, and descent. The final proposed model was validated on real-world datasets, and the results showed that it outperformed other baseline models such as BP, RNN, PSO-LSTM, etc. Among the results, the climbing segment MAE index decreased by more than 40%, the RMSE index decreased by more than 38%, and the R^2 index increased by more than 6%, respectively. The MAE index of the cruise segment decreased by more than 40%, the RMSE index decreased by more than 40%, and the R^2 index increased by more than 5%, respectively. The MAE index of the descending segment decreased by more than 20%, the RMSE index decreased by more than 30%, and the R^2 index increased by more than 5%, respectively. The improved prediction accuracy can be used to implement multi-criteria optimization in flight operations: (1) by quantifying weight–fuel relationships, it supports payload–fuel tradeoff decisions; (2) enhanced phase-specific predictions allow optimized climb/cruise profile selections, balancing time and fuel use; and (3) precise consumption estimates facilitate optimal fuel-loading decisions, minimizing safety margins. The high-precision fuel consumption prediction framework proposed in this study provides actionable insights for airlines to optimize flight operations and design low-carbon route strategies, thereby accelerating the aviation industry's transition toward net-zero emissions.

Keywords: fuel consumption prediction; aircraft traffic management; QAR data; particle swarm arithmetic; LSTM

1. Introduction

With the rapid development of the global economy and the increasing demand for civil aviation transportation, the pollution caused by aircraft fuel emissions is becoming increasingly serious. The emissions caused by aircraft fuel combustion are the largest source of carbon emissions in the aviation industry. According to the International Energy Agency (IEA), the carbon emissions from the aviation industry account for 12% of the global transportation sector's carbon emissions in 2022, with an average annual growth rate of 3.5% [1]. The international community increasingly prioritizes energy conservation and emission reduction in civil aviation, with the promotion of green and sustainable development emerging as a global consensus [2]. The International Civil Aviation Organization (ICAO) passed the "2050 Net Zero Carbon Emissions" resolution in 2022, requiring a minimum of 2.5%/year increase in segment level fuel efficiency through technological innovation and operational optimization [3]. In this context, developing accurate fuel consumption prediction models is crucial for minimizing the gap between actual and theoretical fuel usage, enhancing fuel efficiency through technological innovation, and ultimately reducing carbon emissions—a key pathway toward sustainable aviation [4,5].

Early prediction methods involved calculating aircraft fuel consumption values based on aircraft performance manuals or energy models, which typically included models such as fuel consumption rate, thrust, drag/power ratio, and other parameters closely related to aircraft performance, aerodynamic characteristics, and flight path [6–8]. These parameters come from the performance manual of the aircraft manufacturer and need to be checked with a table. The accuracy is poor and the operation steps are cumbersome. They are not suitable for estimating the fuel consumption of the entire flight trajectory of the aircraft. Therefore, the European Organization for the Safety of Air Navigation (Eurocontrol) has developed the Aircraft Key Performance Database [9,10] and the Base of Aircraft Data (BADA). The BADA performance database provides various theoretical model specifications and related specific performance parameter datasets. It can realistically reproduce the aircraft's motion trajectory throughout the entire flight phase based on existing aircraft performance reference data and then calculate the aircraft's fuel consumption. Kaiser [11] and Dalmau [12] established a fuel consumption model based on the BADA database to avoid the tedious process of consulting static tables. Burzlaff [13] and Pagoni [4] developed a fuel consumption model using the BADA database to calculate the fuel consumption of an aircraft during flight. However, due to the lack of consideration of meteorological parameters, the accuracy of the aircraft fuel consumption model based on the BADA database is often low. To further enhance the practicality of prediction methods, researchers use statistical and mathematical methods to establish the aircraft fuel consumption model. Yutko [14] evaluated typical aircraft operations in order to develop evaluation hypotheses for task-based and instantaneous indicators. Jensen [15] conducted flight by flight analysis using over 200,000 historical operations conducted in 2012, considering the actual wind and temperature experienced by flights, and established an efficiency benefit pool to achieve full system cruise altitude optimization for domestic operations in the United States. However, due to the numerous and complex factors that affect aircraft fuel consumption during flight, there is a significant nonlinear relationship between the influencing factors of each flight segment and aircraft fuel consumption, which is difficult to explain using traditional statistical and econometric models.

This paper aims to dynamically adjust aircraft fuel transportation redundancy by establishing a predictive model to balance economy and safety. In order to build an accurate aircraft fuel consumption prediction model, effective data mining and analysis of performance parameters related to aircraft fuel consumption must be carried out to obtain more accurate aircraft fuel consumption prediction performance parameters. The neural network model, with its unique advantages of distributed storage and parallel processing, can approximate aircraft fuel consumption data with nonlinear performance relationships [16]. Therefore, this paper aims to explore the characteristics of QAR data for climb, cruise, and descent segments, using the CEEMDAN method to decompose aircraft fuel consumption data, fully exploring the changing characteristics of time series, and obtaining more detailed corresponding modal subsequences to provide good feature data for prediction models. A dynamic multidimensional particle swarm optimization algorithm (DMPSO) was developed to address the difficulty of parameter optimization in LSTM network models. The dynamic formula of inertia weight and learning factor was used to construct a DMPSO LSTM full range fuel consumption prediction model, which solves the problem of the particle swarm algorithm easily becoming stuck in local optima and low search accuracy in the solution space optimization process.

The structure of this article is as follows: Section 2 introduces the relevant research on machine learning-based fuel consumption prediction and estimation. Section 3 provides a detailed introduction to the fuel consumption prediction method based on the DMPSO-LSTM model. Section 4 introduces the application and result analysis of the proposed method on a certain airline's A320 aircraft model. Section 5 provides a summary and future work.

2. Literature Review

2.1. Fuel Consumption Prediction

QAR (Quick Access Recorder) is a flight data recording system carried on aircraft, used for real-time acquisition, storage, and transmission of high-frequency flight operating parameters. Its recording range covers aircraft performance parameters (such as airspeed, altitude, attitude), engine status (such as fuel flow, speed, exhaust temperature), environmental conditions (such as temperature, wind speed), and control commands (such as flaps, throttle position), with a sampling frequency typically ranging from 1 to 4 times per second. QAR data are not only the core basis for aviation safety monitoring and accident investigation but also provide high-precision multidimensional data sources for flight performance analysis, fuel efficiency optimization, and prediction model construction. In the field of fuel consumption prediction, QAR data can effectively support data-driven fuel consumption modeling and abnormal pattern recognition by capturing dynamic parameters of the entire flight phase (such as climb rate, cruise Mach number, descent profile), combined with engine operating conditions and environmental variables [17,18]. QAR (Quick Access Recorder) data have become the core data source for aircraft fuel consumption prediction research due to its high-resolution and multidimensional flight parameter acquisition capabilities. In recent years, researchers have proposed various innovative methods by combining statistical modeling, machine learning, and flight performance theory.

Researchers construct a highly interpretable and computationally efficient prediction framework by mining physical correlations in QAR parameters and combining statistical learning techniques. Wang et al. [19]. proposed a hybrid model based on multiple linear regression and principal component analysis (PCA), which utilizes flight phase parameters (such as climb rate and cruise Mach number) and engine performance indicators (such as fuel flow rate and N1 speed) from QAR data to construct fuel consumption prediction equations for flight segments. Research has shown that the 10 key QAR parameters selected through PCA dimensionality reduction can explain 92% of the fuel consumption variance, but this method lacks adaptability to non-steady state flight conditions such as frequent speed regulation in turbulence. To enhance predictive robustness in complex scenarios, Smith and Jones [20] developed a Bayesian Hierarchical Model (BHM) that combines QAR data with aircraft performance benchmarks such as the BADA model. By quantifying the uncertainty of engine performance degradation and environmental wind speed through prior distribution, this model achieved an average absolute percentage error mean absolute percentage error (MAPE) of 3.8% in cross fleet prediction, which is 1.5% lower than traditional regression methods. This achievement highlights the advantages of mathematical statistical methods in integrating prior knowledge of physics with data uncertainty. In addition, García et al. [21] explored a generalized additive model (GAM) driven by QAR data, which fitted the correlation characteristics among flight altitude, airspeed, and fuel consumption through a nonlinear smoothing function. Its research confirms that segmented modeling based on QAR data statistical features (such as distinguishing between climb and cruise phases) can further reduce prediction errors by 12%, but it relies on high-precision flight phase automatic recognition algorithms for support. The International Air Transport Association (IATA) has evaluated the applicability of mathematical statistical methods in QAR data analysis and identified two major challenges that need to be addressed in current research: Firstly, the high dimensionality and multicollinearity of QAR parameters pose a risk of overfitting in traditional regression models. The second challenge is how to effectively identify significant operational variables that affect fuel consumption through statistical hypothesis testing (such as t-test, ANOVA), providing a basis for airlines to formulate fuel-saving strategies [22].

The current research has significantly improved prediction accuracy through physical statistical hybrid modeling and staged parameter optimization, but further efforts are needed to balance the model's generalization ability and real-time performance, while strengthening the standardization process of QAR data quality control and feature engineering. The prediction of aircraft fuel consumption is a very complex multivariate time series fitting problem, and machine learning neural networks are very suitable for nonlinear data fitting budgets. Therefore, using neural network models to predict aircraft fuel consumption has become a research hotspot. This type of method significantly improves the prediction accuracy under complex operating conditions by capturing the nonlinear relationships and spatiotemporal dependencies between flight parameters. Zhang et al. [23] first proposed a framework combining Long Short-Term Memory (LSTM) and Convolutional Neural Network (CNN), which utilizes time series parameters (such as fuel flow rate, airspeed, altitude) and engine state sequences from QAR data to achieve a segment level fuel consumption prediction error of less than 1.5%. This model extracts local features (such as instantaneous fuel consumption fluctuations during the climb phase) through CNN and combines LSTM to model cross stage dependencies, but its computational complexity limits its real-time deployment capability. To optimize the efficiency of feature extraction, Liu et al. [24] designed an attention-based Transformer architecture that embeds QAR parameters (such as N1 RPM, ambient temperature) and flight phase labels (climb, cruise, descent) into multidimensional vectors. The experiment shows that this model reduces the error by 18% compared to traditional LSTM in long haul route prediction and has stronger robustness to data loss. However, the problem of insufficient interpretability has not been solved, making it difficult to quantify the contribution of a single parameter to fuel consumption. In response to the demand for multi-source data fusion, Chen et al. [25] developed a graph neural network (GNN) model that integrates QAR data, radar trajectories, and meteorological grid information to construct a fuel consumption map for the entire flight cycle. This model represents the dynamic fuel consumption correlation between waypoints through graph nodes and improves prediction accuracy by 23% in complex airspace environments such as detours and waiting routes. To further optimize feature engineering, Liu and Wang [26] developed an anomaly detection framework based on QAR data, which identifies fuel efficiency losses caused by non-standard operations (such as non-economic climb rates) by analyzing the synergistic relationship among engine speed (N1), exhaust temperature (EGT), and fuel flow rate. Its research confirms that manipulation features in QAR data, such as flap angle and throttle lever position, can serve as key inputs for predictive models to distinguish the impact of pilot operating preferences on fuel consumption. However, this method has limited modeling capabilities for route structures, such as path deviations caused by air traffic control instructions. In response to the prediction needs in complex airspace environments, Chen et al. [27] proposed a spatiotemporal feature fusion model that combines QAR data with radar trajectories and air traffic control instructions (such as altitude changes and waiting routes) and uses graph neural networks (GNNs) to model the fuel consumption dependency between waypoints. Experiments have shown that this model improves prediction accuracy by more than 15% compared to traditional regression methods in flights with frequent speed adjustments and path extensions. This achievement highlights the potential of collaborative modeling between QAR data and external operating environments but also points out that the current research still needs to improve the balance between real-time data and computational efficiency. Meanwhile, in recent years, researchers have further focused on the collaborative mechanism between fuel consumption prediction and sustainable aviation. Schäfer et al. [28] quantified the reduction effect of route optimization on carbon emission intensity by integrating QAR data with airspace structure parameters. They demonstrated that for every 1% decrease in fuel consumption prediction error during the climb phase, the carbon footprint of the entire flight segment can be reduced by 0.6%. Zhang et al. [29] developed a multi-objective optimization framework based on LSTM, which generates carbon budget schemes that meet CORSIA standards while predicting fuel consumption, but the cross stage dynamic coupling problem has not been solved. In addition, the International Air Transport Association (IATA) has pointed out that existing predictive models have insufficient interpretability in supporting the preparation of ESG (Environmental, Social, and Governance) reports by airlines, and there is an urgent need for a hybrid modeling method that integrates physical mechanisms and data-driven approaches [30]. The "Fuel Efficiency Benchmark 2023" released by Airbus shows that the A350-1000 model reduces additional fuel carrying capacity by 7% to 12% through dynamic prediction and reduces CO_2 emissions by 3200 tons annually [31].

2.2. *Knowledge Gap*

Traditional empirical models, such as the BADA database and energy-based equations, rely on aerodynamic and thermodynamic principles to estimate fuel consumption. While these models provide interpretable results, they often lack adaptability to real-world operational variations (e.g., turbulence, engine degradation). For instance, the BADA model assumes ideal flight conditions and neglects transient effects during phase transitions, leading to average prediction errors of 8–12% in real-world datasets [4,12]. In contrast, data-driven approaches leverage high-resolution QAR data to capture nonlinear relationships between operational variables (e.g., vertical acceleration, wind speed) and fuel consumption, enabling higher accuracy under dynamic conditions. According to current research, there are still significant knowledge gaps at the following theoretical levels:

(1) Efficient decomposition and feature preservation of non-stationary QAR signals. Existing research generally adopts traditional signal decomposition methods (such as EMD and wavelet transform) to process QAR time series data but has not effectively solved

the problems of mode aliasing and non-uniform gaps. This leads to the inability of the decomposed subsequence to accurately characterize the dynamic characteristics of the flight phase (such as instantaneous fuel consumption fluctuations in turbulence), limiting the adaptability of the model to complex operating conditions.

(2) Lightweight characterization and physical interpretability of high-dimensional QAR data. Existing methods such as PCA and linear regression are prone to losing the physical correlation of QAR parameters during dimensionality reduction and have not solved the coupling problem between high-dimensional features and nonlinear fuel consumption relationships. For example, although PCA reduces dimensionality, it cannot preserve the nonlinear mapping between aerodynamic efficiency parameters (such as the lift to drag ratio) and fuel consumption.

(3) Adaptive modeling and collaborative optimization during the dynamic flight phase. The current research focuses on the independent modeling of the climb, cruise, and descent stages, neglecting the dynamic coupling effects across stages (such as the impact of the end of climb on the initial fuel consumption during cruise). In addition, traditional PSO algorithms are prone to becoming stuck in local optima when optimizing LSTM hyperparameters, and do not consider the impact of flight phase differences on the parameter search space.

3. Methodology

3.1. Decomposition of Fuel Consumption Impact Characteristics Based on CEEMDAN

QAR data have non-stationary and nonlinear characteristics, and the characteristics of data in different time periods are different. In response to this situation, the Complete Ensemble Empirical Mode Decomposition with Adaptive Noise (CEEMDAN) method is first used to decompose the aircraft fuel consumption data, fully utilizing the changing characteristics of the time series to obtain a more detailed subset of relevant modes, providing good feature data for the prediction model.

Complementary Ensemble Empirical Mode Decomposition (CEEMDAN) with Adaptive Noise is a further improved decomposition method based on Empirical Mode Decomposition (EMD) and Ensemble Empirical Mode Decomposition (EEMD) [32,33]. The CEEMDAN method first determines the final IMF component of the original signal, then adds white noise and iterates. The decomposition process is as follows:

(1) Preprocess the signal to avoid extreme values or monotonic trends.

(2) Construct an adaptive noise sequence and add it to the original signal to generate multiple sets of random experimental signals.

(3) Perform CEEMDAN on each experimental signal to obtain a set of IMFs.

(4) Combine and weight each group of IMFs to generate a set of total IMFs.

(5) Perform CEEMDAN on the total IMF to obtain a new set of IMFs.

(6) Check whether the newly generated IMF meets the convergence condition. If not, return to step (2) to adjust the adaptive noise.

(7) After satisfying the convergence condition, iterate repeatedly according to steps 3–6 until the number of IMFs obtained no longer increase.

In the CEEMDAN process, different IMFs have different fluctuation characteristics, resulting in the original data having different fluctuation characteristics at different time intervals. CEEMDAN captures the instability, randomness, and variability of the original time series data through intrinsic mode functions (IMFs), effectively highlighting localized features across different temporal scales. The flowchart of the CEEMDAN feature decomposition algorithm is shown in Figure 1.

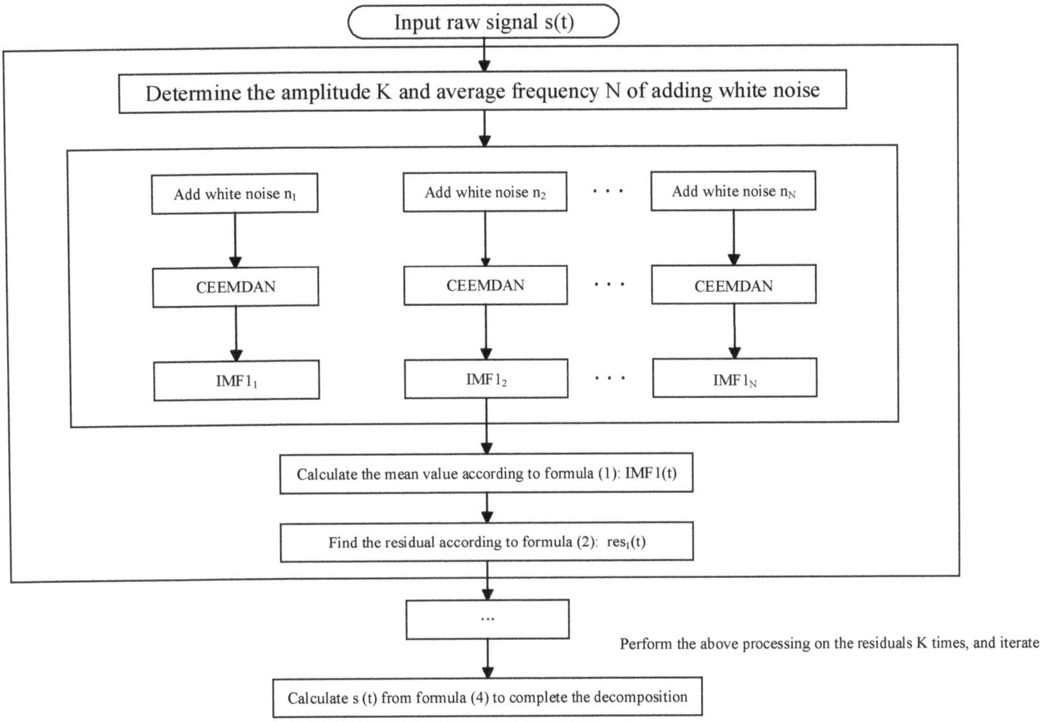

Figure 1. CEEMDAN algorithm decomposition flowchart.

The specific decomposition process is as follows:

(1) If the original data sequence of factors affecting aircraft fuel consumption for each flight segment (represented by *s(t)*) adds noise with *a* mean of 0 (represented by $\varphi_j(t)$) before decomposition, then $s(t)+\alpha_0\varphi_j(t)$ is the new original decomposition sequence. On the other hand, parameter α_0 is used to ensure that the signal-to-noise ratio of the added noise to the original signal remains within an acceptable range and can be adjusted accordingly. The standard deviation of $\varphi_j(t)$ is k_ε, k is the coefficient of the noise fluctuation amplitude value, and ε is the standard deviation of the sequence.

(2) Perform CEEMDAN on the $s(t) + \alpha_0\varphi_j(t)$ sequence.

(3) Repeat steps (1) and (2) N times, adding different noise sequences $\varphi_j(t)$ in each execution, where the first modal component is the average of the components in N executions, calculated as shown in Equation (1), and the remaining $res_1(t)$ are the differences between the initial data and the modal components, calculated as shown in Equation (2).

$$\widetilde{IMF_1}(t) = \frac{1}{N}\sum_{j=1}^{N} IMF_1^j(t) \tag{1}$$

$$res_1(t) = s(t) - \widetilde{IMF_1}(t) \tag{2}$$

(4) Continuing to introduce white noise and residuals to eliminate errors caused by adding noise, the new signal to be decomposed is $res_1(t) + \alpha_1 EMD_1(\varphi_j(t))$. The calculation formula for the second modal component is shown in Equation (3).

$$\widetilde{IMF_2}(t) = \frac{1}{N}\sum_{j=1}^{N} EMD(res_1(t) + \alpha_1 EMD_1(\varphi_j(t))) \tag{3}$$

(5) Repeat step (4) until the residual $res_i(t)$ satisfies the termination condition, obtaining M specific modal components and the final residual $res(t)$. The calculation method for the final initial signal sequence is shown in Equation (4).

$$s(t) = \sum_{i=1}^{M} \widetilde{IMF_1}(t) + res(t) \tag{4}$$

3.2. Dimension Reduction in Aircraft Fuel Consumption Impact Characteristics Based on KPCA

The data sequence obtained through CEEMDAN highlights the fluctuation characteristics of the original data at different time scales, but the dimensionality of the input data variables also increases accordingly. To enhance computational efficiency and mitigate overfitting, the kernel principal component analysis (KPCA) algorithm is employed for dimensionality reduction in input variables, thereby improving the computational efficiency and accuracy of the model while retaining the effectiveness and representativeness of information. KPCA introduces the idea of kernel functions, which map data from low dimensions to high dimensions by performing inner product operations in space. Data extraction can be achieved without specifying a specific mapping function, and the introduction of kernel functions further reduces the computational complexity of the entire system. The implementation steps are as follows:

(1) For a given $m * n$ -dimensional original sample set X with a total of m samples, the covariance matrix is first calculated using the method shown in Equation (5).

$$\overline{C} = \frac{1}{N} \sum_{i=1}^{N} \varnothing(X_i)\varnothing(X_i)^T \tag{5}$$

Among them, $\varnothing(X_i)$ is the nonlinear mapping function in the conversion process, and $\sum\limits_{i=1}^{N} \varnothing(X_i) = 0$.

(2) According to the covariance matrix, calculate the corresponding decomposition eigenvalue V, and corresponding eigenvector V, as shown in Equations (6) and (7).

$$\lambda V = \overline{C}V = \frac{1}{N} \sum_{i=1}^{N} \left\langle \varnothing(X_i)^T, V \right\rangle \varnothing(X_i)\lambda V \tag{6}$$

$$V = \sum_{i=1}^{N} \varnothing(X_i) \tag{7}$$

(3) By introducing the kernel function matrix into the nonlinear mapping function, p sets of kernel principal elements can be obtained, where the calculation method for the *k-th* principal element is shown in Equation (8).

$$t_k = \langle V_k, \varnothing(x) \rangle = \sum_{i=1}^{N} \alpha_i^k \langle \varnothing(X_i), \varnothing(x) \rangle \;\; (k = 1, 2, \ldots, p) \tag{8}$$

Gaussian radial basis functions are chosen as the kernel functions because they have fewer parameters and are semi-positively defined symmetric functions. The calculation method is shown in Equation (9).

$$k(x, y) = e^{\frac{|x-y|^2}{2\sigma^2}} \tag{9}$$

(4) According to the set of main elements, statistical measures reflecting the degree of data model fitting and data trend can be obtained. The calculation method of the KPCA model statistical measures is shown in Equation (10).

$$T^2 = [t_1, \ldots, t_p] \Lambda^{-1} [t_1, \ldots, t_p]^T \tag{10}$$

The stability of KPCA data is often evaluated using the T^2 statistic, which represents the distance between the sample data and the mean of the variable data in the high-dimensional KPCA space. This reflects the degree to which the aircraft fuel consumption data deviates from the stable operating state in terms of trend and amplitude. In the equation, $[t_1, \ldots, t_p]$ is the principal component of the aircraft fuel consumption data extracted by KPCA, which can be calculated according to Equation (8). Λ^{-1} is the diagonal inverse matrix composed of the eigenvalues corresponding to each principal component.

3.3. *Aircraft Fuel Consumption Prediction Model Based on Improved PSO-LSTM*

3.3.1. DMPSO Algorithm

Particle swarm optimization (PSO) [34] is a random parallel search algorithm that simulates the foraging behavior of birds and fish in nature. It consists of a set of particles, where one particle moves in the search space based on its best position (*pbest*) and the best position reached by any particle entering the particle swarm (*gbest*). All particles have velocities, positions, and corresponding fitness values that determine their direction of motion. The particles will follow the particle with the best historical fitness value and search for a solution in space until they work together in a group to find the optimal solution and corresponding fitness value for the problem to be optimized. The process of PSO is shown in Figure 2, and the specific steps are as follows.

(1) In the D-dimensional solution space, assuming the number of particles is m, the inertia weight is w, the acceleration constants are c_1 and c_2, and the random numbers are $rand_{1d}$ and $rand_{2d}$.

(2) Randomly initialize the initial position x and initial velocity v of each particle and evaluate the fitness value of each particle to obtain the current individual optimal position solution *pbest* and the global optimal position solution *gbest*.

(3) Update the particle position and velocity according to the following Formulas (11) and (12):

$$v_{id} = v_{id} + c_1 rand_{1d}(pbest_{id} - x_{id}) + c_2 rand_{2d}(gbest_{id} - x_{id}) \tag{11}$$

$$x_{id} = x_{id} + v_{id} \tag{12}$$

Among them, $pbest_{id}$ and $gbest_{id}$ respectively refer to the optimal position searched by the *i-th* particle and the entire particle swarm in d-dimensional space; x_{id} and v_{id} respectively refer to the position and velocity of the particle corresponding to the *d-th* variable.

(4) Calculate the fitness value of each particle's function and update the individual optimal position solution *pbest* and global optimal position solution *gbest* of the particle.

(5) Continuously iterate until the specified termination condition is met, then stop searching and output the optimal result, otherwise return to step (3).

However, the main problem with traditional PSO algorithms is that they are prone to becoming stuck in local optima during the optimization process, which can lead to lower search accuracy. In order to accelerate convergence speed and avoid becoming stuck in local optima, this paper constructs a dynamic multidimensional particle swarm optimization algorithm (DMPSO). This algorithm utilizes a dynamically changing formula adaptively adjusted by inertia weights [34,35] and learning factors [36,37] to transform fixed parameters into nonlinear transformation parameters, in order to improve the accuracy and speed of convergence. The specific details are as follows.

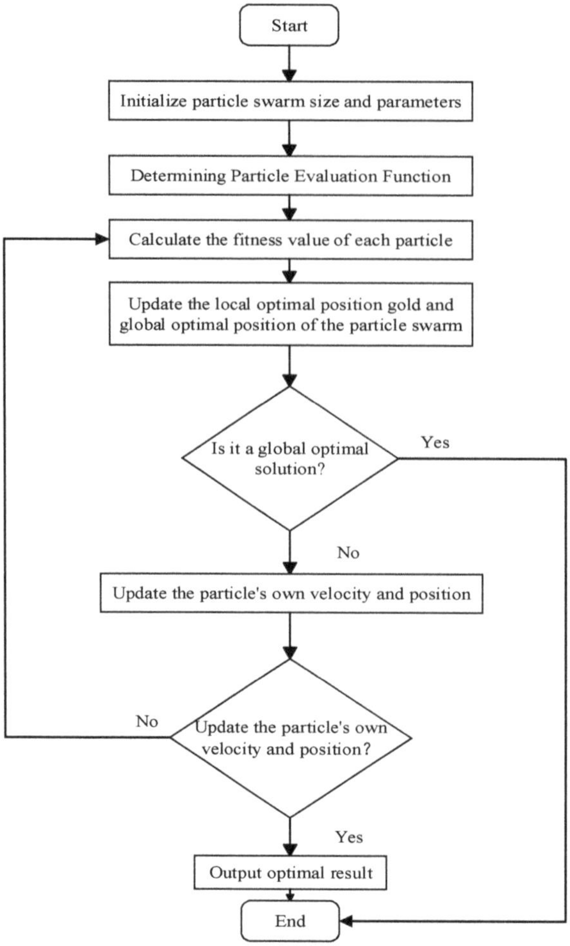

Figure 2. PSO algorithm flowchart.

(1) Adaptive optimization of inertia weight ω. The inertia weight not only determines the influence of the particle's historical flight state on the current flight speed but also balances the global search ability of the particle swarm algorithm. In order to improve the optimization performance of particle swarm optimization algorithm, in the early optimization stage, the global search ability of each particle should be expanded as much as possible, while in the later optimization stage, it is more desirable to have higher fine search ability. This article introduces a nonlinear decreasing inertia weight adaptive adjustment method, which not only gradually reduces the weight value as the number of runs increases, thereby achieving the goal of transforming the particle swarm algorithm from early global search to later fine search mode, but also effectively overcomes the problem of poor local search ability in the later stage of linear decreasing weight. The improved calculation method is shown in Equation (13).

$$w = a - (a - b)\sqrt{\frac{t}{N}} \tag{13}$$

In the formula, a and b are the maximum and minimum values of the inertia weight, t is the current number of iterations, and N is the maximum number of iterations.

(2) Adaptive optimization learning factors c_1 and c_2. The learning factor c_1 is the expression of particle self-awareness, used for the influence of self-awareness on flight trajectories, and c_2 is the expression of particle group consciousness, used for the influence of social cognition on flight trajectories. In the early stages of optimization, the focus must be on the individual's self-awareness, while in the later stages, the focus must be on the individual's cognitive ability to acquire social information. Therefore, in order to balance the influence of learning factors on flight trajectories, an optimization method was introduced where the learning factors adaptively change with inertia weights. The improved calculation formulas are shown in Equations (14) and (15), respectively.

$$c_1 = e_1 - (e_1 - e_2)cos(w) \tag{14}$$

$$c_2 = f_{1+}(f_2 - f_1)\ cos(w) \tag{15}$$

In the formula, w is the inertia weight, e_1 and f_1 are the initial values of the learning factor, and e_2 and f_2 are the termination values for the learning factor.

3.3.2. DMPSO-LSTM Model

Long Short-Term Memory Network (LSTM) is a special type of RNN with memory properties [38]. LSTM introduces a control input/output unit structure consisting of forget gates, input gates, and output gates. During information transmission, due to the unique internal structure of the storage unit, this structure can maintain a stable error flow, ensuring that the entire network does not cause gradient explosion or vanishing during training. It also has excellent memory performance and is very suitable for classifying, processing, and predicting time series data. The structure diagram of the LSTM network model is shown in Figure 3. Among them, f is the forget gate, i is the input gate, a is the output gate, C_t is the update process of the cell state, x_t is the input at the current time point, h_t is the hidden state at the current time point, h_{t-1} is the hidden state at the previous time point, and O is the output after function processing.

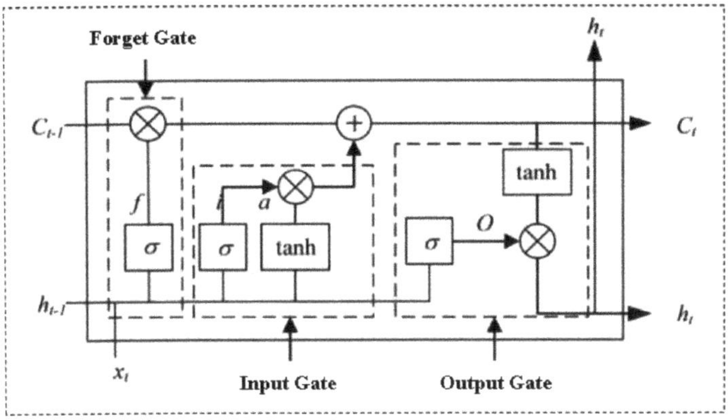

Figure 3. Schematic diagram of LSTM network model structure.

The values of parameters such as learning rate, step size, and iteration times in the LSTM model can affect the prediction results of the model. In order to avoid subjective factors in the parameter selection process and explore more reasonable and effective methods for predicting aircraft fuel consumption, a dynamic multidimensional particle swarm optimization algorithm (DMPSO) is proposed to optimize the learning parameters in the LSTM neural network, that is, to establish a DMPSO-LSTM model to predict the time series of aircraft fuel consumption.

The DMPSO-LSTM model takes the batch size, number of hidden layer units, time window size, and other related parameters of the LSTM model as optimization objects and uses an improved PSO algorithm to adjust and optimize its parameters. The flowchart for optimizing the LSTM model with DMPSO is shown in Figure 4.

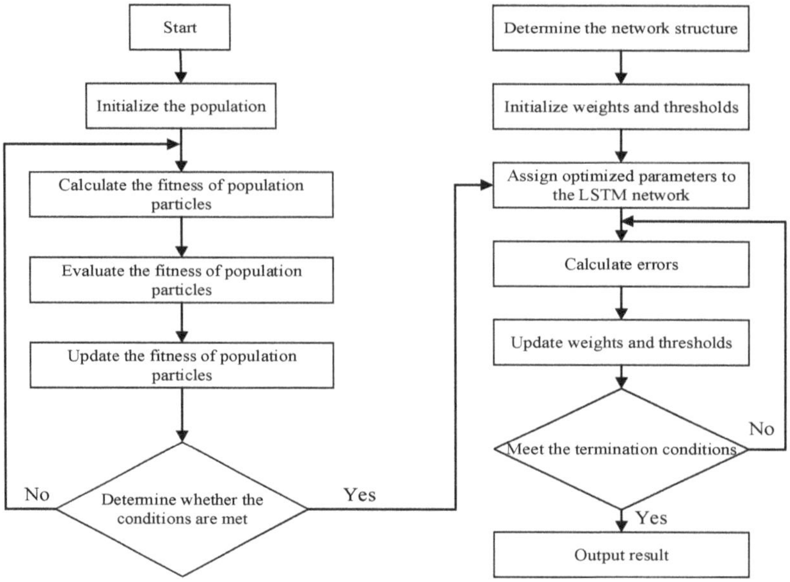

Figure 4. DMPSO optimization LSTM flowchart.

The specific process is as follows:

(1) PSO algorithm parameter initialization encoding. Initialize particle swarm parameters such as inertia weights, population size, iteration times, learning factors, and parameter solving intervals based on the LSTM parameters to be optimized.

(2) Determine the optimization parameters and range. Initialize the position x and velocity v of the particle swarm with the batch size, number of hidden layer units, and time window size of the LSTM model as the parameters to be solved and ensure that x and v are within the solution interval of the parameters.

(3) Calculate and evaluate the fitness function. Select the loss function of LSTM on the training set as the individual fitness function of the algorithm and initialize the global optimal particle position solution *gbest* and the local optimal particle position solution *pbest* based on the principle of minimizing fitness values.

(4) Update the fitness of the particle population. Calculate the weights and learning factors of the particle swarm based on Equations (13)–(15), and update *pbest* and *gbest* in real-time.

(5) Determine termination conditions. If satisfied, terminate the iteration and return the optimal parameter value, otherwise return to step (3).

(6) Output the result. Substitute the parameters into the LSTM neural network for training and output the prediction results through the test dataset.

The DMPSO-LSTM model structure consists of an input layer, two LSTM layers, and an output layer. The optimal parameter time window size of the adaptive particle swarm model is set to 10, the learning factors c_1 and c_2 are 1.5, the batch size is 60, the number of hidden layer units in the first layer is 12, the number of hidden layer units in the second layer is 22, the learning rate is set to 0.001, and the particle swarm population size is set to 30. The maximum iteration number is 500, and the velocity inertia weight w is 0.8. The Adam algorithm is used to optimize the model training process, and the loss function uses mean square error. The initial parameter settings for the DMPSO LSTM aircraft fuel consumption prediction model is shown in Table 1.

Table 1. DMPSO-LSTM aircraft fuel consumption prediction model parameter settings.

Parameter	Value
Population size	30
Time window	10
Learning factor	1.5
Inertia weight	0.8
Number of iterations	500
Learning rate	0.001
Batch size	60
Number of neurons in the first hidden layer	12
Number of neurons in the second hidden layer	22

4. Case Study

The QAR data used in this article are selected from the QAR data of a reference airline's A320 aircraft model as the experimental source, including 27,015 flight records. The QAR system captures a large amount of flight parameters and operational data from onboard systems, including time, takeoff airport, landing airport, aircraft type, takeoff runway, altitude, speed, heading, fuel consumption, and other data. Table 2 shows some sample data.

Table 2. Examples of QAR datasets.

Fuel Volume	Mach Number	Vertical Velocity	Takeoff Weight	Barometric Altitude	...	Vertical Acceleration	Longitudinal Acceleration	Airspeed	Ground Speed	Wind Speed	Bank Angle
5060	0.743	−128	64.084	31,196	...	0.93	0.0195	273.75	463	14	−0.44
4812	0.743	−208	64.084	31,192	...	0.93	0.0195	273.88	463	89	−0.09
4696	0.744	−304	64.084	31,188	...	0.93	0.0156	273.88	463	17	−2.64
4500	0.744	−400	64.084	31,188	...	0.93	0.0156	274	463	53	−1.58
...	...	...	...	...	...	...	...	...	...	51	−1.76
3976	0.744	−576	64.084	31,168	...	0.93	0.0078	274	462	...	...
3816	0.744	−656	64.084	31,156	...	0.93	0.0078	274.25	462	6	−2.72
3612	0.745	−720	64.084	31,148	...	0.938	0.0039	274.13	462	6	−2.46
3456	0.744	−784	64.084	31,128	...	0.949	0	274.63	462	6	−1.9
3380	0.744	−832	64.084	31,120	...	0.949	0	274.38	462	5	−2.72

The influencing factors for the final selection of the climb, cruise, and descent segments are shown in Table 3. The overall framework of the DMPSO-LSTM aircraft fuel consumption prediction model is shown in Figure 5. The overall process steps are as follows:

(1) Perform CEEMDAN on the fuel consumption influencing factors of each flight segment after data preprocessing to reduce its volatility.

(2) Use the kernel principal component analysis (KPCA) method to reduce the dimensionality of decomposed data with fewer input parameter dimensions.

(3) Use the DMPSO algorithm to optimize the optimal values of the LSTM network parameters and reduce the random errors caused by the manually set LSTM network parameters.

(4) The LSTM network optimized using the DMPSO algorithm is used to make preliminary predictions for the decomposed reconstructed sequence and residual sequence, respectively.

Table 3. List of influencing factors for each flight segment.

Flight Segment Name	Influence Factor
Climbing phase	Mach number, vertical velocity, barometric altitude, longitudinal acceleration, airspeed, ground speed, tilt angle
Cruise phase	Vertical speed, aircraft weight, barometric altitude, longitudinal acceleration, airspeed, tilt angle, wind speed, wind direction
Descending phase	Mach number, vertical velocity, barometric altitude, longitudinal acceleration, aircraft weight, tilt angle, wind speed, wind direction

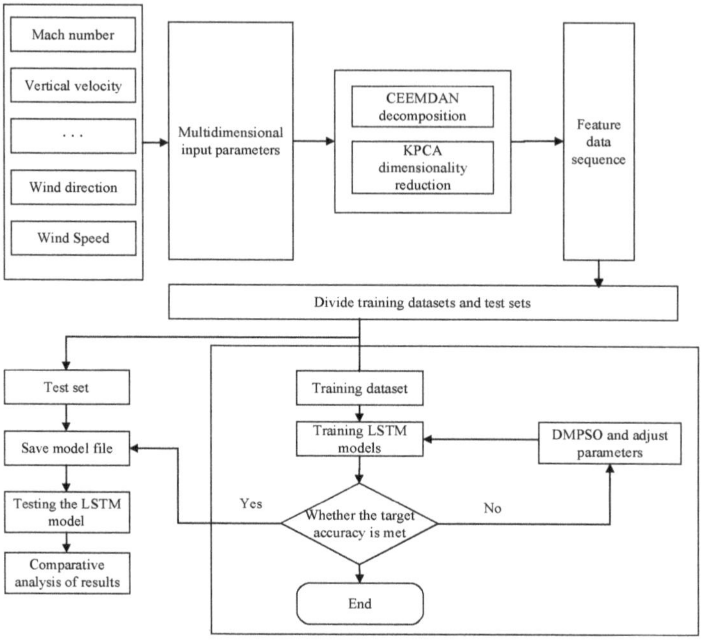

Figure 5. Overall framework diagram of CEEMDAN-KPCA-DMPSO-LSTM aircraft fuel consumption prediction model.

4.1. Data Preprocessing

This article will use methods such as missing value completion, outlier removal, moving average smoothing, and data normalization to preprocess QAR data, in order to construct a dataset for aircraft fuel consumption prediction based on neural networks. The dataset generation process is shown in Figure 6.

This paper uses the mean interpolation method to complete missing values. The mean interpolation method interpolates data based on the average value of the sample data or the average value of multiple samples. The calculation method is shown in Equation (16).

$$\overline{y} = \frac{\sum_{i=1}^{n} \beta_i y_i}{n_i} \tag{16}$$

Among them, β_i is a descriptive symbol indicating whether to answer, $\beta_i = 1$ represents "yes", $\beta_i = 0$ means "no", and n_i is a number.

Due to sensor malfunctions or errors, abnormal data collection and transmission, as well as flight operations and operating conditions, outliers are prone to occur in the QAR dataset. Figure 7 is a box plot of the selected QAR parameter data.

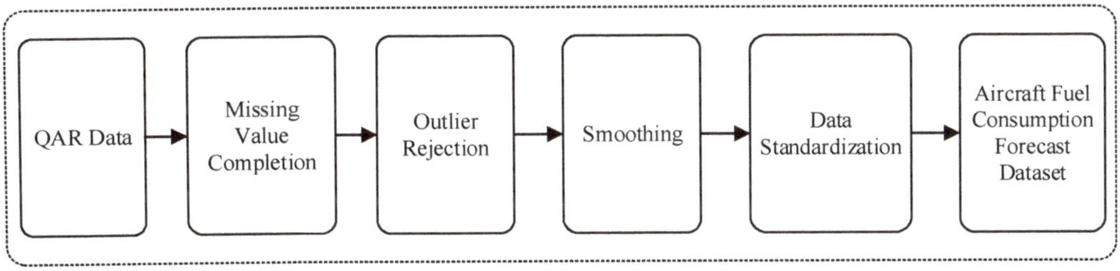

Figure 6. Dataset generation process.

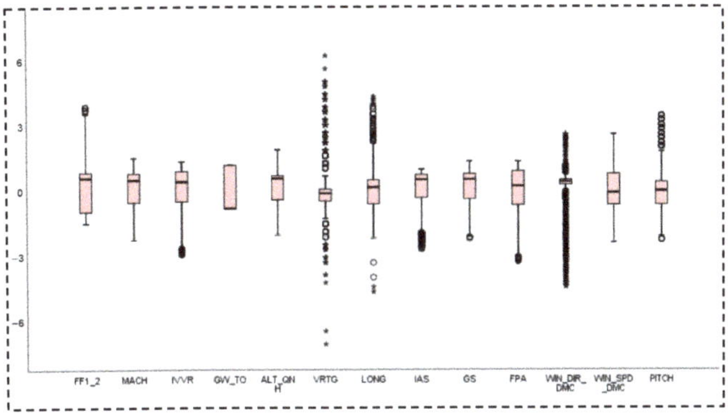

Figure 7. Selected parameter QAR data box plot.

From Figure 7, it can be seen that in the QAR data of the selected parameters, there are no outliers in the Mach number (MACH), aircraft weight (GW_TO), and barometric altitude (ALT_QNH) parameter columns. However, there are outliers in the remaining total fuel flow rate (FF1_2), vertical velocity (IVVR), vertical acceleration (VRTG), longitudinal acceleration (LONG), airspeed (IAS), ground speed (GS), tilt angle (FPA), wind speed (WIN_SPD_DMC), wind direction (WIN_DIR_DMC), and pitch angle (PITCH). This paper uses box plot analysis to remove outliers. The principle of removing outliers in box plots is based on the statistical method of interquartile range (IQR). The standard for removing outliers from box plots is to use the quartiles and interquartile range of the box plot to experimentally test the data. If the monitored data appears outside the box plot, it is abnormal data and needs to be removed. Circles represent ordinary outliers and asterisks indicate extreme outliers or multiple overlapping outliers.

This paper uses the moving average method to process noisy data. It is a method of smoothing the window by moving the average over the data to denoise the data. The algorithm not only has the advantage of low complexity but also can remove short-term fluctuations in the time series, making the data smoother and eliminating differences between various sample data of aircraft fuel consumption. Figure 8 shows the implementation results of the moving average method for smoothing different windows. From Figure 8, it can be seen that the setting of different windows in the moving average method has a significant impact on the effectiveness of data smoothing. When the window size is set to 3, the smoothed QAR data almost retain the local variation characteristics of the original aircraft fuel consumption data. As the window size increases, the smoothed QAR data curve becomes smoother but also lags more. When the window size is set to 9, the smoothed QAR data are too smooth to present the changes between adjacent time series. To preserve as much local variation as possible in the original aircraft fuel consumption QAR data, remove unnecessary noise, and obtain relatively smooth aircraft fuel consumption data, this section chooses to set the window size for smoothing processing to 5 in order to prepare for further analysis of aircraft fuel consumption characteristics and model construction.

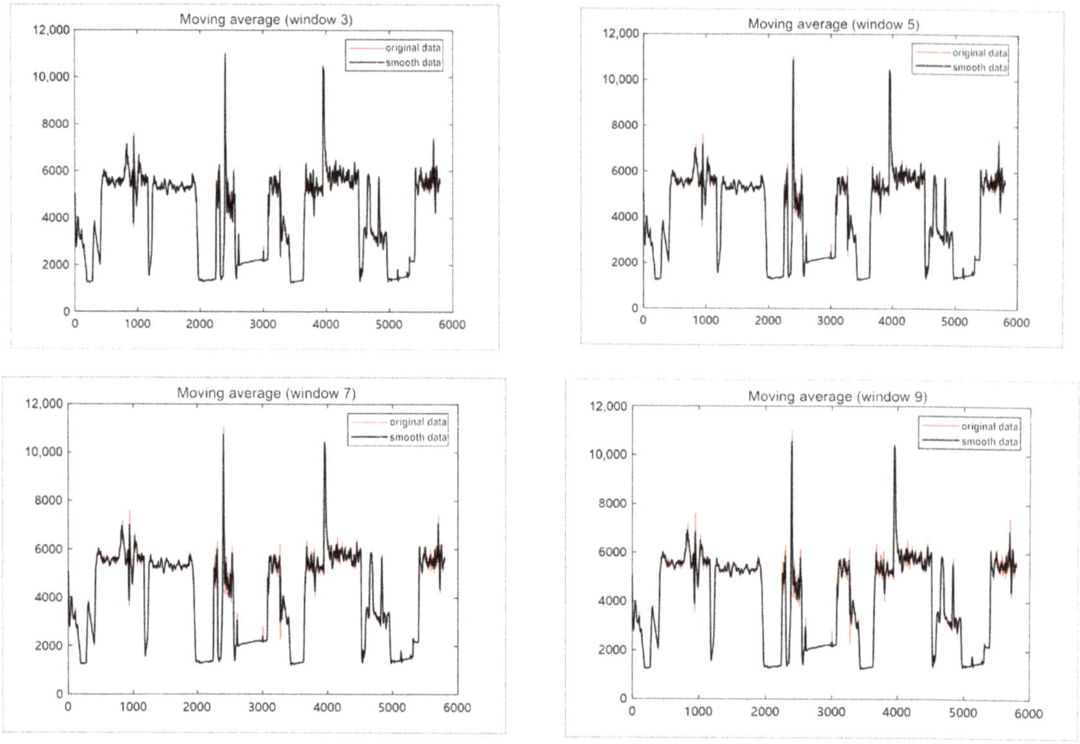

Figure 8. Schematic comparison of different window smoothing processes of moving average method.

Due to the different dimensions and units of QAR data parameters, in order to eliminate the influence of different parameter data ranges on prediction results, it is necessary to use the same dimensions or standardized data for the data before establishing the model. This article uses normalization to process the data, converting the processed data parameters into dimensionless data between [0, 1], thereby eliminating errors and improving prediction accuracy. The conversion method is shown in Equation (17).

$$\dot{X_{tn}} = \frac{X_{tn} - X_{tmin}}{X_{tmax} - X_{tmin}} \quad (17)$$

where $\dot{X_{tn}}$ is the normalized value of the n-th value in the t-th parameter, X_{tn} is the value before normalization of the n-th value in the t-th parameter, X_{tmax} is the maximum value in the t-th parameter, and X_{tmin} is the minimum value in the t-th parameter. The partial data after normalization are shown in Table 4, where the unit of fuel volume is kg.

Table 4. Example of partially normalized dataset.

Ground Speed	Longitudinal Acceleration	Vertical Acceleration	Wind Speed	...	Bank Angle	Airspeed	Pressure Altitude	Fuel Volume
0.25	0.16	−0.37	−0.75	...	0.36	0.67	−0.67	0.44
0.92	0.29	−0.53	0.98	...	0.53	0.99	0.99	−1.00
0.04	0.45	−0.60	−0.68	...	−0.70	0.34	−0.69	0.47
0.57	0.24	−0.53	0.15	...	−0.19	0.88	0.26	−0.55
0.58	0.22	−0.53	0.10	...	−0.28	0.88	0.24	−0.54
...	...	...	...	...	...	...	...	...
−0.93	0.03	−0.49	−0.93	...	−0.74	−0.05	−0.99	1.00
−0.93	0.24	−0.49	−0.93	...	−0.62	−0.08	−0.99	0.99
−0.7	0.22	−0.49	−0.93	...	−0.35	0.89	−0.99	0.98
−0.34	0.28	−0.29	−0.95	...	−0.74	−0.05	−0.99	0.92

4.2. Decomposition and Dimensionality Reduction in Factors Affecting Aircraft Fuel Consumption

4.2.1. Decomposition of Factors Influencing Aircraft Fuel Consumption Characteristics

The CEEMDAN algorithm is used to decompose the data sequence of factors affecting aircraft fuel consumption, and 500 sets of noise pairs are added to the decomposition algorithm with a noise amplitude set to 0.2. The input data of this algorithm is the data sequence of factors affecting aircraft fuel consumption in each flight segment, and the output is the IMF component and residual component (res). According to the adaptive decomposition performance of the algorithm, the Mach number data sequence was decomposed into seven sets of IMF components of different scales and one set of residual components by CEEMDAN. The decomposition results are shown in Figure 9.

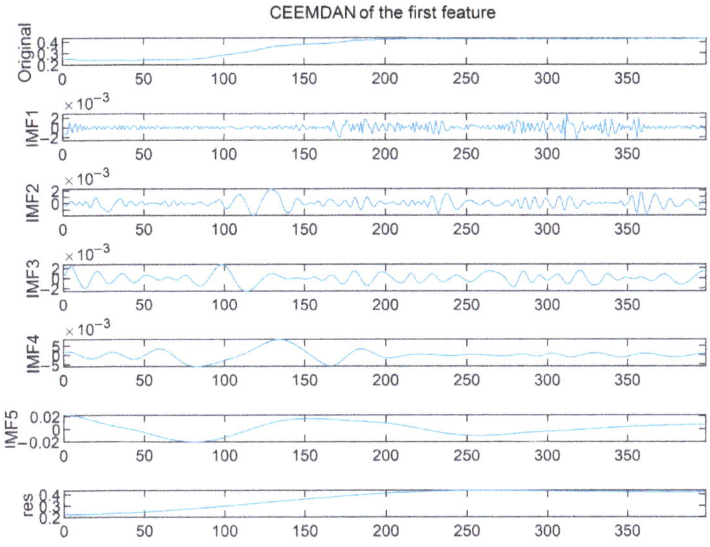

Figure 9. CEEMDAN results of Mach number data series.

The number of IMF components and remaining components obtained by decomposing the factors affecting aircraft fuel consumption during the climb, cruise, and descent segments using CEEMDAN is shown in Tables 5–7.

Table 5. IMF components and remaining components obtained by decomposition of series of influencing factors in climbing segment.

Factors Affecting of the Climbing Phase	IMF Number of Components/Piece	Remaining Number of Components/Piece
Mach number	10	1
Vertical velocity	11	1
Pressure altitude	10	1
Longitudinal acceleration	10	1
Airspeed	10	1
Ground speed	11	1
Bank angle	9	1
Total	71	7

Table 6. IMF components and remaining components obtained by decomposition of sequence of influencing factors in cruise segment.

Factors Affecting Cruise Phase	IMF number of Components/Piece	Remaining Number of Components/Piece
Vertical velocity	9	1
Airplane weight	8	1
Pressure altitude	9	1
Longitudinal acceleration	10	1
Airspeed	8	1
Bank angle	9	1
Wind speed	10	1
Wind direction	8	1
Total	71	8

Table 7. IMF components and remaining components obtained by decomposition of influencing factor series in descending segment.

Factors Influencing the Decline Stage	IMF Number of Components/Piece	Remaining Number of Components/Piece
Mach number	7	1
Vertical velocity	8	1
Pressure altitude	7	1
Longitudinal acceleration	8	1
Airplane weight	7	1
Bank angle	8	1
Wind speed	7	1
Wind direction	8	1
Total	60	8

4.2.2. Dimensionality Reduction in Aircraft Fuel Consumption Impact Characteristics

Use KPCA algorithm to perform dimensionality reduction on the data sequence obtained through CEEMDAN, with kernel function parameters set to 10, total input dimension set to 2, and cumulative contribution rate set to principal components greater than 95%. The individual feature contribution rates of the climb segment, cruise segment, and descent segment after KPCA dimensionality reduction are shown in Figures 10–12.

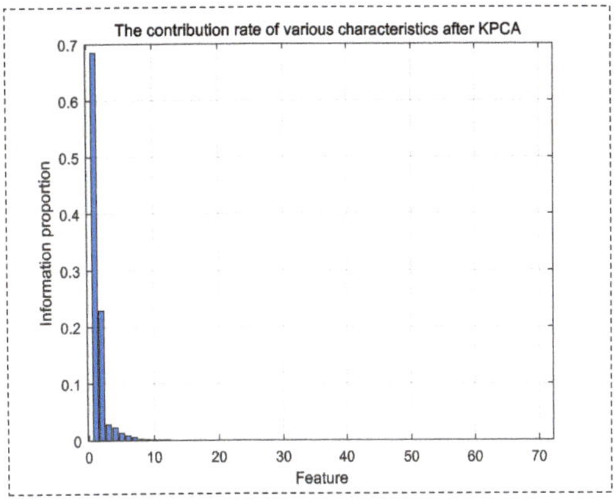

Figure 10. Contribution rate of each feature after KPCA dimensionality reduction in climbing segment.

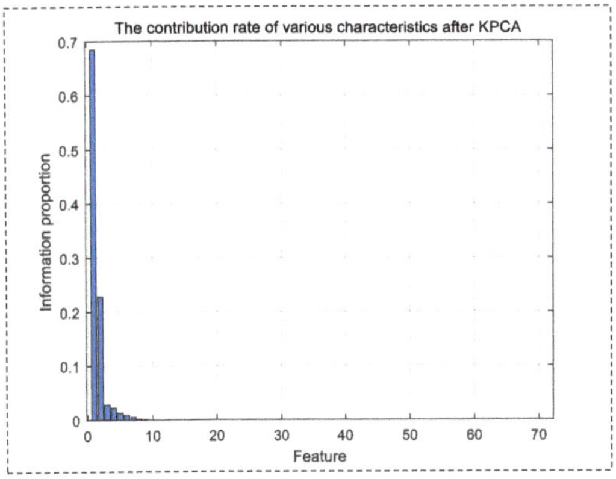

Figure 11. Contribution rate of each feature after KPCA dimensionality reduction in cruise segment.

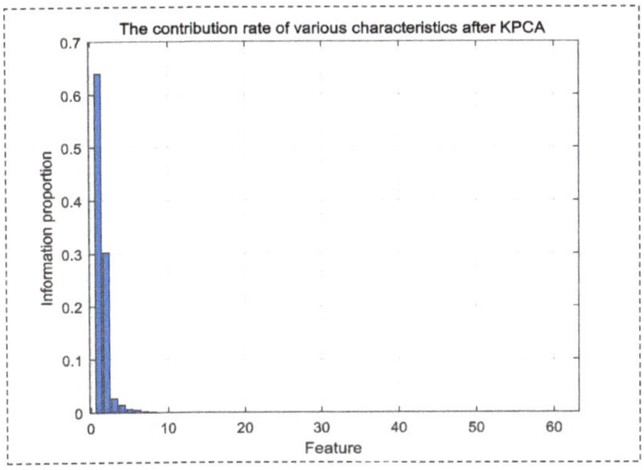

Figure 12. Contribution rate of each feature after KPCA dimensionality reduction in decline segment.

From the above experimental results, it can be seen that after KPCA dimensionality reduction, the contribution rates of the principal component numbers of the decomposed feature sequence are sorted from high to low, and the principal component numbers with relatively high contribution rates are selected based on a cumulative contribution rate of 98%. As a result, the factors affecting aircraft fuel consumption during the climb segment decreased from 71 dimensions to 10 dimensions, the factors affecting aircraft fuel consumption during the cruise segment decreased from 71 dimensions to 9 dimensions, and the factors affecting aircraft fuel consumption during the descent segment decreased from 60 dimensions to 9 dimensions.

4.3. Experimental Results

To verify the effectiveness of the proposed model, this paper compared the prediction results of four models: LSTM, CEEMDAN-PSO-LSTM, CEEMDAN-KPCA-PSO-LSTM, and CEEMDAN-KPCA-DPSO-LSTM. The comparison of aircraft fuel consumption prediction results for the climb, cruise, and descent segments is shown in Figures 13–15, and the corresponding prediction errors for each segment are shown in Figures 16–18. It can be seen from the figure that the CEEMDAN-KPCA-DMPSO-LSTM aircraft fuel consumption prediction model proposed in this paper has higher prediction accuracy than the single model that has not been optimized, whether in the climb leg, cruise leg, or descent leg. The change trend of the prediction results is more consistent with the actual change in the real value, and the prediction error of the model is smaller.

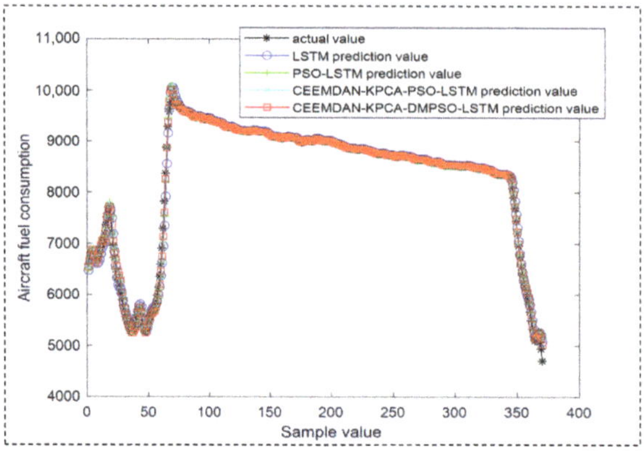

Figure 13. Forecast of aircraft fuel consumption during climb segment.

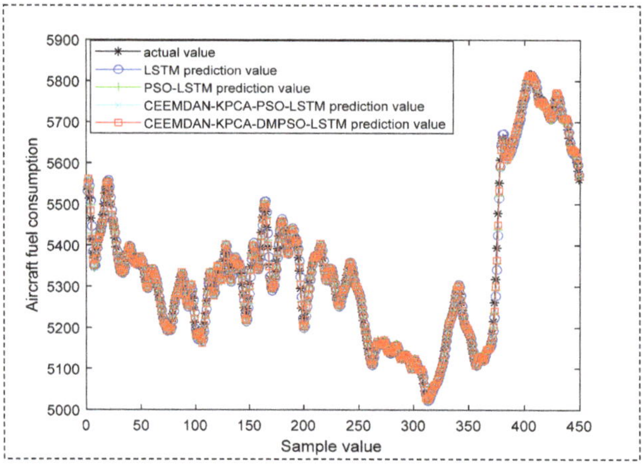

Figure 14. Forecast of aircraft fuel consumption during cruise segment.

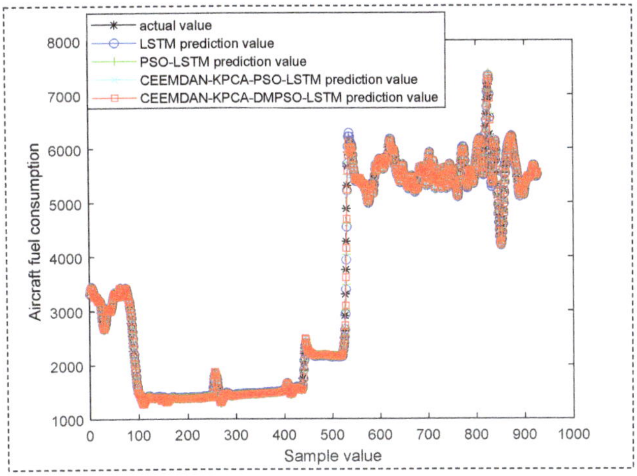

Figure 15. Forecast of aircraft fuel consumption in descent segment.

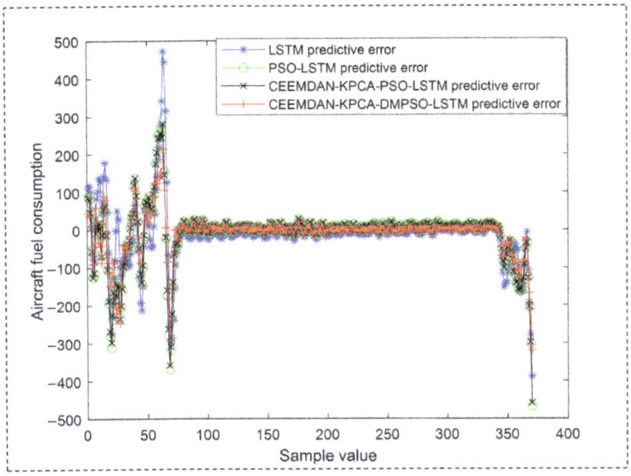

Figure 16. Prediction error of aircraft fuel consumption during climb segment.

The evaluation indicators for the prediction accuracy of the climb, cruise, and descent segments are shown in Tables 8–10. Compared with the three models of LSTM, PSO-LSTM, and CEEMDAN-KPCA-PSO-LSTM, the CEEMDAN-KPCA-DPSO-LSTM aircraft fuel consumption prediction model proposed in this paper achieved the best predictive performance in the climb, cruise, and descent segments. Compared with the other three models, the MAE index of the climbing segment decreased by 32.44%, 30.55%, and 18.48%, respectively, the RMSE index decreased by 33.34%, 30.21%, and 26.78%, respectively, and the goodness of fit error R^2 index increased by 2.27%, 2.12%, and 0.44%, respectively. Compared with the other three models, the MAE index of the cruise segment decreased by 30.06%, 26.69%, and 16.26%, respectively, the RMSE index decreased by 30.60%, 29.05%, and 22.34%, respectively, and the goodness of fit error R^2 index increased by 1.42%, 1.22%, and 0.43%, respectively. Compared with the other three models, the MAE index of the descending segment decreased by 14.25%, 12.09%, and 1.74%, the RMSE index decreased by 19.87%, 18.87%, and 11.93%, and the goodness of fit error R^2 index increased by 2.07%, 1.73%, and 1.00%, respectively.

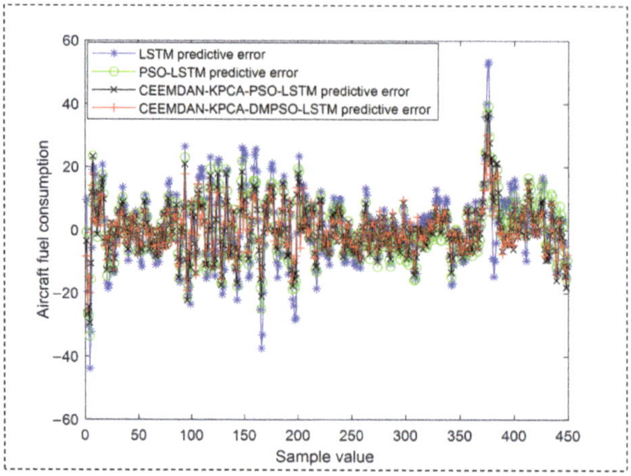

Figure 17. Aircraft fuel consumption prediction error during cruise segment.

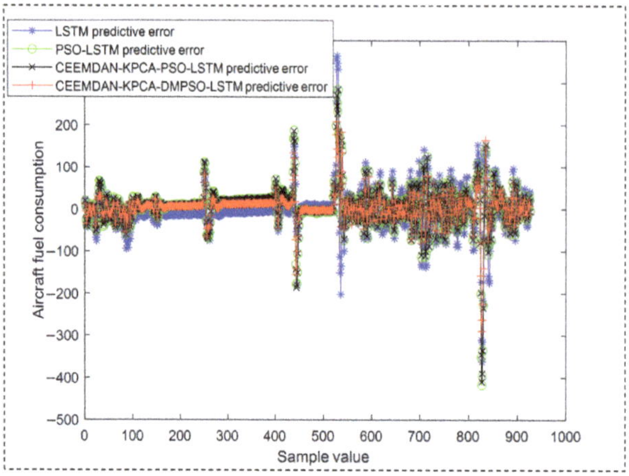

Figure 18. Aircraft fuel consumption prediction error in descent segment.

Table 8. Comparison of aircraft fuel consumption prediction performance during climb segment.

Model	MAE	RMSE	R^2
LSTM	20.9576	25.9763	0.95001
PSO-LSTM	20.3879	25.3159	0.95146
CEEMDAN-KPCA-PSO-LSTM	17.3672	23.6483	0.96731
CEEMDAN-KPCA-DMPSO-LSTM	14.1574	17.3145	0.97165

Table 9. Comparison of aircraft fuel consumption prediction performance in cruising segment.

Model	MAE	RMSE	R^2
LSTM	21.9853	26.4680	0.95942
PSO-LSTM	20.9731	25.8921	0.96132
CEEMDAN-KPCA-PSO-LSTM	18.3626	23.6528	0.96891
CEEMDAN-KPCA-DMPSO-LSTM	15.3752	18.3682	0.97312

Table 10. Comparison of aircraft fuel consumption prediction performance in descending segment.

Model	MAE	RMSE	R^2
LSTM	22.9428	24.0616	0.95998
PSO-LSTM	22.3781	23.7652	0.96316
CEEMDAN-KPCA-PSO-LSTM	20.0219	21.8924	0.97012
CEEMDAN-KPCA-DMPSO-LSTM	19.6724	19.2792	0.97987

In order to further validate the predictive performance of the model, this paper compared it with the BP and RNN models for verification. We conducted a systematic comparison with two baseline neural networks (BP and RNN) across all flight phases. The comprehensive evaluation metrics in Tables 11–13 demonstrate substantial performance improvements through three critical dimensions: prediction accuracy (MAE), error stability (RMSE), and model explanatory power (R^2). Notably, the proposed model achieved a remarkable 36.5–45.4% reduction in MAE compared to the BP networks and 13.9–31.0% improvement over RNNs across different flight phases. The consistent superiority in RMSE metrics (16.6–42.4% lower than BP, 24.0–35.3% better than RNNs) further confirmed the enhanced prediction stability. Particularly significant is the R^2 enhancement to the 0.971–0.980 range, approaching ideal fitting status ($R^2 = 1$), which represents 6.1–8.3% improvement over BP and 4.7–6.9% advancement beyond RNN baselines. Table 11 reveals that during the climb phase, our model reduces fuel consumption prediction errors by 41.2% (MAE) and 45.0% (RMSE) compared to BP, while maintaining 97.2% data variance explanation capability. The cruising phase analysis in Table 12 shows even greater stability improvements, with 45.4% MAE reduction from BP baseline. Table 13 demonstrates particular advantages in descent phase modeling, where the complex aerodynamic characteristics are effectively captured through our feature decomposition and optimization mechanisms, yielding 97.99% R^2, which significantly outperforms both benchmarks.

Table 11. Comparison of predicted performance of aircraft fuel consumption of different models in climb segment.

Model	MAE	RMSE	R^2
BP	24.0872	31.4789	0.89685
RNN	22.3257	28.3695	0.91465
CEEMDAN-KPCA-DMPSO-LSTM	14.1574	17.3145	0.97165

Table 12. Comparison of fuel consumption prediction performance of different models in cruising segment.

Model	MAE	RMSE	R^2
BP	28.1542	34.5732	0.90347
RNN	25.4379	31.4635	0.92578
CEEMDAN-KPCA-DMPSO-LSTM	15.3752	18.3682	0.97312

Table 13. Comparison of fuel consumption prediction performance of different models in descent segment.

Model	MAE	RMSE	R^2
BP	26.3580	33.4637	0.91615
RNN	24.4732	27.3468	0.93575
CEEMDAN-KPCA-DMPSO-LSTM	19.6724	19.2792	0.97987

5. Conclusions

With the increasing global attention to the sustainable development of the aviation industry, reducing carbon emissions and improving fuel efficiency have become key challenges facing the aviation industry. The paper constructs an aircraft fuel consumption prediction model based on DMPSO-LSTM and verifies it through simulation using QAR data of an A320 aircraft model from a certain airline. The results show that the proposed aircraft fuel consumption prediction model has good predictive performance in the climb, cruise, and descent stages. The enhanced prediction accuracy achieved in this paper directly contributes to fuel consumption optimization through three main mechanisms: (1) enabling precise weight–fuel consumption mapping for optimal payload distribution, (2) supporting multi-objective optimization of flight trajectories considering both temporal and energetic efficiency, and (3) facilitating dynamic fuel loading strategies that reduce unnecessary safety margins while maintaining operational safety. The main contributions of this article are as follows:

(1) The CEEMDAN method was used to decompose aircraft fuel consumption data, obtaining time series data that better highlight local features and solve the problems of modal aliasing and uneven modal gaps that may occur in traditional decomposition methods when processing non-stationary signals.

(2) By using the KPCA method to reduce the dimensionality of the decomposed data while maintaining information validity and representativeness, the computational efficiency and prediction accuracy of the model have been improved.

(3) By constructing a dynamic multi-swarm particle swarm optimization algorithm (DMPSO), the problem of the traditional PSO algorithm easily becoming stuck in local optima and having low search accuracy in the solution space optimization process is solved.

The limitation of this article is that it does not consider the influence of complex external environments, such as aircraft health monitoring data, aircraft maintenance, etc., which may affect the prediction accuracy of the model in special circumstances. While the current model captures weight-related impacts through mass-dependent parameters, future work will explicitly incorporate aircraft weight optimization models considering both structural limits and operational requirements. Meanwhile, the causal relationship between the predicted results and actual flight operations has not been established, and there is a lack of diagnostic ability for abnormal fuel consumption. Moreover, traditional machine learning may face decision risks due to data noise. In the future, this article will consider introducing multi-source heterogeneous data such as meteorological data, maintenance records, and aircraft health monitoring data, constructing a spatiotemporal physical fusion feature set, and introducing an engine performance degradation equation to develop a domain adaptive cross aircraft transfer learning framework to reduce the model reconstruction cost of heterogeneous fleets. At the same time, SHAP (Shapley Additive explant) values are introduced to quantify the contribution of features and conduct intuitive root cause analysis of fuel consumption anomalies.

Author Contributions: Conceptualization, J.X. and C.Z.; methodology, C.Z. and J.X.; formal analysis, Y.S. and J.X.; investigation, C.Z. and G.Y.; resources, J.X.; data curation, Y.W. and C.Z.; verification, C.Z., Y.S. and J.X.; writing—original draft preparation, J.X., C.Z. and Y.W.; writing—review and editing, Y.S., G.Y. and C.Z.; visualization, G.Y. and C.Z.; supervision, J.X.; project administration, Y.W. and J.X. All authors have read and agreed to the published version of the manuscript.

Funding: This research was supported by the Ministry of Industry and Information Technology of China (No. 2022-A04R-1-1), the Chinese National Nature Science Foundation Project (No. 72471133), and the Natural Science Foundation of Shanghai Municipality (No. 21ZR1423800).

Institutional Review Board Statement: Not applicable.

Informed Consent Statement: Not applicable.

Data Availability Statement: The data are confidential and were obtained from a specific airline.

Conflicts of Interest: Chunling Zou was employed by the company Scania China Manufacturing Co., Ltd. Yongbing Wan was employed by the company Scania Shanghai Shentong Metro Group Co., Ltd. Technical Center. The remaining authors declare that the research was conducted in the absence of any commercial or financial relationships that could be construed as a potential conflict of interest.

References

1. IEA. *CO2 Emissions in 2022*; International Energy Agency: Paris, France, 2023.
2. Siavash, K.; Eetu, R.; Dmitrii, B.; Christian, B. Global Transportation demand development with impacts on the energy demand and greenhouse gas emissions in a climate-constrained world. *Energies* **2019**, *12*, 3870. [CrossRef]
3. ICAO. *2022 Environmental Report: Aviation's Roadmap to Net-Zero Carbon Emissions by 2050*; ICAO: Montreal, QC, Canada, 2022.
4. Pagoni, I.; Psaraki-Kalouptsidi, V. Calculation of aircraft fuel consumption and CO_2 emissions based on path profile estimation by clustering and registration. *Transp. Res. Part D Transp.* **2017**, *54*, 172–190. [CrossRef]
5. Baklacioglu, T. Modeling the fuel flow-rate of transport aircraft during flight phases using genetic algorithm-optimized neural networks. *Aerosp. Sci. Technol.* **2016**, *49*, 52–62. [CrossRef]
6. Nikoleris, T.; Gupta, G.; Kistler, M. Detailed estimation of fuel consumption and emissions during aircraft taxi operations at Dallas/Fort Worth International Airport. *Transp. Res. Part D Transp. Environ.* **2011**, *16*, 302–308. [CrossRef]
7. Yanto, J.; Liem, R.P. Aircraft fuel burn performance study: A data-enhanced modeling approach. *Transp. Res.* **2018**, *65*, 574–595. [CrossRef]
8. Zhang, M.; Huang, Q.; Liu, S.; Zhang, Y. Fuel Consumption Model of the Climbing Phase of Departure Aircraft Based on Flight Data Analysis. *Sustainability* **2019**, *11*, 4362. [CrossRef]
9. Nuic, A.; Poles, D.; Mouillet, V. BADA: An advanced aircraft performance model for present and future ATM systems. *Int. J. Adapt. Control Signal Process.* **2022**, *24*, 850–866. [CrossRef]
10. Khan, W.A.; Ma, H.-L.; Ouyang, X.; Mo, D.Y. Prediction of aircraft trajectory and the associated fuel consumption using covariance bidirectional extreme learning machines. *Transp. Res. Part E Logist. Transp. Rev.* **2021**, *145*, 102189. [CrossRef]
11. Kaiser, M.; Schultz, M.; Fricke, H. Enhanced jet performance model for high precision 4D flight path prediction. In Proceedings of the 1st International Conference on Application and Theory of Automation in Command and Control Systems, Barcelona, Spain, 26–27 May 2011; pp. 33–40.
12. Dalmau, R.; Prats, X. Fuel and time savings by flying continuous cruise climbs. *Transp. Res. Part D Transp. Environ.* **2019**, *35*, 62–71. [CrossRef]
13. Burzlaff, M. Aircraft Fuel Consumption-Estimation and Visualization. Ph.D. Thesis, Hamburg University of Applied Science, Hamburg, Germany, 2017.

14. Yutko, B.M. Approaches to Representing Aircraft Fuel Efficiency Performance for the Purpose of a Commercial Aircraft Certification Standard. Ph.D. Thesis, Massachusetts Institute of Technology, Cambridge, MA, USA, 2021.
15. Jensen, L.; Hansman, R.J.; Venuti, J.C.; Reynolds, T. Commercial airline speed optimization strategies for reduced cruise fuel consumption. In Proceedings of the 2013 Aviation Technology, Integration, and Operations Conference, Los Angeles, CA, USA, 12–14 August 2013; p. 4289.
16. Trani, A.; Wing-Ho, F.; Schilling, G.; Baik, H.; Seshadri, A. A Neural Network Model to Estimate Aircraft Fuel Consumption. *Aiaa J.* **2024**, *10*, 61–68.
17. Wang, Z.M.; Xue, D.B.; Wu, L.X.; Yan, R. A reliable predict-then-optimize approach for minimizing aircraft fuel consumption. *Transp. Res. Part D Transp.* **2025**, *142*, 104693. [CrossRef]
18. Wang, B.; Zou, R.Y.; Mao, J.F.; Wu, C.L.; Xue, D.B. Developing an aircraft takeoff mass estimation model based on the hybrid KMI-DNN-BI model using quick access recorder (QAR) data. *Aerosp. Sci. Technol.* **2025**, *158*, 109918. [CrossRef]
19. Wang, F.; Li, X.; Zhang, K. Multivariate regression with PCA for aircraft fuel consumption prediction: A case study of QAR data analytics. *Transp. Res. Part D Transp. Environ.* **2022**, *108*, 103315.
20. Smith, T.; Jones, P. Bayesian hierarchical modeling of aircraft fuel burn integrating QAR data and performance benchmarks. *Aerospace* **2021**, *8*, 256.
21. García, M.; Pérez, A.; López, R. Generalized additive models for fuel consumption prediction: A QAR data-driven approach with flight phase segmentation. *J. Air Transp. Manag.* **2023**, *107*, 102335.
22. International Air Transport Association [IATA]. *Statistical Methods for Aviation Fuel Efficiency Analysis Using Flight Data;* IATA Technical Report No. 2022-ATFR-15; IATA: Montreal, QC, Canada, 2022.
23. Zhang, Y.; Zhou, M.; Chen, H.; Wang, L. Hybrid CNN-LSTM model for aircraft fuel burn prediction using high-frequency QAR data. *Transp. Res. Part C Emerg. Technol.* **2021**, *132*, 103405.
24. Liu, Y.; Wang, H.; Zhang, T. A transformer-based approach for aircraft fuel consumption prediction with attention to QAR data spatiotemporal features. *Expert Syst. Appl.* **2022**, *207*, 117952.
25. Chen, X.; Li, Z.; Zhang, Q. Graph neural networks for integrated fuel prediction: Combining QAR data, radar tracks, and weather impacts. *IEEE Trans. Aerosp. Electron. Syst.* **2023**, *59*, 1450–1463.
26. Liu, H.; Wang, J. Data-driven fuel efficiency optimization in commercial aviation: A QAR-based anomaly detection framework. *Aerosp. Sci. Technol.* **2022**, *128*, 107754.
27. Chen, X.; Li, Z.; Zhang, Q.; Liu, Y. Spatiotemporal feature fusion for fuel prediction: Integrating QAR data and air traffic control constraints. *IEEE Trans. Intell. Transp. Syst.* **2023**, *24*, 2876–2888.
28. Schäfer, A.W.; Yadav, P.; Stolarski, D.G. Fuel consumption prediction and trajectory optimization for climate-efficient aviation. *Transp. Res. Part D Transp. Environ.* **2022**, *109*, 103335.
29. Zhang, Y.; Li, Q.; Wang, J. A multi-objective deep learning framework for fuel prediction and emission budgeting in sustainable aviation. *Appl. Energy* **2023**, *334*, 120678.
30. IATA. *Machine Learning for ESG Reporting in Aviation;* IATA: Geneva, Switzerland, 2023.
31. Airbus. *Global Market Forecast: Navigating the Future;* Airbus: Leiden, The Netherlands, 2023.
32. Zhang, Y.; Li, H.; Chen, Z. An enhanced CEEMDAN method with adaptive noise amplitude selection for rotating machinery fault diagnosis. *Mech. Syst. Signal Process.* **2023**, *189*, 110432. [CrossRef]
33. Wang, L.; Liu, J.; Yang, Q. CEEMDAN-based hybrid deep learning model for ECG signal denoising and arrhythmia classification. *IEEE Trans. Instrum. Meas.* **2022**, *71*, 1–12.
34. Li, X.; Zhang, Y.; Wang, H. An adaptive particle swarm optimization with dynamic inertia weight and learning factors for high-dimensional optimization. *Expert Syst. Appl.* **2023**, *213*, 119213.
35. Chen, R.; Li, W.; Zhang, T. A dynamic multi-strategy particle swarm optimizer with adaptive parameters for engineering design optimization. *Swarm Intell.* **2024**, *18*, 45–67.
36. Gupta, S.; Deep, K. A hybrid PSO algorithm with adaptive learning mechanism for complex optimization problems. *Swarm Evol. Comput.* **2022**, *71*, 101090.
37. Fu, W.; Wang, B.; Li, X.; Liu, L.; Wang, Y.l. Particle swarm optimization algorithm of learning factors and time factor adjusting to weights. *Appl. Res. Comput.* **2014**, *31*, 3291–3294.
38. Sudhir-Kumar, P.; Narendra, B.C.; Mukul, B. Long short-term memory. *Big Data Min. Anal.* **2021**, *4*, 279–297.

energies

MDPI

Article

A Data Imputation Approach for Missing Power Consumption Measurements in Water-Cooled Centrifugal Chillers

Sung Won Kim [1] and Young Il Kim [2,*]

[1] Department of Architectural Engineering, Graduate School, Seoul National University of Science & Technology, Seoul 01811, Republic of Korea; rogun@koies.or.kr
[2] School of Architectural, Seoul National University of Science & Technology, Seoul 01811, Republic of Korea
* Correspondence: yikim@seoultech.ac.kr

Abstract: In the process of collecting operational data for the performance analysis of water-cooled centrifugal chillers, missing values are inevitable due to various factors such as sensor errors, data transmission failures, and failure of the measurement system. When a substantial amount of missing data is present, the reliability of data analysis decreases, leading to potential distortions in the results. To address this issue, it is necessary to either minimize missing occurrences by utilizing high-precision measurement equipment or apply reliable imputation techniques to compensate for missing values. This study focuses on two water-cooled turbo chillers installed in Tower A, Seoul, collecting a total of 118,464 data points over 3 years and 4 months. The dataset includes chilled water inlet and outlet temperatures (T_1 and T_2) and flow rate ($\dot{V}_1$) and cooling water inlet and outlet temperatures (T_3 and T_4) and flow rate ($\dot{V}_3$), as well as chiller power consumption ($\dot{W}_c$). To evaluate the performance of various imputation techniques, we introduced missing values at a rate of 10–30% under the assumption of a missing-at-random (MAR) mechanism. Seven different imputation methods—mean, median, linear interpolation, multiple imputation, simple random imputation, k-nearest neighbors (KNN), and the dynamically clustered KNN (DC-KNN)—were applied, and their imputation performance was validated using MAPE and CVRMSE metrics. The DC-KNN method, developed in this study, improves upon conventional KNN imputation by integrating clustering and dynamic weighting mechanisms. The results indicate that DC-KNN achieved the highest predictive performance, with MAPE ranging from 9.74% to 10.30% and CVRMSE ranging from 12.19% to 13.43%. Finally, for the missing data recorded in July 2023, we applied the most effective DC-KNN method to generate imputed values that reflect the characteristics of the studied site, which employs an ice thermal energy storage system.

Keywords: centrifugal chiller; CVRSME; data imputation; DC-KNN; MAPE; performance analysis

1. Introduction

1.1. Research Background and Objectives

Water-cooled chillers play a crucial role in maintaining stable indoor environments in large-scale cooling systems. They primarily handle the cooling demands of large buildings and, as a result, account for a significant proportion of total energy consumption [1]. In recent years, improving the efficiency of chillers has become a critical challenge, particularly in response to climate change. Notably, the summers of 2023 and 2024 experienced record-breaking heat waves, along with abnormal autumnal high temperatures, leading to a sharp increase in cooling demand. Consequently, it has become essential to manage chillers efficiently to maintain a stable indoor climate while simultaneously reducing energy consumption [2]. The control of the chilled water outlet temperature in water-cooled chillers directly influences the coefficient of performance (COP) and overall power consumption. This indicates that optimizing operational conditions can significantly reduce building energy usage [3]. To assess this relationship, it is necessary to collect and analyze data over an extended period, including chilled water and cooling water temperatures, flow rates, and power consumption. However, collected data often contain missing values due to sensor failures, data transmission errors, or environmental factors. If missing values occur in sensor data during cooling operations, the predictive accuracy of analytical models decreases. Missing values can introduce errors in model training or reduce the dataset size when removed, ultimately leading to a decline in model performance [4]. Thus, an effective strategy for imputing missing data is crucial to ensure data quality. This study applies and evaluates various imputation techniques to enhance the quality of chiller operational data. A total of 118,464 real-world data points were subjected to a missing-at-random (MAR) mechanism. The missing data were then imputed using seven different techniques, and their performance was assessed. Finally, the most effective imputation method was applied to replace missing values in the dataset for July 2023.

1.2. Limitations of Previous Studies

Previous studies have utilized a variety of statistical techniques and machine learning algorithms for missing data imputation. However, several recurring limitations have been observed:

A. Many prior works applied a single imputation method without conducting a thorough analysis of the missing data mechanism, such as missing completely at random (MCAR) or missing at random (MAR). As a result, the selected methods often failed to reflect the true nature of the data, leading to reduced imputation quality [5];
B. In high-dimensional and dynamically changing environments, such as time-series sensor data, traditional studies have predominantly relied on simplistic statistical approaches such as mean substitution, linear interpolation, or regression-based imputation. While computationally efficient, these methods are limited in their ability to capture complex temporal patterns and inter-cluster heterogeneity [6];
C. Conventional data-driven methods such as k-nearest neighbors (KNNs) rely on fixed-distance similarity measures, which fail to consider the existence of clustered structures or local variations. As a result, their performance tends to degrade under varying operational conditions or device-specific patterns [7];
D. There is a lack of advanced imputation techniques tailored to real-world sensor-based environments. Particularly in physical systems such as water-cooled chillers, various types of missing data may occur due to environmental disturbances, sensor malfunctions, or communication failures. However, case studies addressing such practical scenarios remain limited [8].

To overcome these challenges, this study proposes a novel algorithm—dynamically clustered k-nearest neighbors (DC-KNNs)—which reflects inherent data clusters through unsupervised clustering and applies dynamic weighting schemes within each cluster. By improving on the limitations of conventional KNNs, the proposed method provides more precise imputations that respect the structural complexity of time-series sensor data. This advancement not only enhances imputation precision but also significantly contributes to more reliable and accurate forecasting in intelligent energy systems [9].

1.3. Research Objectives and Scope

This study aims to identify missing values within a dataset consisting of 118,464 operational records from two water-cooled turbo chillers over a three-year period. The research systematically compares and analyzes various statistical imputation techniques to determine the most effective method. Missing values were imputed using mean imputation, median imputation, linear interpolation, multiple imputation, simple random imputation, and KNN imputation. The most effective approach was then further enhanced by proposing a DC-KNN imputation method. The DC-KNN method enhances traditional KNN imputation by addressing two core limitations: ① equal weighting of variables regardless of their predictive power and ② blind selection of neighbors solely based on global distance metrics without considering local data structures.

A. Variable-specific importance weighting

Unlike conventional KNNs, where all variables contribute equally to the distance computation, the DC-KNN method assigns a tailored weight to each variable based on its statistical importance in predicting the target variable. For instance, in the context of chiller operation data, features such as flow rate or evaporator outlet temperature may exhibit stronger correlations with the missing variable than others (e.g., ambient temperature). These weights are either of the following:

- Computed via correlation coefficients, mutual information, or domain-specific feature ranking;
- Integrated directly into the distance function, allowing more relevant variables to exert greater influence.

This mechanism improves sensitivity to domain-relevant features, enhancing imputation precision in complex physical systems like heating, ventilation, and air conditioning (HVAC) operations.

B. Cluster-based neighbor selection

The standard KNN method selects the k-nearest neighbors globally, which risks drawing in contextually dissimilar instances, especially in heterogeneous datasets with multiple operational modes or patterns.

The DC-KNN method first performs unsupervised clustering based on data similarity profiles and then restricts neighbor selection to the cluster that the instance belongs to. This allows the method to achieve the following:

- Respect local data topology;
- Reduce contamination from out-of-distribution points;
- Preserve the semantic integrity of imputed values.

For example, a chiller operating under partial load conditions will seek imputation references only from similar partial-load samples rather than from full-load or standby modes.

1.4. Key Contributions of This Study

This study proposes the DC-KNN imputation method to overcome the limitations of existing techniques and enhance the reliability of chiller operation data. By improving the accuracy of COP (Coefficient of Performance) analysis, this method is expected to contribute to improving data quality within Building Energy Management Systems (BEMSs). The key contributions of this study are as follows:

A. Introduction of a novel approach to address missing data in chiller operation data:

  - Analyzes the limitations of commonly used imputation methods, such as mean imputation, median imputation, linear interpolation, and KNN imputation;

- Develops a new DC-KNN imputation method that improves upon conventional KNNs by incorporating clustering-based filtering and dynamic weighting, leading to more precise missing data imputation.

B. Demonstration of practical applicability using real operational data:

- The dataset consists of 118,464 chiller sensor records collected over three years in a real-world operational environment, ensuring high practical applicability;
- The analysis is conducted based on long-term observed data, enhancing the reliability of the imputed results by utilizing an extensive dataset.

C. Analysis of the optimal combination of missing data rates and imputation techniques:

- This study compares and analyzes various missing data imputation techniques to identify the optimal imputation method based on different missing rates (10%, 20%, and 30%). This research provides valuable insights into data-driven chiller performance optimization and offers a robust guideline for handling missing data in HVAC system analysis.

2. Theoretical Review

2.1. Outlier Detection

2.1.1. Thermodynamic Characteristics of Chillers

The thermodynamic characteristics of water-cooled chillers are defined by the interaction between heat and work within the refrigeration cycle. Heat and work represent the two fundamental forms of energy transfer in thermodynamics, each governed by distinct principles and mechanisms. Heat transfer occurs due to temperature differences, whereas work is the transfer of energy resulting from a force applied over a distance. Unlike stored energy, such as internal energy, heat and work exist only when energy moves across system boundaries. This distinction highlights the transient nature of heat and work, where energy can enter or leave a system in these forms but cannot be stored as heat or work itself.

The relationship between heat and work is formalized by the first law of thermodynamics, which states that energy cannot be created or destroyed but can only be converted from one form to another. This principle plays a crucial role in refrigeration systems, where mechanical work applied by the compressor enables heat transfer from a low-temperature heat source (evaporator) to a high-temperature heat sink (condenser). During this process, the refrigerant absorbs heat in the evaporator, requiring work input to circulate the refrigerant, and subsequently releases heat in the condenser.

One of the key concepts in the interaction between heat and work is entropy. While heat transfer always involves entropy transfer, work transfer does not. This distinction indicates that heat contains a certain level of disorder, whereas work represents a more structured form of energy transfer. Consequently, the efficiency of a refrigeration cycle depends not only on the amount of energy transferred but also on the mode of transfer and its impact on system entropy.

Understanding the interaction between heat and work in practical refrigeration applications is essential for optimizing system performance. The refrigeration cycle consists of four primary processes: evaporation, compression, condensation, and expansion. These processes are maintained through a controlled interaction of heat absorption, mechanical work input, and heat release. Effective management of these energy transfers improves chiller efficiency, reduces energy consumption, and ensures optimal cooling performance.

The refrigerant absorbs heat in the evaporator, releases heat in the condenser, and consumes energy in the compressor to circulate throughout the system. In the evaporator, the refrigerant utilizes the temperature difference between the inlet and outlet chilled water to achieve cooling, thereby providing a cooling effect. The system's components, operating mechanisms, and thermal storage processes are illustrated in Figure 1. The energy transfer process in the chiller is expressed by the following heat transfer equation [10]:

$$\dot{Q} = \rho \dot{V} c \Delta T \tag{1}$$

$$\dot{Q}_e = \rho_b \dot{V}_1 c_b (T_1 - T_2) \tag{2}$$

$$\dot{Q}_c = \rho_w \dot{V}_3 c_w (T_4 - T_3) \tag{3}$$

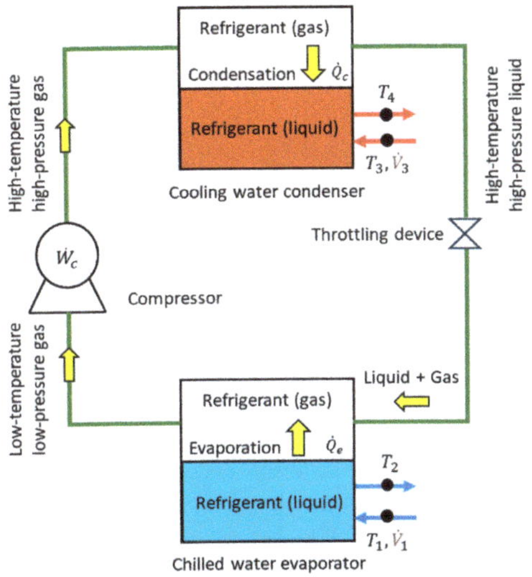

Figure 1. Schematic of a vapor-compression water chiller.

2.1.2. Energy Balance Error (EB)

In this study, reliable data were selected based on the first law of thermodynamics by utilizing the energy balance error (EB). $\dot{Q}_e$ and $\dot{Q}_c$ were calculated to assess whether the data satisfied the first law of thermodynamics. Data with an EB within ±5% were considered reliable, whereas data outside this range were likely affected by sensor measurement errors, refrigerant flow fluctuations, or partial load operations [11,12].

As a result, only data within an EB of ±5% were deemed reliable in this study. This approach ensured data quality, significantly enhancing the model's reliability and predictive accuracy. The EB is calculated as follows: In Equation (4), the numerator $\left(\dot{Q}_e + \dot{W}_c\right) - \dot{Q}_c$ represents the deviation in energy balance and dividing this by $\dot{Q}_c$ yields the relative error (ratio). Consequently, the EB value is a dimensionless quantity, allowing the system's balance to be expressed as a percentage (%):

$$\mathrm{EB}(\pm 5\%) = \frac{\left(\dot{Q}_e + \dot{W}_c\right) - \dot{Q}_c}{\dot{Q}_c} \tag{4}$$

2.1.3. Coefficient of Performance (COP)

COP is an indicator of how efficiently a chiller operates. Since the primary objective of a chiller is to minimize the energy consumed in the process of heat removal or transfer, the coefficient of performance serves as a critical metric for evaluating the energy efficiency of a chiller system. COP is expressed by the following equation [13]:

$$\mathrm{COP} = \frac{\dot{Q}_e}{\dot{W}_c} \quad (5)$$

2.2. *Imputation Methods*

Statistical methods for missing data imputation have been widely used for an extended period due to their computational simplicity and ease of application. In this study, various representative statistical imputation techniques were compared and analyzed to determine the most suitable method.

2.2.1. Mean Imputation

Mean imputation replaces missing values with the mean of the corresponding variable, effectively preserving the overall average trend of the dataset. However, this method is highly sensitive to outliers and may introduce distortions in the data distribution. Despite these limitations, mean imputation is computationally simple and widely used, particularly in the initial stages of data preprocessing [14,15]:

$$y_{\text{imputed}} = \frac{1}{N}\sum_{i=1}^{N} y_i \quad (6)$$

2.2.2. Median Imputation

Median imputation replaces missing values with the median of the corresponding variable, making it suitable for mitigating the influence of non-normal distributions and outliers. This method provides stable imputation results, especially for datasets with extreme values. The median is determined by sorting the data in ascending order and selecting the central value, effectively minimizing the impact of outliers when imputing missing data [16,17]:

$$y_{\text{imputed}} = Median(y) \quad (7)$$

2.2.3. Linear Interpolation

Linear interpolation estimates missing values in time-series data by interpolating between adjacent values based on a linear relationship. This method is useful for maintaining the natural trend of the data and offers computational simplicity and high interpretability. However, it has limitations in capturing abrupt fluctuations or nonlinear relationships within the data [18,19]:

$$y_{\text{imputed}} = y_{i-1} + (y_{i+1} - y_{i-1})\frac{t_i - t_{i-1}}{t_{i+1} - t_{i-1}} \quad (8)$$

2.2.4. Multiple Imputation

Multiple imputation replaces missing values multiple times and calculates the final imputed value as the average of these imputed results, thereby reflecting the uncertainty associated with missing data. This method provides reliable results and is particularly effective for handling missing-at-random (MAR) data [20,21]:

$$y_{\text{imputed}} = \frac{1}{m}\sum_{j=1}^{m} Y_{i,j} \quad (9)$$

2.2.5. Simple Random Imputation

Simple random imputation replaces missing values with randomly generated values based on the existing distribution of the variable. While this approach aims to maintain the probabilistic characteristics of the original data, it has the drawback of introducing significant variability in the imputed results [22,23]:

$$y_{\text{imputed}} = y_{obs} + \epsilon \tag{10}$$

2.2.6. KNN Imputation

K-nearest neighbor (KNN) imputation replaces missing values with either the mean or median of the K most similar neighboring data points. This method is particularly useful for datasets containing highly correlated variables, as it can capture complex multidimensional relationships while preserving data structure and patterns. However, KNN imputation is computationally expensive, and processing time increases as the dataset size grows [24,25]:

$$y_{\text{imputed}} = \frac{1}{K}\sum_{i=1}^{K} y_{\text{neighbor}} \tag{11}$$

2.2.7. Euclidean Distance

Euclidean distance is suitable for continuous data, assuming that closer data points exhibit higher similarity. This method calculates the straight-line distance between two points. Given two data points, x and y, the Euclidean distance is defined as follows [26]:

$$d_{\text{Euc}}(y_i, y_j) = \sqrt{\sum_{i=1}^{n}\left(y_{ik} - y_{jk}\right)^2} \tag{12}$$

2.2.8. Cosine Similarity

Cosine similarity focuses on the direction rather than the magnitude of the data and is particularly effective for normalized datasets. This method calculates similarity by measuring the angle between two vectors, with values ranging from -1 to 1. Cosine similarity is defined as follows [27]:

$$S_{\text{COS}}(y_i, y_j) = \frac{y_i \cdot y_j}{||y_i||\ ||y_j||} \tag{13}$$

2.2.9. DC-KNN Imputation

The DC-KNN imputation method is an enhanced imputation approach developed in this study. It first clusters data with similar patterns to the missing-value-containing records using the K-means algorithm and then applies KNNs within each cluster to perform imputation. This method improves upon the limitations of conventional KNN imputation, which relies solely on distance-based neighbor selection, by integrating clustering-based filtering and variable-specific dynamic weighting into an extended KNN approach. Initially, missing-value-containing records are grouped into clusters with similar patterns based on operational variables such as chilled water temperatures and flow rates. Among various clustering techniques, K-means is explicitly employed in this study due to its suitability for partitioning chiller operation data by temperature and load range. Although other methods, such as DBSCAN, are referenced in the literature for noise removal in sensor data with abrupt variations [14,28], only K-means is used in the implementation of DC-KNNs for this research.

K-means is a distance-based unsupervised clustering algorithm that partitions the dataset into K clusters by minimizing the within-cluster sum of squared distances (WCSS). It iteratively assigns each data point to the nearest cluster centroid and updates the centroids until convergence. This approach is particularly well-suited for time-series sensor data with continuous, normalized variables—such as temperature and flow rate—where Euclidean distance is a meaningful similarity metric.

On the other hand, density-based clustering techniques, like DBSCAN, are advantageous for handling nonlinear cluster shapes or identifying noise points in unstructured data. However, they are less suitable for structured operational datasets, especially in scenarios where a fixed number of interpretable clusters is required. DBSCAN's sensitivity to parameter settings (e.g., epsilon and minimum samples) and lack of control over the number of clusters make it less compatible with the structured nature of chiller operation data, particularly in the presence of sparse or periodic missingness.

The number of clusters, K = 5, was empirically determined using the elbow method. The WCSS was computed across a range of cluster numbers (from 3 to 8), and K = 5 was selected as the point where marginal gains in intra-cluster variance reduction began to plateau. In addition to statistical justification, this configuration aligned well with distinguishable operational modes of the chiller system—such as full-load cooling, partial-load cooling, nighttime storage, transition periods, and idle states—making the clusters both interpretable and physically meaningful. This selection of K enhanced both the contextual relevance and computational efficiency of the DC-KNN imputation process.

In the actual application, seven physical variables were used: T_1, T_2, T_3, T_4, $\dot{V}_1$, $\dot{V}_3$, and $\dot{W}_C$. The first six variables represent inlet–outlet temperatures and flow rates of chilled and cooling water and were used for clustering. The final variable, power consumption ($\dot{W}_C$), was the target for imputation. To group similar operational patterns, K-means clustering was applied with K = 5 clusters. This number was empirically determined and applied separately for each chiller. Clustering enables KNNs to operate locally, increasing the contextual accuracy of imputation.

K-means is an unsupervised clustering algorithm that partitions a dataset into K clusters $\{C_1, C_2, \ldots, C_K\}$ by minimizing the within-cluster sum of squared distances (WCSS). The objective function of K-means is defined as:

$$\min_{\{C_k\}} \sum_{k=1}^{K} \sum_{x_i \in C_k} \|x_i - \mu_k\|^2 \tag{14}$$

where μ_k is the centroid of cluster C_K, computed as:

$$\mu_k = \frac{1}{|C_k|} \sum_{x_i \in C_k} x_i \tag{15}$$

After clustering, the dataset is represented as:

$$C = \{C_1, C_2, \ldots, C_n\} \tag{16}$$

Each missing value X_{miss} identifies K-nearest neighbors within the same cluster C_j. The weights w_1 and w_2 are dynamically assigned based on data characteristics, satisfying the constraint:

$$w_1 + w_2 = 1 \tag{17}$$

When applying KNNs, the similarity between samples is defined solely using Euclidean distance, as it more appropriately reflects absolute numerical differences in physical variables such as temperature and flow rate:

$$d(y_i, y_j) = \|y_i - y_j\|_2 \tag{18}$$

The final DC-KNN imputed value is computed using a weighted average of the nearest neighbors [14]:

$$y_{\text{imputed}} = \frac{\sum_{i=1}^{K} w_{\text{i}}, y_{\text{neighbor},i}}{\sum_{i=1}^{K} w_{\text{i}}} \tag{19}$$

where w_{i} represents the weight assigned to the i-th nearest neighbor, with higher weights applied to closer neighbors:

$$w_{\text{i}} = \frac{1}{d(y_{miss}, y_i)} \tag{20}$$

The DC-KNN imputation method refines the conventional KNN approach by incorporating K-means-based clustering to better capture data patterns relevant to operational context. As a result, it improves the reliability of missing value imputation, maintains data quality, and achieves optimal predictive performance.

The detailed procedure of the proposed DC-KNN method is summarized in Algorithm 1, providing a step-by-step workflow for implementation and reproducibility.

Algorithm 1. The step-by-step procedure of the proposed DC-KNN imputation method.

DC-KNN Imputation
Input: D (dataset with missing values), X (observed variables), K (number of clusters), and k (neighbors).

1. Normalize input variables X;
2. Apply K-means clustering to group data into K clusters;
3. For each missing value Y_{i}:
   a. Find cluster C_{i} containing Y_{j};
   b. In C_{i}, find the k-nearest neighbors based on Euclidean distance;
   c. Assign weights: $w_{\text{i}} = 1/\text{distance}$;
   d. Compute the imputed value:
      $\hat{Y}_j = \Sigma(w_{\text{i}} * Y_{\text{i}})/\Sigma(w_{\text{i}})$
4. Return the imputed dataset.

2.3. Validation Methods

2.3.1. MAPE

The mean absolute percentage error (MAPE) expresses the error between predicted and actual values as a percentage and calculates the mean of these percentages. It is commonly used to evaluate the accuracy of predictive models, providing an intuitive measure of how much the model deviates from actual values [29,30]. The MAPE is defined as follows:

$$\text{MAPE} = \frac{1}{n}\sum_{i=1}^{n}\left|\frac{y_i - \hat{y}_i}{y_i}\right| \times 100 \tag{21}$$

2.3.2. CVRMSE

The coefficient of variation of the root mean squared error (CVRMSE) is a normalized metric of the root mean squared error (RMSE) used to assess the relative accuracy of predicted or imputed values. A CVRMSE value below 10% indicates a high-accuracy model, while a value between 10% and 20% suggests moderate accuracy. Values exceeding 20% indicate the need for further model improvement [31,32]. The CVRMSE is defined as follows:

$$\mathrm{CVRMSE} = \frac{\left[\frac{\sum(y_i - \hat{y}_i)^2}{(n-1)}\right]^{1/2}}{\overline{y}} \times 100 \tag{22}$$

3. Research Methodology

3.1. Measurement Target and Period

This study was conducted on two 450 USRT-class turbo chillers installed on the third basement floor of Tower A in Seoul. Tower A is a 33-story commercial building with a total floor area of approximately 70,000 m^2, equipped with a central cooling system that operates under various conditions. Data were collected in real time through the building energy management system (BEMS) from March 2021 to July 2024. Due to the characteristics of cooling loads, measurements were taken only during the cooling season, from May to October.

During the daytime, the chiller operates under partial load while simultaneously utilizing the thermal energy storage (TES) system to handle the cooling load. At night, the system operates in a mode where ice is stored using capsule-type storage media. The cooling system in the study site operates under four primary modes:

A. Ice-making mode: The process of freezing capsule-type media inside the thermal storage tank. The chiller is designed to operate at an average temperature of −4.5 °C. Ice storage primarily occurs during late-night hours, and the system stops after 10 h of operation or when the storage reaches 100% capacity;
B. TES-only operation: The cooling load is handled solely by the thermal storage tank without operating the chiller. This mode is mainly used during transitional seasons or at night when the cooling demand is low;
C. Parallel operation of the chiller and the TES: The chiller and the TES tank operate together to meet peak cooling loads. In this mode, the chiller operates at ambient conditions;
D. Chiller-only operation: The chiller independently handles the cooling load without utilizing the TES system. This mode is used when the TES system malfunctions or when the thermal storage is completely depleted.

The collected data were processed by removing missing values and outliers from the original raw data. A partially missing-at-random (MAR) dataset with 10–30% missing values was then generated, and six imputation methods, including mean imputation, along with the DC-KNN imputation method, were applied to fill in the missing values. The imputed values were then compared with the actual values to identify the most effective imputation method. The final imputation accuracy was validated using the MAPE and CVRMSE, and the best-performing imputation method (DC-KNNs) was ultimately used to predict the actual missing values. Figure 2 illustrates the overall process [33].

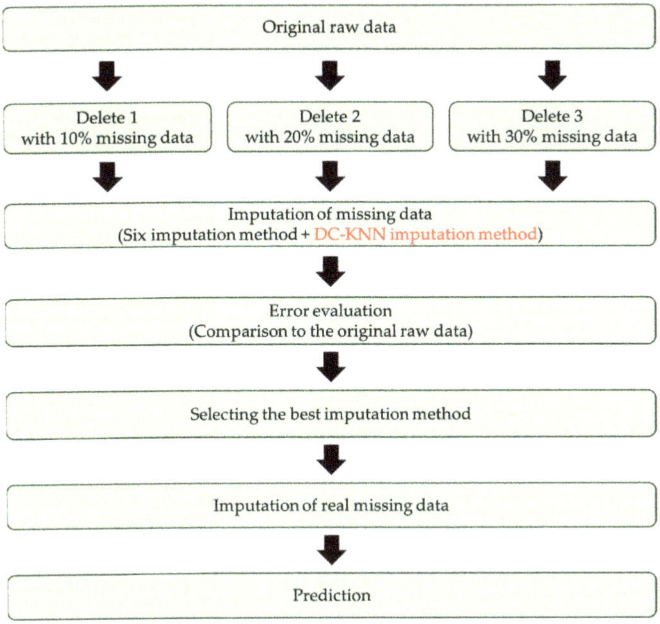

Figure 2. Flowchart of the methodology used in this study.

3.2. Data Measurement

3.2.1. Data Measurement Method

For the performance analysis of the chillers, the study measured and recorded the flow rate and temperature of the chilled and cooling water, as well as the chiller's power consumption. To measure the flow rate, an AUDS-100M ultrasonic flowmeter from AutoFLO (Bucheon-si, Gyeonggi-do, Republic of Korea) was used. This device is a time-difference ultrasonic flowmeter specifically designed for BEMS applications. The flowmeter uses a clamp-on sensor, allowing it to be attached externally without modifying the piping structure or cutting the pipe.

The temperature of the chilled and cooling water was measured using an NT-320G insertion-type temperature sensor from Hankyung Electric (Seoul, Republic of Korea). This sensor has a measurement range of −50 °C to 150 °C with an accuracy of ±0.5%, ensuring precise temperature data essential for chiller system performance analysis.

The chiller's power consumption was measured using a GEMS3512 AC power meter from BMT (Yangsan-si, Republic of Korea). The collected data were transmitted via Modbus TCP communication to a direct digital controller (DDC), where they were later integrated and analyzed within the BEMS. Data for flow rate, temperature, and power consumption were transmitted in real time through RS485 communication and stored on the server at 15 min intervals. The types and key characteristics of the measurement equipment are summarized in Table 1 [33].

Table 1. Types and characteristics of measuring instruments.

Category	Ultrasonic Flow Meter	Temperature Sensor	AC Power Meter
Manufacturer	AutoFLO	Hankyung electric	BMT
Model	AUDS-100M	NT-320G	GEMS3512
Measurement principle	Transit time	Insertion type	CT or Rogowski coil
Measurement range	±0.02~15 m/s	−50~150 °C	0~415 VAC
Error	±1%	±0.5%	±0.5%

3.2.2. Uncertainty Analysis

To reliably evaluate the COP of the chiller, it is essential to consider the measurement uncertainty of the instruments used. In this study, the uncertainty of the measurement equipment listed in Table 1 was incorporated into the analysis. Generally, when a function R is determined by multiple measured variables $x_{1,}x_{2,} \cdots x_n$, the total uncertainty W_R of the result R is calculated as follows [4]:

$$W_R = \sqrt{\sum \left(\frac{\partial R}{\partial x_i} W_{xi} \right)^2} \tag{23}$$

During daytime operation, the COP was 4.04, and the total uncertainty W_R at a 95% confidence level was determined to be ± 0.073. In contrast, during nighttime operation, the COP was 2.26, with a corresponding W_R ± 0.036 at the same confidence level. These uncertainty values reflect the errors of the measurement instruments, and the analysis confirmed that the temperature sensors and power meters were the most significant contributors to COP uncertainty.

In addition to the uncertainties derived from instrument specifications, practical sources of error during data collection were also considered. The original power consumption data were measured using BEMS sensors with a rated accuracy of $\pm 0.5\%$. To minimize the influence of abnormal values or sensor drift, EB filtering with a threshold of $\pm 5\%$ was applied.

Furthermore, the dataset was recorded at hourly intervals, which is appropriate for capturing the dynamic characteristics of chiller operation. The data spanned from May to October, encompassing a wide range of operational load conditions during the cooling season. By addressing these sources of uncertainty, the reliability of the imputation performance evaluation was further strengthened.

3.2.3. Initial Review of Collected Power Consumption Data

A total of 59,232 data points were collected from chiller #1 and another 59,232 from chiller #2, resulting in a total of 118,464 data points. At the study site, chillers operate using both direct cooling and the TES during the daytime (09:00–21:00), while at night (21:00–09:00), the chillers store ice. The collected data included chiller power consumption ($\dot{W}_c$), which was analyzed based on the ratio of actual values to design values, as shown in Table 2. The formula for each series is as follows:

$$R_W = \frac{\dot{W}_o}{\dot{W}_a} \tag{24}$$

Table 2. Measured R_W of chiller #1 and chiller #2.

Parameters	Chiller #1			Chiller #2		
	59,232 Data Doints			59,232 Data Points		
	Min	Mean	Max	Min	Mean	Max
R_W	0.01	0.94	42.10	0.01	0.89	51.70

R_W is defined as the ratio of $\dot{W}_a$ to $\dot{W}_o$. This ratio serves as a critical indicator for evaluating how well the chiller operates relative to its designed capacity under the current operating conditions. The interpretation of the ratio is as follows [3]:

- $R_W > 1$: The actual capacity exceeds the nominal capacity, indicating that the chiller is operating beyond its design conditions. This situation is typically caused by high load conditions or unexpected external environmental changes;

- $R_W = 1$: The actual capacity matches the nominal capacity, meaning that the system is operating as per the design conditions;
- $R_W < 1$: The actual capacity is lower than the nominal capacity, suggesting that the chiller is operating under partial load conditions. This state frequently occurs in energy-saving modes or low-load conditions.

From the 118,464 collected data points (chiller #1: 59,232, chiller #2: 59,232), the missing values and outliers were identified and separated (Table 3). Outliers were defined as data points falling outside the ±5% energy balance error (EB) range [11,12].

Table 3. The missing data for $\dot{W}_c$ in chiller #1 and chiller #2.

Parameters	Chiller #1		Chiller #2	
	59,232 Data Points		59,232 Data Points	
	Daytime	Nighttime	Daytime	Nighttime
Missing data	26,838	16,460	24,727	14,789
Outlier data	1474	8769	3504	12,493

To analyze the distribution of $\dot{W}_c$, the hourly variation of R_W was examined using the raw dataset. As shown in Figure 3, valid data points appeared only between May and October, reflecting the seasonal operation of the chiller system. This figure serves as a baseline reference to define the effective data period for subsequent preprocessing and experimental scenario construction.

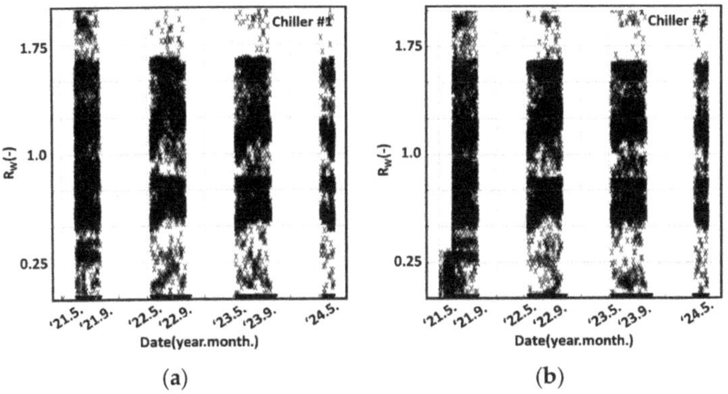

Figure 3. (**a**) Linear graph of the initially measured 59,232 R_W values and missing data for chiller #1; (**b**) Linear graph of the initially measured 59,232 R_W values and missing data for chiller #2.

3.2.4. Review of Valid Data

The valid data preprocessed based on the EB (±5%) consists of 3236 entries for chiller #1 and 524 entries for chiller #2 (Table 4).

Table 4. Appropriate R_W for chiller #1–#2.

Parameters	Chiller #1			Chiller #2		
	3236 Data Points			524 Data Points		
	Min	Mean	Max	Min	Mean	Max
R_W	0.52	1.21	1.71	0.21	0.64	1.88

The time-based variation of R_W was re-examined after applying missing value and outlier removal, as shown in Figure 4. Compared to the raw data in Figure 3, a substantial reduction in data points is observed, reflecting the impact of preprocessing. This figure serves to verify the effectiveness of the EB (±5%) filtering process and define the cleaned dataset used as the baseline for subsequent imputation experiments.

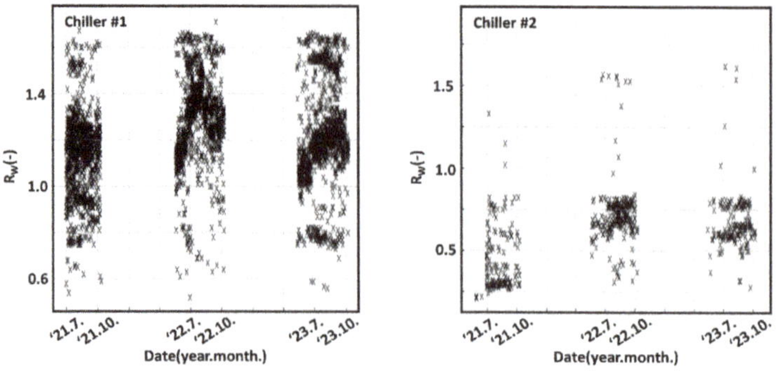

(a) (b)

Figure 4. (**a**) Linear graph of the 3236 valid R_W data points for chiller #1; (**b**) Linear graph of the 524 valid R_W data points for chiller #2.

3.3. Partially MAR Data

A total of 3236 values of R_W from chiller #1 and 524 from chiller #2 were artificially removed using the missing-at-random (MAR) assumption, based on the preprocessed dataset filtered by the EB (±5%). To evaluate the robustness of the imputation algorithms under increasing levels of data loss, the missingness was categorized into 10% (Figure 5), 20% (Figure 6), and 30% (Figure 7) scenarios. These figures visualize the spatial and temporal distribution of the missing values introduced at each level, serving as references for the experimental conditions defined in Table 5.

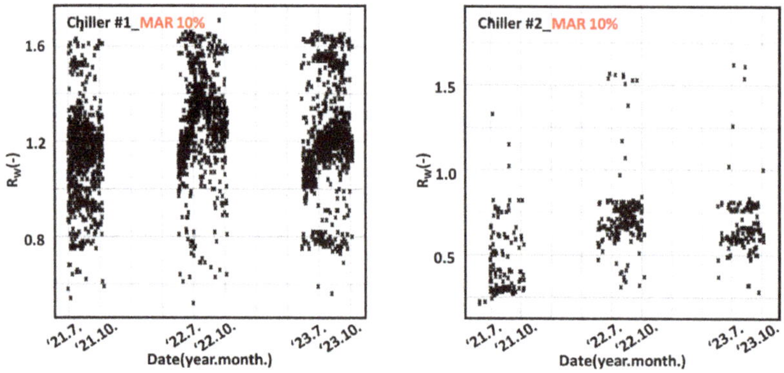

(a) (b)

Figure 5. (**a**) R_W data points for chiller #1 after MAR 10%; (**b**) R_W data points for chiller #2 after MAR 10%.

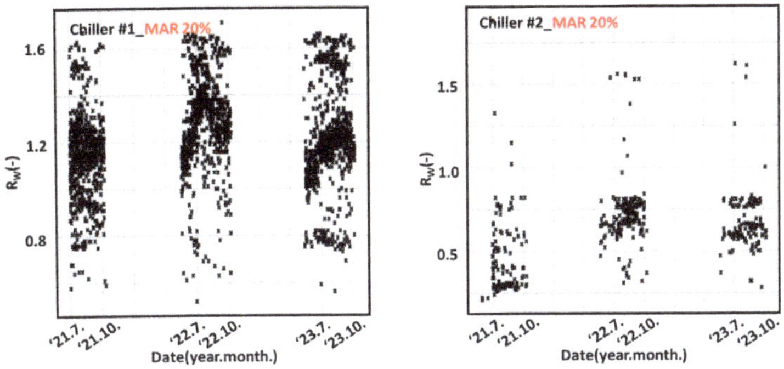

(a) (b)

Figure 6. (**a**) R_W data points for chiller #1 after MAR 20%; (**b**) R_W data points for chiller #2 after MAR 20%.

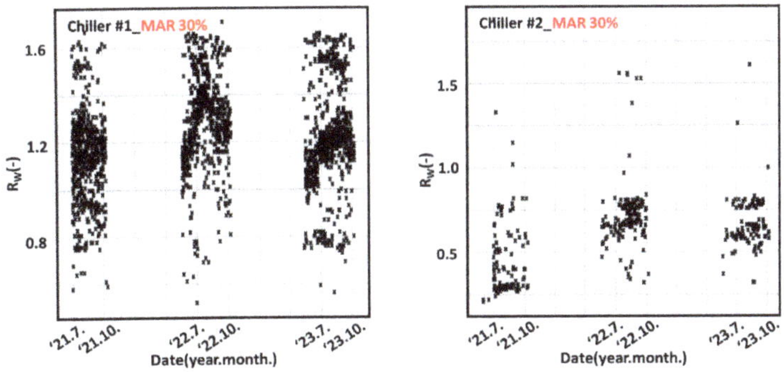

(a) (b)

Figure 7. (**a**) R_W data points for chiller #1 after MAR 30%; (**b**) R_W data points for chiller #2 after MAR 30%.

Table 5. The number of missing data under MAR 10–30% conditions for chillers #1–#2.

Parameters	Chiller #1		Chiller #2	
	3236 Data Points		524 Data Points	
	Remain	Missing	Remain	Missing
MAR 10%	2913	323	472	52
MAR 20%	2589	647	420	104
MAR 30%	2266	970	367	157

To implement the MAR condition, missing values were applied to the power consumption variable $\dot{W}_c$ with conditional dependence on the observed variables T_1, T_2, T_3, T_4, $\dot{V}_1$ and $\dot{V}_3$. This ensures that the probability of a missing value occurring in $\dot{W}_c$ depends on other observed values in accordance with the MAR framework. The missing data were introduced in stages: an initial 10% of records were randomly selected based on these conditions, followed by an additional 10% (cumulative 20%) and a final 10% (cumulative 30%), each applied to the remaining observed data. This incremental approach preserved the MAR logic while controlling the overall missingness.

According to the MAR framework, missing values are assumed to be dependent on other observed variables but not on the missing values themselves, allowing for statistically valid imputations based on the available data [31]. The corresponding mathematical expression is presented below:

$$P(R|Y_{obs}, Y_{mis}) = P(R|Y_{obs}) \quad (25)$$

4. Experimental Results and Discussion

4.1. Results of Replacing Partially Missing-at-Random Data

4.1.1. Mean Imputation

In terms of the MAPE, chiller #1 showed a stable performance across the MAR levels, with values of 14.62% at 10%, 14.28% at 20%, and 14.69% at 30%. Chiller #2 followed a similar trend with 13.76%, 13.52%, and 13.86%, respectively. The error rates remained relatively consistent, indicating that mean imputation is resilient to increasing missingness in terms of average accuracy.

For the CVRMSE, chiller #1 recorded 17.43%, 17.33%, and 17.65%, while chiller #2 showed slightly lower variability at 16.91%, 16.66%, and 17.01% across the same MAR levels. This suggests that the mean imputation maintains moderate predictive stability and is particularly effective when the goal is to minimize model complexity and preserve central tendencies (Table 6).

Table 6. Results of MAR imputation validation.

Method	MAR	Chiller #1		Chiller #2	
		MAPE	CVRMSE	MAPE	CVRMSE
Mean	10%	14.62%	17.43%	13.76%	16.91%
	20%	14.28%	17.33%	13.52%	16.66%
	30%	14.69%	17.65%	13.86%	17.01%
Median	10%	14.96%	17.66%	14.06%	17.13%
	20%	14.71%	17.63%	13.87%	16.94%
	30%	15.10%	17.94%	14.22%	17.27%
Linear	10%	14.57%	19.39%	14.44%	18.98%
	20%	15.08%	19.81%	14.26%	18.79%
	30%	15.37%	20.06%	14.61%	19.08%
Multiple	10%	14.62%	17.43%	13.76%	16.91%
	20%	14.28%	17.33%	13.52%	16.66%
	30%	14.69%	17.65%	13.86%	17.01%
Simple	10%	21.51%	25.78%	20.94%	24.98%
	20%	20.27%	23.95%	20.14%	23.81%
	30%	21.15%	25.02%	20.53%	24.46%
KNNs	10%	9.92%	12.37%	10.04%	12.87%
	20%	10.23%	13.20%	10.42%	13.41%
	30%	10.35%	13.23%	10.58%	13.62%
DC-KNNs	10%	9.86%	12.19%	9.74%	12.65%
	20%	10.30%	13.19%	10.17%	13.32%
	30%	10.08%	13.06%	10.29%	13.43%

Mean imputation replaces all missing entries with a single average value, resulting in a horizontally aligned distribution. While this approach is computationally efficient and robust against outliers, it reduces natural variance in the data and can mask underlying patterns. Thus, it is best suited for datasets with low missingness and when preserving structural variation is not critical (Figures 8–10).

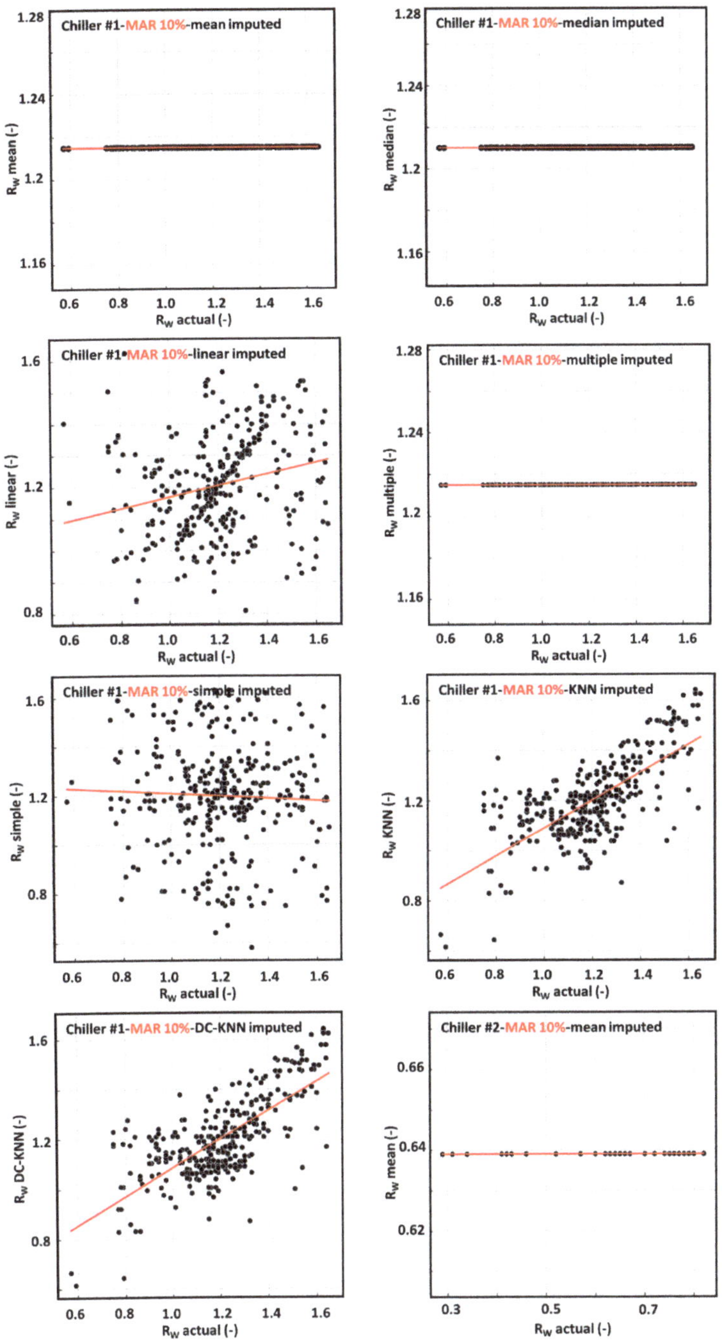

Figure 8. *Cont.*

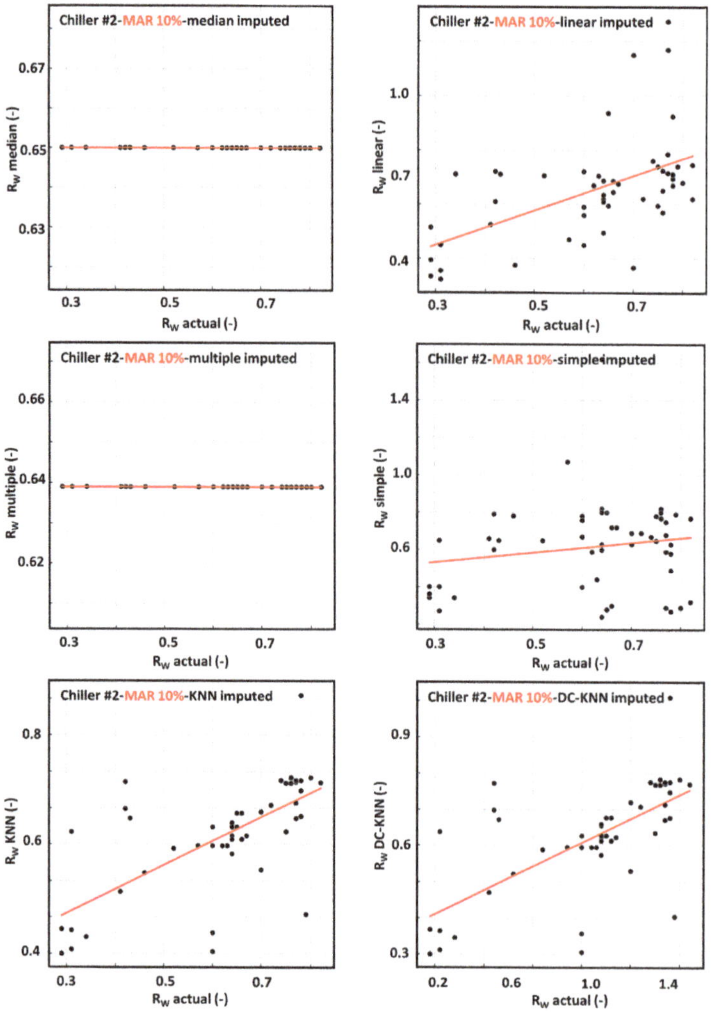

Figure 8. Imputation of R_W missing values using seven methods following 10% MAR in chiller #1-2.

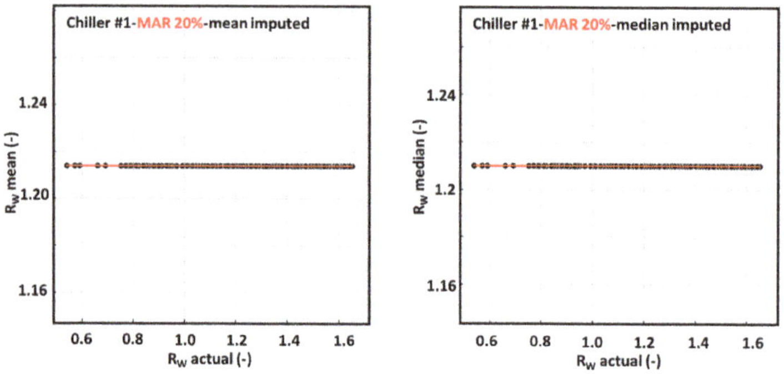

Figure 9. *Cont.*

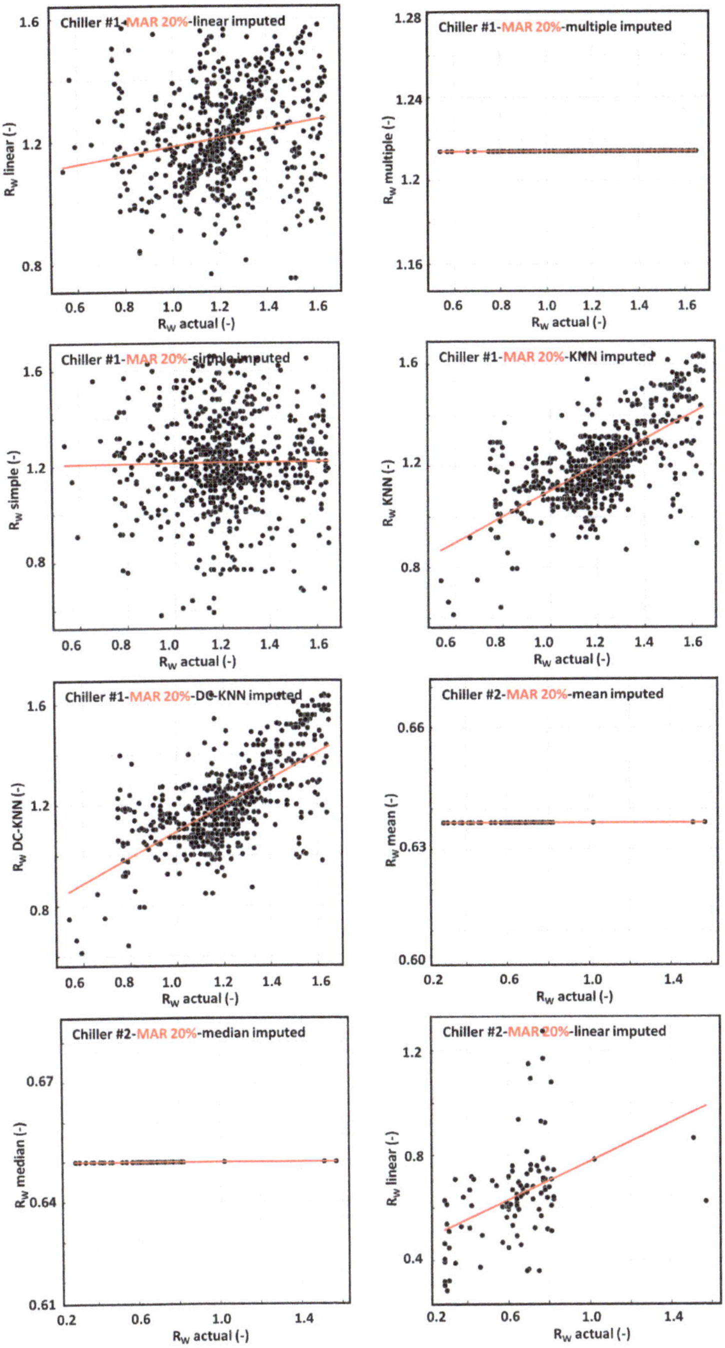

Figure 9. *Cont.*

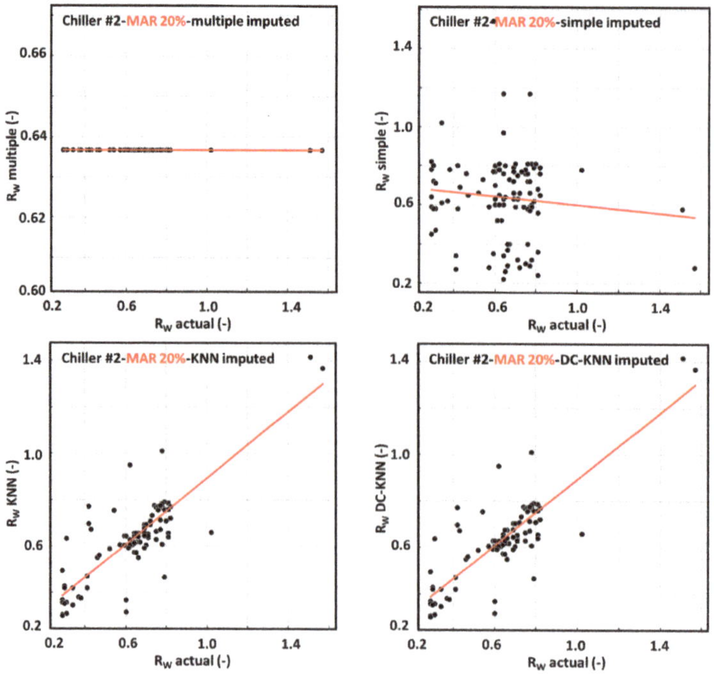

Figure 9. Imputation of R_W missing values using seven methods following 20% MAR in chiller #1-2.

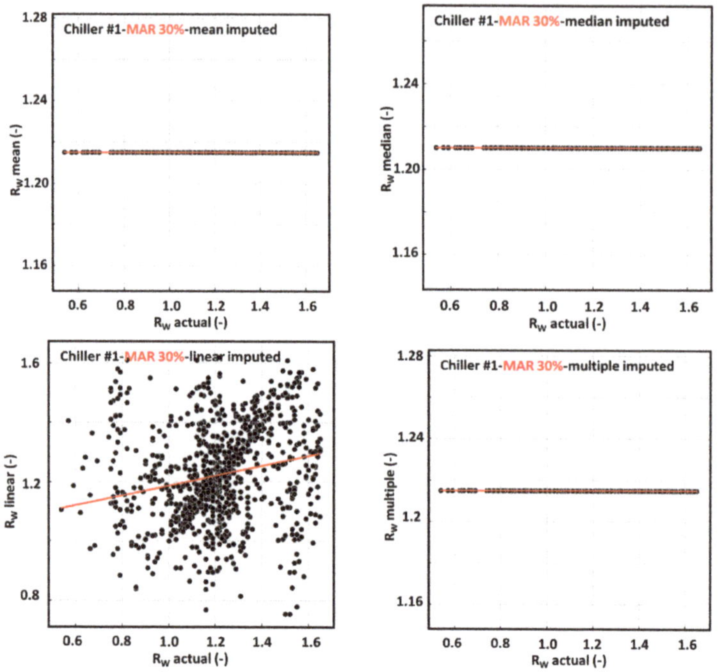

Figure 10. *Cont.*

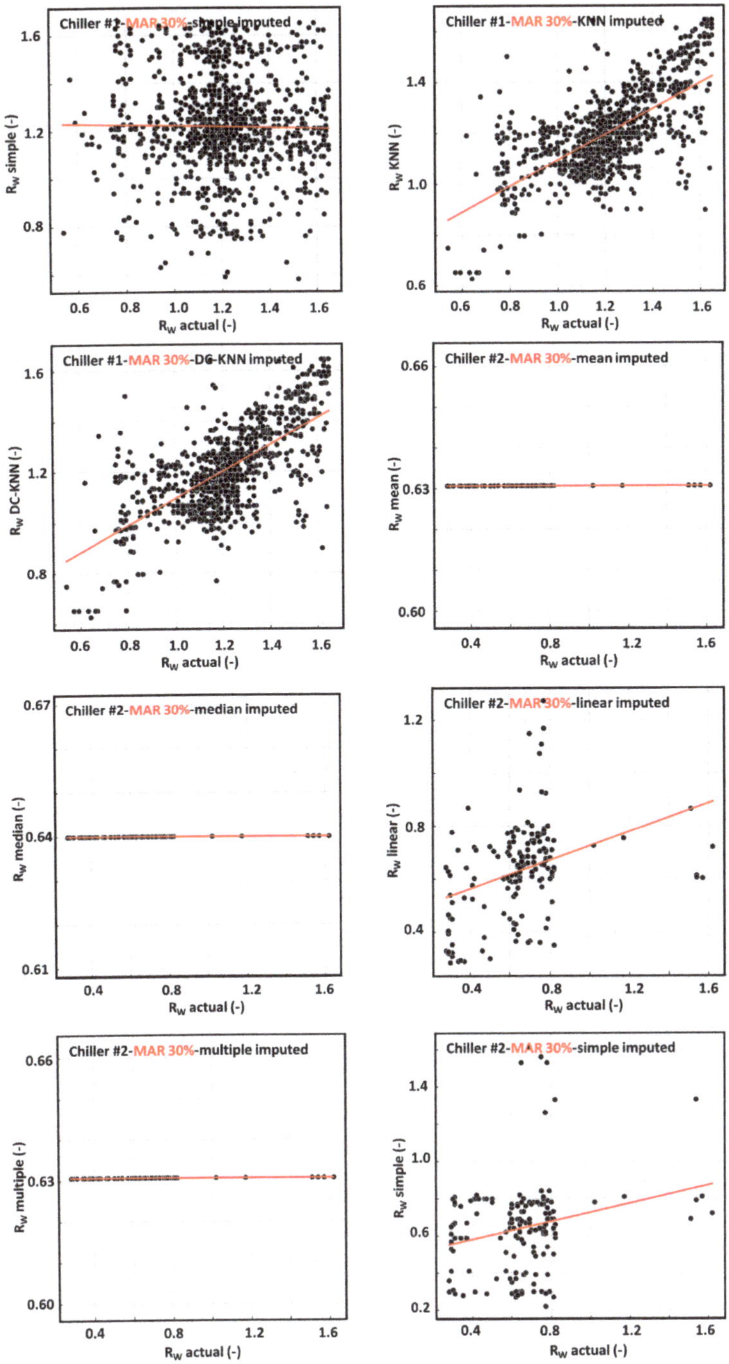

Figure 10. *Cont.*

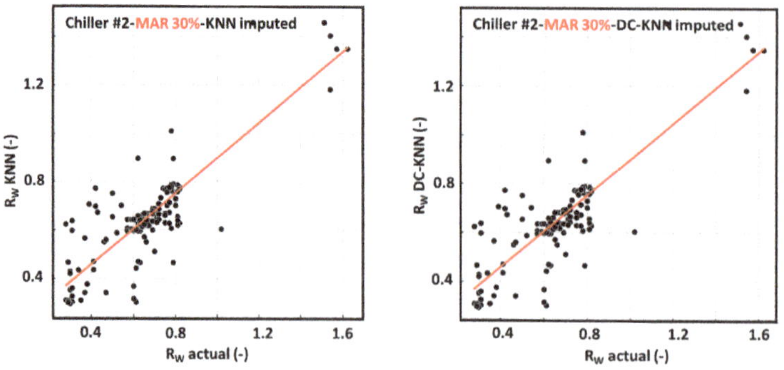

Figure 10. Imputation of R_W missing values using seven methods following 30% MAR in chiller #1-2.

4.1.2. Median Imputation

For the MAPE, chiller #1 recorded 14.96%, 14.71%, and 15.10% at MAR levels of 10%, 20%, and 30%, respectively, indicating a slight degradation in accuracy as the missing rate increased. Chiller #2 displayed comparable results with 14.06%, 13.87%, and 14.22%, suggesting stable and consistent imputation performance across varying levels of missingness.

In terms of the CVRMSE, chiller #1 maintained values of 17.66%, 17.63%, and 17.94%, while chiller #2 showed slightly lower variation with 17.13%, 16.94%, and 17.27%, respectively. These results confirm that median imputation offers reliable performance in terms of both accuracy and variance, especially in datasets with modest outlier presence (Table 6).

Similar to mean imputation, median imputation substitutes missing values with a single central tendency value. However, unlike the mean, the median is robust to extreme outliers, making it preferable in skewed distributions or datasets with anomalies. Although it preserves the central location of the data more effectively than the mean, it still reduces variability and fails to capture the dynamics of temporal or correlated variables. Therefore, it is most appropriate for non-time-series data with a relatively low to moderate rate of missingness and the presence of outliers (Figures 8–10).

4.1.3. Linear Interpolation

For the MAPE, chiller #1 showed a slight increase across missingness levels, with 14.57%, 15.08%, and 15.37% at MAR 10%, 20%, and 30%, respectively. Chiller #2 followed a similar trend with values of 14.44%, 14.26%, and 14.61%. These results suggest that while linear interpolation maintains moderate accuracy at lower missing rates, its performance degrades slightly as the extent of missingness increases.

Regarding the CVRMSE, chiller #1 exhibited relatively higher variability compared to other methods, with 19.39%, 19.81%, and 20.06% at MAR 10%, 20%, and 30%. Chiller #2 also recorded increasing variability, with 18.98%, 18.79%, and 19.08%, indicating that linear interpolation is more susceptible to data pattern distortion, particularly in highly dynamic time-series environments (Table 6).

Linear interpolation estimates missing values by connecting known values on either side with a straight line, thereby assuming smooth transitions between points. This results in a diagonally aligned imputation structure, which is effective in preserving local continuity and trends in sequential datasets. However, it lacks the flexibility to capture nonlinear fluctuations or abrupt shifts, especially when large consecutive gaps exist. Thus, while linear interpolation is suitable for temporally continuous data with mild variability, its performance may be limited in systems with sharp transitions or complex dynamics (Figures 8–10).

4.1.4. Multiple Imputation

For the MAPE, chiller #1 showed consistent accuracy across missingness levels, recording 14.62%, 14.28%, and 14.69% at MAR 10%, 20%, and 30%, respectively. Chiller #2 exhibited very similar results with 13.76%, 13.52%, and 13.86%, indicating that multiple imputation maintains stable predictive accuracy regardless of the proportion of missing data. This trend suggests that multiple imputation is robust to increased missingness, particularly when the data distribution remains relatively consistent.

In terms of the CVRMSE, chiller #1 maintained steady variability with values of 17.43%, 17.33%, and 17.65%, showing minimal deviation across the MAR levels. Chiller #2 also demonstrated low and stable variability, with 16.91%, 16.66%, and 17.01%. These results indicate that multiple imputation effectively preserves the variance structure of the original dataset across both chillers (Table 6).

Unlike single-value replacement methods such as mean or median imputation, multiple imputation involves generating multiple plausible values based on probabilistic modeling. These imputed values are then aggregated to form a more realistic estimate that reflects the underlying distribution. This approach helps to retain natural variability and improves inference quality.

The strengths of multiple imputation include enhanced statistical validity, the ability to handle uncertainty, and improved performance in downstream analysis. However, it requires more computational resources and methodological rigor and is sensitive to the assumptions underlying the imputation model. Overall, it is well-suited for cases where maintaining distributional integrity and prediction robustness is critical, especially in the context of machine learning model preparation or inferential studies (Figures 8–10).

4.1.5. Simple Random Imputation

For the MAPE, chiller #1 recorded 21.51%, 20.27%, and 21.15% at MAR levels of 10%, 20%, and 30%, respectively, indicating high and inconsistent error rates across different missingness levels. Similarly, chiller #2 exhibited substantial errors with MAPE values of 20.94%, 20.14%, and 20.53%. These results demonstrate that simple random imputation offers low accuracy and limited robustness, particularly in structured, time-dependent datasets, such as chiller sensor records.

In terms of the CVRMSE, chiller #1 recorded 25.78%, 23.95%, and 25.02%, while chiller #2 showed 24.98%, 23.81%, and 24.46% at MAR 10%, 20%, and 30%, respectively. These values are the highest among all tested methods, highlighting the poor ability of simple random imputation to preserve data variability and predictive fidelity (Table 6).

Simple random imputation involves replacing missing values with randomly sampled values from the observed dataset without considering time order, contextual similarity, or feature relationships. While this method may minimize systematic bias, it introduces high variance and disrupts temporal or structural dependencies within the data. This often leads to noisy imputations and degraded model performance in both analytical and predictive tasks.

Due to its disregard for pattern integrity and sequence continuity, simple random imputation is generally unsuitable for time-series or sensor data. Its application is discouraged except in exploratory contexts or for benchmarking purposes where simplicity is prioritized over accuracy (Figures 8–10).

4.1.6. KNN Imputation

For the MAPE, chiller #1 showed 9.92%, 10.23%, and 10.35% at MAR levels of 10%, 20%, and 30%, respectively, while chiller #2 recorded 10.04%, 10.42%, and 10.58% for the same missingness levels. These results indicate that KNN imputation maintains relatively consistent and moderate error rates, even as the proportion of missing data increases, suggesting robustness across varying levels of data incompleteness.

In terms of the CVRMSE, chiller #1 yielded 12.37%, 13.20%, and 13.23%, and chiller #2 exhibited 12.87%, 13.41%, and 13.62% across the respective MAR levels. While not the lowest among the tested methods, these values indicate stable variability control, particularly when compared to more simplistic imputation techniques (Table 6).

KNN imputation identifies the K most similar complete records (neighbors) based on a defined distance metric—commonly the Euclidean distance—and imputes missing values using a weighted or unweighted aggregation of those neighbors. This method preserves local data structure, capturing both linear and nonlinear relationships, making it particularly effective in datasets with complex inter-feature interactions.

However, its performance is sensitive to the choice of hyperparameters such as K and the distance metric. Additionally, KNNs can suffer from high computational cost and diminished accuracy when dealing with sparse datasets or high proportions of missingness. Despite these limitations, the KNN method offers a balanced trade-off between performance and flexibility, making it a strong candidate for structured sensor datasets where local contextual information is important (Figures 8–10).

4.1.7. DC-KNN Imputation

For the MAPE, chiller #1 recorded 9.86%, 10.30%, and 10.08% at MAR levels of 10%, 20%, and 30%, respectively, while chiller #2 showed 9.74%, 10.17%, and 10.29%. These results indicate that DC-KNN imputation maintains stable and competitive accuracy, with slightly improved performance over standard KNNs, particularly at lower MAR levels.

In terms of the CVRMSE, chiller #1 yielded 12.19%, 13.19%, and 13.06%, while chiller #2 showed 12.65%, 13.32%, and 13.43%. The consistent and controlled variability across missingness rates suggests that DC-KNN imputation is robust in preserving underlying data characteristics, even under increased data sparsity (Table 6).

DC-KNN imputation improves upon conventional KNNs by integrating dynamic weighting of variables and cluster-based neighbor selection. Rather than selecting neighbors solely based on global distance, the DC-KNN method first identifies clusters of data with similar patterns—such as those defined by temperature or flow rate levels—and restricts neighbor search within the same cluster. This structure-sensitive approach increases the contextual relevance of the selected neighbors.

Furthermore, the DC-KNN method introduces variable-specific weighting that balances between Euclidean distance and cosine similarity depending on feature characteristics, allowing more flexible modeling of both magnitude and directional relationships. This enhances the fidelity of the imputed values while mitigating distortion caused by irrelevant neighbors.

While the method incurs higher computational cost and complexity, its advantages lie in its ability to maintain structural consistency and increase imputation realism, particularly in multidimensional and time-series sensor datasets. It is especially suitable for preprocessing stages in high-stakes predictive modeling, where data integrity and precision are essential (Figures 8–10).

4.1.8. Visual Comparison and Interpretation of Algorithm Performance

Figures 8–10 present a comparative visualization of the prediction performance of seven imputation methods under three MAR conditions (10%, 20%, and 30%) using MAPE as the evaluation metric. Each figure illustrates how the accuracy of each method varies across chiller #1 and #2, providing a clear depiction of how performance differences evolve as the level of missingness increases.

The proposed DC-KNN method consistently achieved the lowest or near-lowest MAPE values across all missingness levels, demonstrating strong accuracy and robustness. In particular, it outperformed conventional KNNs under higher missing rates, indicating that its cluster-based neighbor selection and dynamic weighting mechanisms effectively capture the underlying structure of the data.

By separating the figures by missingness level, the layout enables an assessment not only of absolute performance but also of the stability and scalability of each method under increasing data sparsity. This visual evidence supports the conclusion that the DC-KNN method is not only precise but also structurally superior in preserving contextual integrity in time-series sensor environments.

4.2. Selection of Optimal Imputation Methods

4.2.1. Optimal Imputation Method Based on MAPE

DC-KNNs consistently achieved the lowest MAPE across all MAR levels for both chiller #1 and chiller #2, demonstrating the highest level of accuracy among all evaluated methods. Specifically, DC-KNNs recorded 9.86%, 10.30%, and 10.08% for chiller #1, and 9.74%, 10.17%, and 10.29% for chiller #2 at MAR 10%, 20%, and 30%, respectively. These results reflect DC-KNNs' superior ability to minimize absolute predictive error. Although KNNs also maintained competitive performance, it was consistently outperformed by DC-KNNs, particularly at lower MAR levels. In contrast, simple random imputation recorded the highest MAPE across all conditions, indicating poor performance and high volatility. These findings affirm that DC-KNNs provide the most accurate imputation across various degrees of data sparsity (Figure 11).

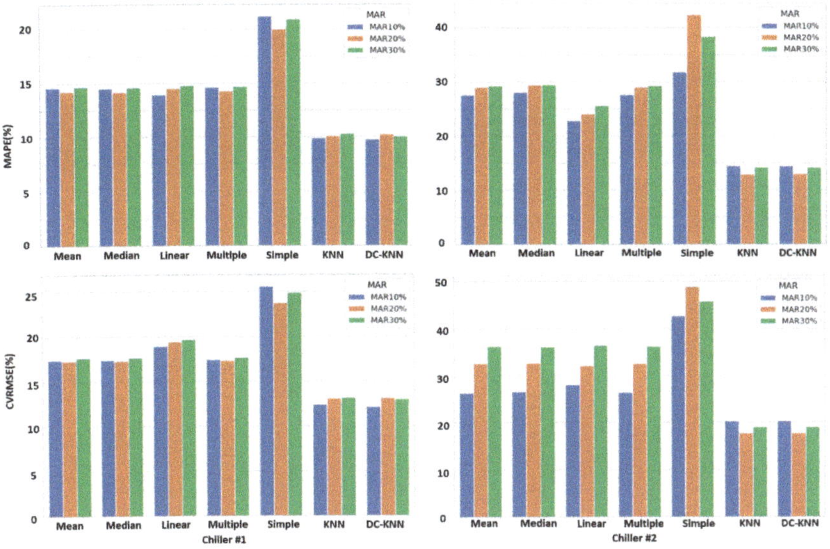

Figure 11. Comparison graph of MAPE and CVRMSE for each imputation method in chiller #1-2.

4.2.2. Optimal Imputation Method Based on CVRMSE

For the CVRMSE, the DC-KNN method again demonstrated strong performance, especially for chiller #2, where it recorded the lowest variability at all MAR levels (12.65%, 13.32%, 13.43%). For chiller #1, while mean, median, and multiple imputation performed slightly better at MAR 10%, DC-KNNs remained highly competitive and recorded lower error rates than KNNs and all other methods as the MAR increased. By contrast, simple random imputation produced the highest CVRMSE values for both chillers, highlighting its limited reliability for variance-sensitive applications. These results emphasize the robustness of DC-KNNs in preserving not only accuracy but also the variance structure of the data (Figure 11).

4.2.3. Changes According to MAR Levels

All imputation methods exhibited increasing MAPE and CVRMSE values as the MAR level rose, reflecting the inherent challenge of accurately recovering missing information under high data sparsity. Notably, simple random imputation, linear interpolation, and KNNs showed steep performance degradation at MAR 30%, demonstrating sensitivity to missing data proportions. In contrast, DC-KNNs and multiple imputation maintained stable performance across all MAR levels. Their resilience under high missingness suggests suitability for robust imputation in real-world applications where complete data availability cannot be guaranteed (Figure 11).

4.2.4. Derivation of the Optimal Imputation Method

Considering both error metrics—the MAPE and the CVRMSE—the DC-KNN imputation was identified as the most reliable and versatile method. Its hybrid architecture, which integrates clustering, dynamic weighting, and distance-based neighbor selection, allows it to adapt effectively to the structural characteristics of chiller sensor data. While mean, median, and multiple imputation methods provided reasonable performance at lower MAR levels and may be suitable for simple scenarios, their limitations become more apparent as the missing rate increases. Linear interpolation, although preserving continuity, struggled under high missingness, and simple random imputation proved inadequate across all conditions. Therefore, the DC-KNN approach is recommended as the optimal imputation method for both accuracy and robustness in high-dimensional, time-dependent sensor datasets (Figure 11).

4.3. DC-KNN Imputation for Actual Missing Data

Using raw data spanning a total of three years, the missing values of R_W for chiller #1 and chiller #2 in July 2023 were imputed using the DC-KNN method. Given the site characteristics, the power consumption of the chiller can be significantly low when utilizing thermal energy storage, and the imputed values appear to reflect this characteristic. The black points represent actual values, while the red points indicate imputed values (Figure 12).

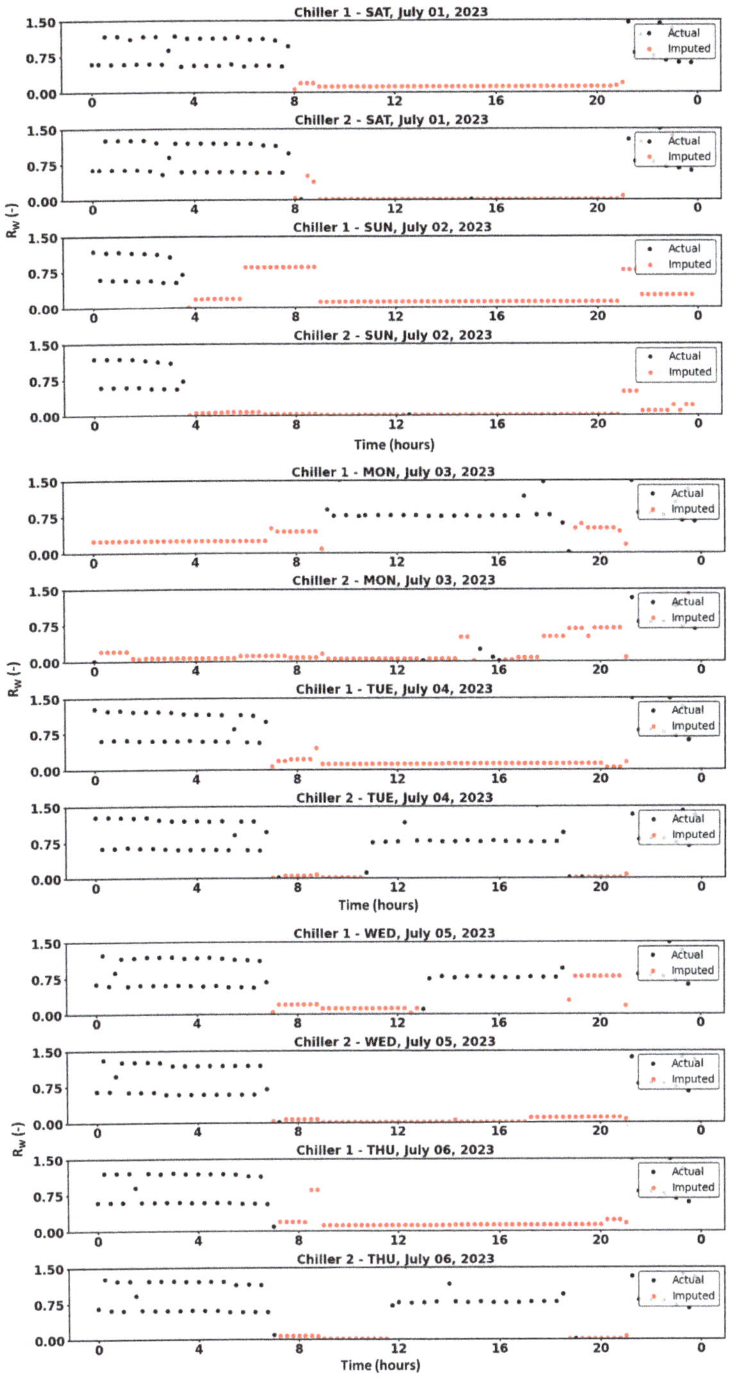

Figure 12. *Cont.*

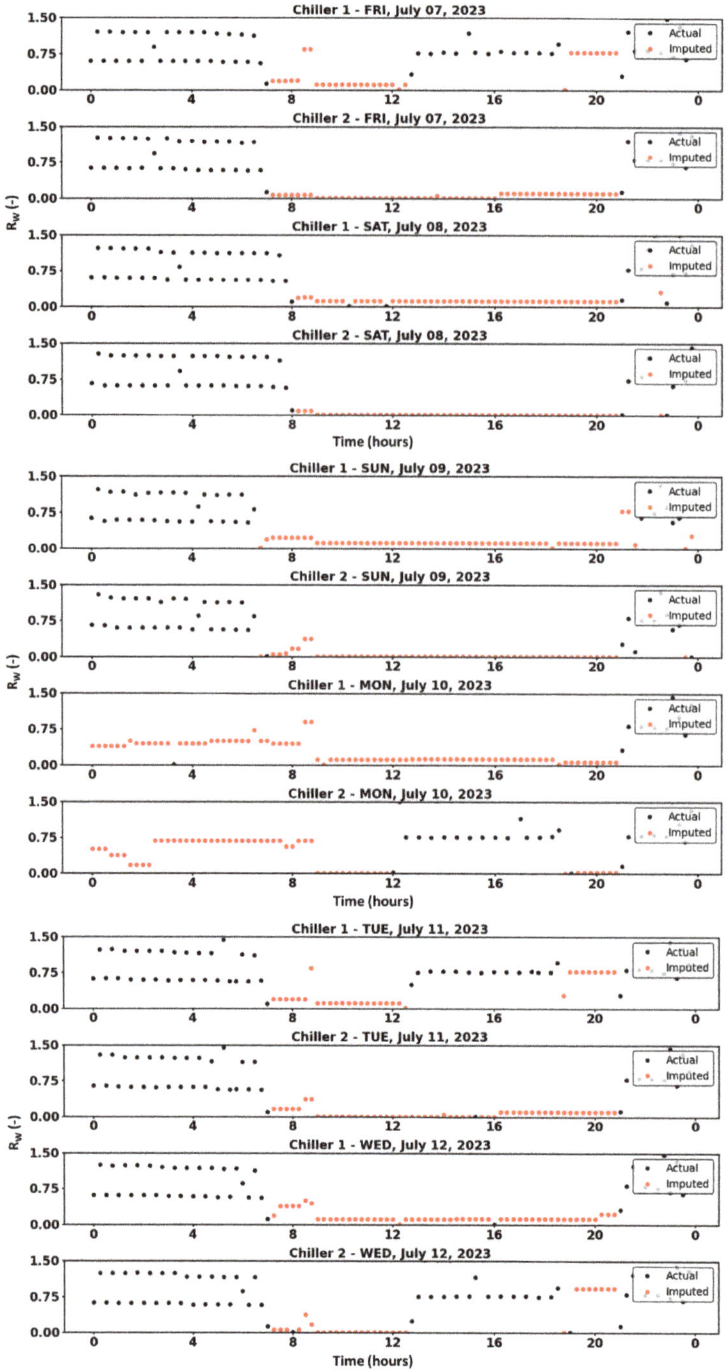

Figure 12. *Cont.*

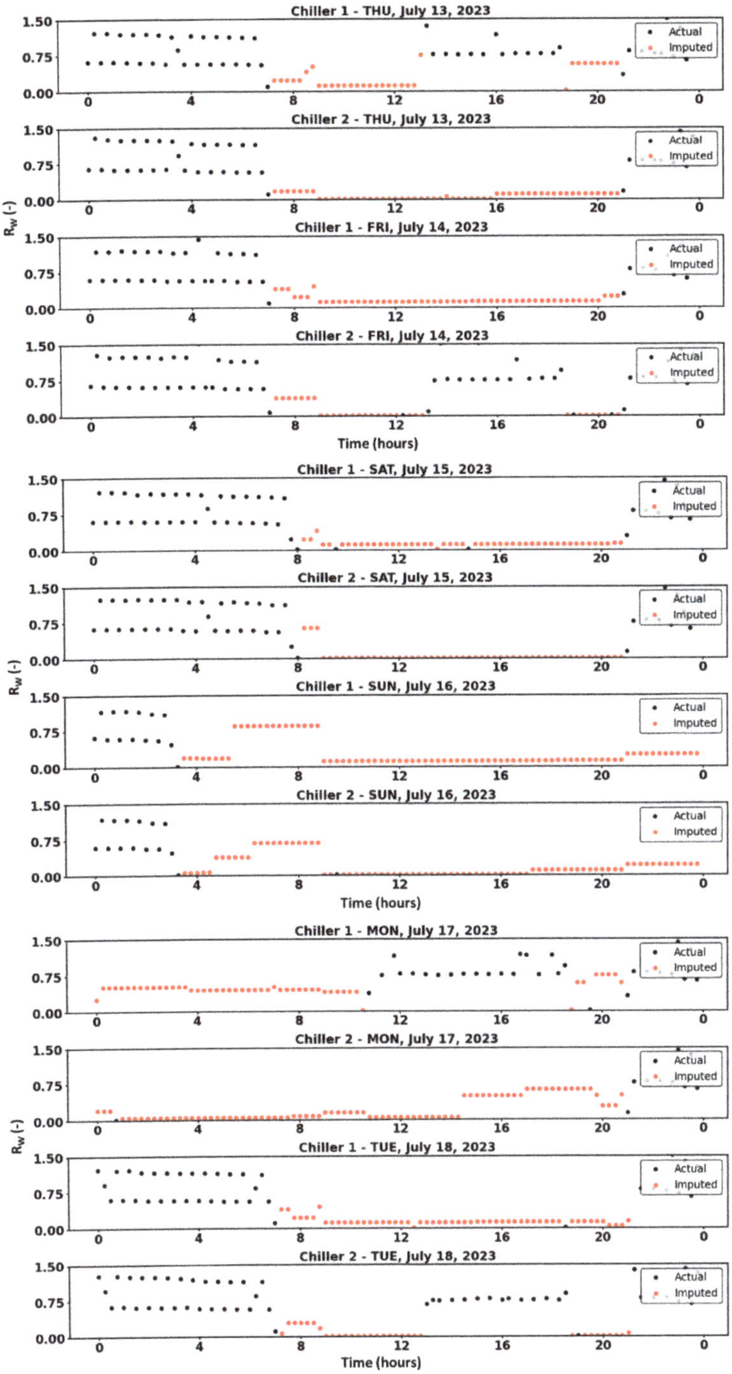

Figure 12. *Cont.*

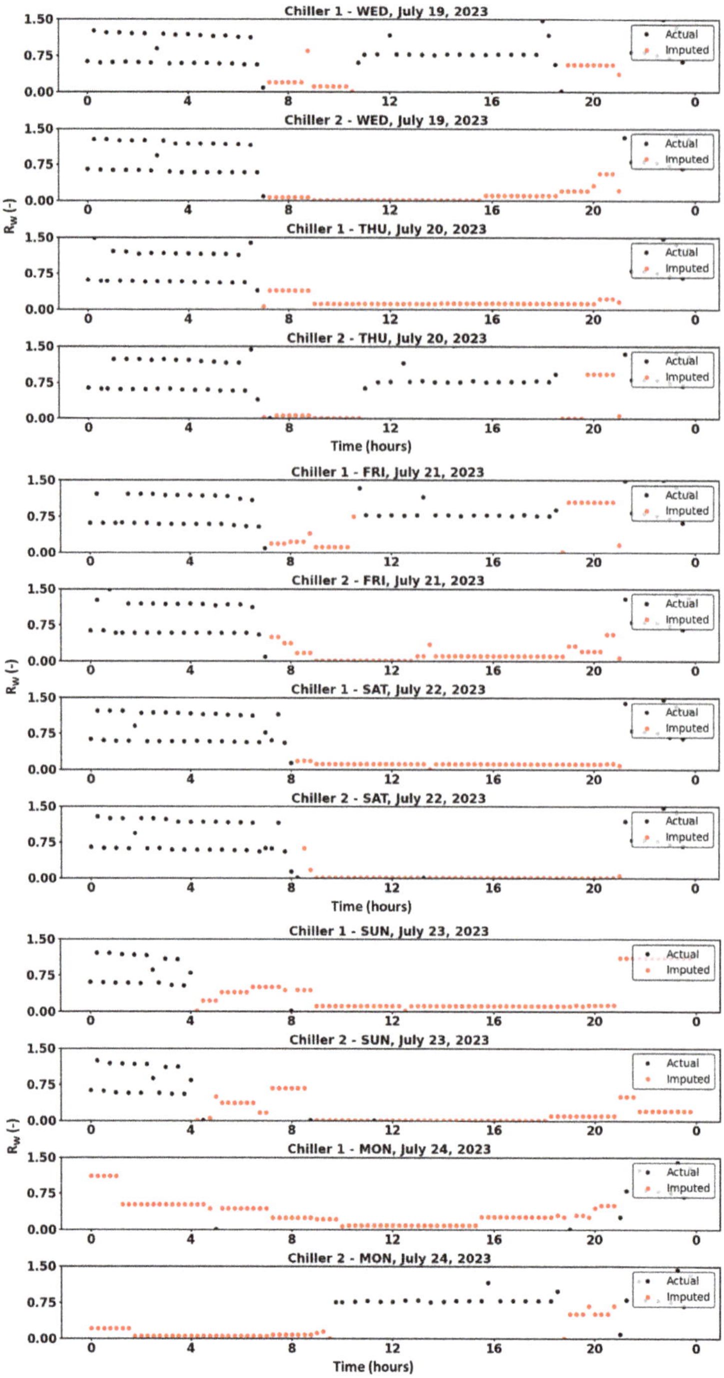

Figure 12. *Cont.*

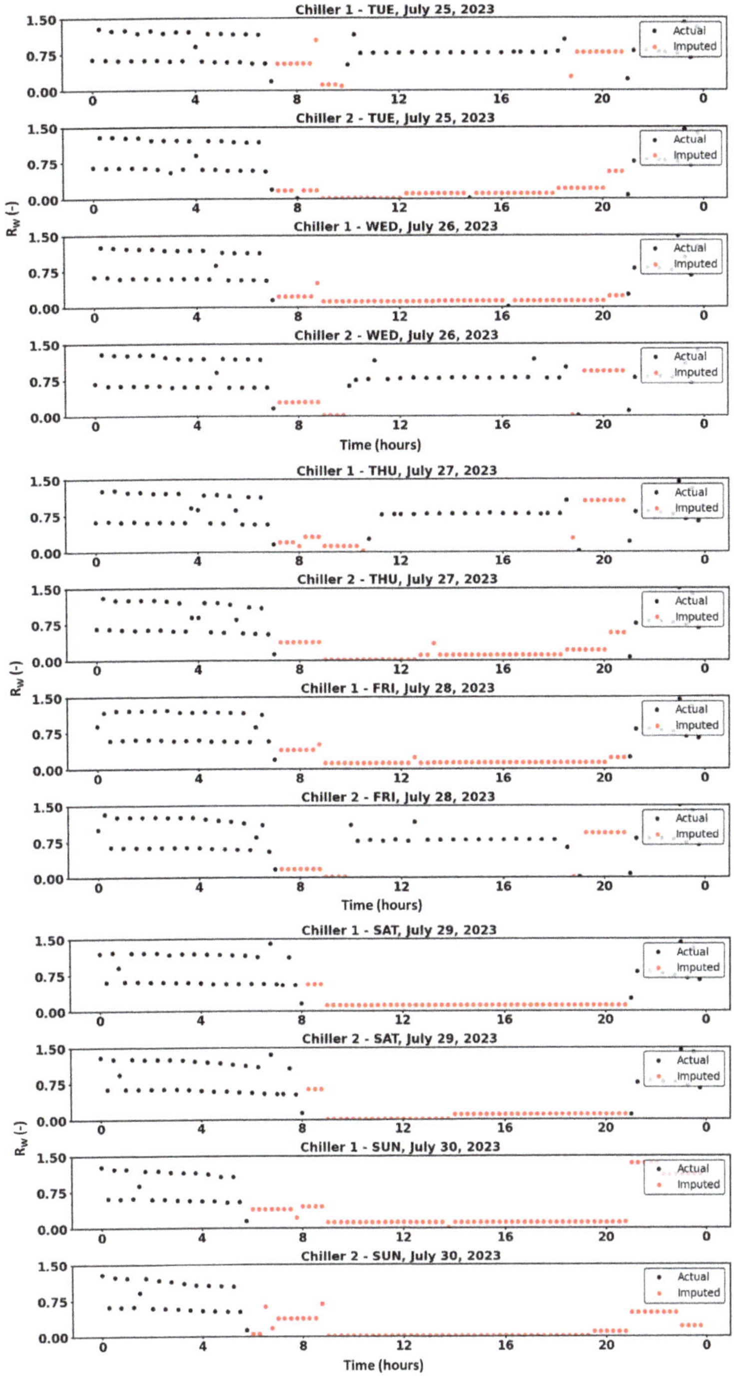

Figure 12. *Cont.*

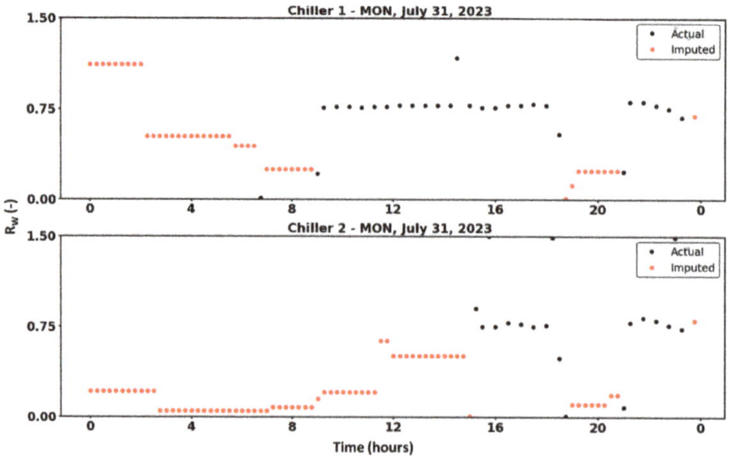

Figure 12. Hourly scatter plot of R_W in July 2023 for chiller #1-2 with missing values imputed using the DC-KNN method.

5. Discussion

Various missing data imputation techniques were compared and analyzed for the power consumption records of water-cooled centrifugal chillers, and the performance of the proposed DC-KNN method was evaluated accordingly. The DC-KNN method was designed to overcome the limitations of conventional KNN-based imputation by integrating K-means clustering to guide neighbor selection within localized operational contexts and applying dynamic weights based on feature-specific characteristics. This approach enables a more precise reflection of the physical behavior of the data.

The use of Euclidean distance for similarity measurement, cluster-based neighbor selection, and inverse-distance weighted averaging proved effective in preserving the structural consistency of the chiller operation data. When applied under missing-at-random (MAR) conditions with missing rates ranging from 10% to 30%, the DC-KNN method consistently outperformed conventional approaches—including mean, median, linear interpolation, multiple imputation, simple random imputation, and standard KNNs—in terms of both MAPE and CVRMSE.

Specifically, the DC-KNN method maintained a stable performance with MAPE values below 10% across all missingness levels for both chiller #1 and chiller #2 and demonstrated lower variability in the CVRMSE compared to other methods. These results suggest that the DC-KNN approach effectively restores underlying patterns and variance structures, even under high missingness conditions. In contrast, methods such as simple random imputation and linear interpolation showed significant performance degradation as the missing rate increased, whereas the DC-KNN method maintained robustness, indicating its high practical applicability.

However, the performance gap among imputation methods was not statistically large in all cases; particularly under low missingness, multiple imputation and even mean substitution also exhibited a relatively strong performance. Thus, in practical settings, a flexible choice among various validated imputation methods—including DC-KNNs—can be made based on data characteristics and the nature of missingness. Applying advanced methods, such as DC-KNNs, may further enhance data quality where complex or structured missing patterns are present.

It is also noted that the DC-KNN method may be sensitive to certain hyperparameters, such as the number of clusters (K) and distance-based weighting schemes, and its computational cost may increase with large-scale datasets. Future work should explore automated cluster optimization and algorithmic improvements to enhance scalability and efficiency.

On the other hand, although the DC-KNN method demonstrated superior performance in terms of imputation accuracy due to its cluster-based local interpolation mechanism, the algorithm inherently requires repeated execution of K-means clustering and KNN searches within each cluster. This structure can lead to increased computational time, particularly in large-scale datasets or real-time streaming environments.

While this study was conducted using a static dataset and did not include measurements of real-time computational load, future research should consider lightweight adaptations—such as online clustering, approximate KNNs, or parallel processing-based implementations—to improve computational efficiency.

In practical applications such as building energy management systems (BEMSs) or digital twin platforms, real-time processing capability is as critical as accuracy. Therefore, adapting the structure of the DC-KNN method to meet the requirements of real-time control environments will be an essential direction for its practical advancement.

While deep learning-based imputation methods, such as BiLSTM and autoencoders, have recently gained attention in time-series recovery tasks, this study focused on interpretable and practically applicable methods based on statistical and tree-based models. Deep learning-based approaches are currently under investigation in a separate follow-up study, with a focus on sequence learning, nonlinear dynamic pattern recognition, and model generalizability for energy forecasting and real-time control in chiller systems.

Beyond technical improvements in imputation accuracy, the reliability of restored sensor data plays a critical role in enabling data-driven decision-making in intelligent energy systems. Similarly, Es-sakali et al. [34] developed a predictive maintenance strategy based on real-time monitoring of a VRF system installed in an actual building, highlighting the importance of comprehensive data preprocessing—including missing value imputation, outlier removal, and noise filtering—for accurate fault diagnosis and energy efficiency optimization.

Their comparative analysis of multiple machine learning models demonstrated that data quality is a decisive factor in the effectiveness of predictive maintenance. These findings support the practical applicability and technical value of the proposed DC-KNN method, especially in scenarios where restored data serve as the foundation for high-stakes decision-making in intelligent HVAC control systems [34].

Prior studies in soft science fields have emphasized the importance of trustworthy data in supporting predictive maintenance, operational optimization, and adaptive energy control strategies in smart building environments [35]. The DC-KNN method proposed in this study preserves interpretability while enhancing accuracy under high-missingness conditions, making it a promising candidate for integration into real-time control architectures and context-aware decision frameworks.

In sensor-based energy management systems, the reliability and completeness of input data are critical for enabling effective control strategies and achieving measurable energy savings. Previous studies have demonstrated the impact of sensor-integrated room control systems on reducing energy consumption in real-world building environments [36], as well as the importance of stable and high-quality data in performance evaluations of advanced energy storage applications [37]. In this context, the DC-KNN method enhances the robustness of time-series datasets, thereby reinforcing the data foundation required for intelligent forecasting, control, and decision-making in smart energy systems [36,37].

6. Conclusions

This study proposed the DC-KNN method for improving missing data imputation accuracy in time-series chiller operation data. The proposed method integrates KNNs with K-means clustering, enabling context-aware neighbor selection and maintaining thermodynamic consistency during imputation. Unlike cosine similarity, which was not adopted in this study, Euclidean distance was exclusively used to preserve the sensitivity to absolute differences in physical variables, such as temperature and flow rate.

Across all tested MAR levels (10%, 20%, and 30%), the DC-KNN method achieved the lowest MAPE and CVRMSE scores in most scenarios, demonstrating high accuracy and robustness even under increased missingness. In contrast, traditional techniques—particularly at MAR 30%—showed substantial degradation in performance. The DC-KNN method was successfully applied to actual chiller data from July 2023 and preserved operational characteristics while maintaining the thermodynamic relationship between variables.

These results imply the strong applicability of the DC-KNN method as an automated imputation module in real-world systems such as BEMSs, where accurate data recovery is crucial for control optimization and fault detection. Furthermore, the chiller system analyzed in this study operates under a thermal energy storage (TES) scheme, which induces a clear separation between daytime cooling and nighttime ice-making modes. These load-shifting behaviors result in heterogeneous operational profiles and introduce structured missing data patterns. The DC-KNN method addresses this complexity by selecting neighbors within operationally similar clusters, thereby enhancing the reliability and contextual accuracy of the imputation, even under dynamic system conditions.

Future research may extend this approach to other HVAC systems and assess its effectiveness under real-time and streaming conditions.

Moreover, the imputed high-quality data can be further utilized for fault detection, energy consumption prediction, and digital twin-based system simulations in intelligent HVAC applications. These downstream applications critically rely on the integrity of time-series sensor data, making robust imputation a foundational step.

Additionally, while this study focused on MAR (missing-at-random) scenarios, future work may explore extending the DC-KNN approach to handle missing not at random (MNAR) conditions, where the probability of missingness depends on the unobserved values themselves. Integrating probabilistic models or iterative estimation strategies—such as expectation–maximization (EM)—could enhance its generalizability to more complex data environments. The applicability of DC-KNNs to multivariate time-series forecasting tasks also presents a promising avenue for future research.

Author Contributions: Conceptualization, S.W.K.; methodology, S.W.K.; experiment, S.W.K.; software, S.W.K. and Y.I.K.; verification, Y.I.K.; formal analysis, S.W.K.; investigation, S.W.K.; resources, S.W.K.; data Curation, S.W.K. and Y.I.K.; writing—original draft preparation, S.W.K. and Y.I.K.; writing—review and editing, Y.I.K.; visualization, S.W.K.; director, Y.I.K.; project management, Y.I.K.; funding, S.W.K. All authors have read and agreed to the published version of the manuscript.

Funding: This study was supported by research programs funded by Seoul National University of Science and Technology and the Korea Institute of Educational Facility Safety.

Data Availability Statement: The original contributions presented in the study are included in the article, further inquiries can be directed to the corresponding author.

Conflicts of Interest: The authors declare no conflicts of interest.

Nomenclature

COP	Coefficient of performance (-)
COP_a	Actual COP (-)
COP_o	Nominal COP (-)
COP_p	Predicted daytime COP (-)
COP_m	Mean COP (-)
$y_{imputed}$	Imputed value
y_i	Observed value
y_{miss}	Missing value
$y_{neighbor}$	Neighboring value in KNNs
$\hat{y}_i$	Predicted value
$d\left(y_i \cdot y_j\right)$	Distance function
w_i	Weight in KNNs
n	Number of data points
$\overline{y}$	Mean of value
$d_{Euc}\left(y_i \cdot y_j\right)$	Euclidean distance
$S_{COS}\left(y_i \cdot y_j\right)$	Cosine similarity
EB	Energy balance (%)
BEMS	Building energy management system
RSS	Residual sum of squares
MSE	Mean squared error
RMSE	Root mean square error
CVRMSE	Coefficient of variation of the root mean squared error
R^2	Coefficient of determination
R_Q	Heat transfer ratio (-)
R_W	Power consumption ratio (-)
R_{COP}	COP ratio (-)
R_T	Water temperature ratio (-)
R_V	Circulation flow ratio (-)
s	Standard deviation
$\dot{Q}$	Rate of heat transfer (kW)
$\dot{Q}_e$	Rate of heat transfer on the evaporator side (kW)
$\dot{Q}_c$	Rate of heat transfer on the condenser side (kW)
$\dot{Q}_a$	Actual heat transfer (kW)
$\dot{Q}_o$	Nominal heat transfer (kW)
$\dot{W}_c$	Power consumed by the compressor (kW)
$\dot{W}_o$	Actual power consumption (kW)
$\dot{W}_a$	Nominal power consumption (kW)
Y_{obs}	Observed data
Y_{mis}	Missing data
$P(R\|Y_{obs})$	Probability of missingness considering both observed and missing data
ρ	Fluid density (kg/m^3)
c_p	Specific heat of fluid (kJ/kg°C)
c_w	Specific heat of water (kJ/kg°C)
T_a	Actual water temperature (°C)
T_o	Nominal water temperature (°C)
c_b	Specific heat of brine (kJ/kg°C)
$\dot{V}$	Volumetric flow rate (L/min)
$\dot{V}_1$	Chilled water volumetric flow rate (L/min)
$\dot{V}_3$	Cooling water volumetric flow rate (L/min)
$\dot{V}_{a1}$	Actual chilled water circulation volumetric flow (L/min)
$\dot{V}_{o1}$	Nominal chilled water circulation volumetric flow (L/min)

$\dot{V}_{a3}$	Actual cooling water circulation volumetric flow (L/min)
$\dot{V}_{o3}$	Nominal cooling water circulation volumetric flow (L/min)
ΔT	Fluid temperature difference (°C)
T_1	Chilled water inlet temperature (°C)
T_2	Chilled water outlet temperature (°C)
T_3	Cooling water inlet temperature (°C)
T_4	Cooling water outlet temperature (°C)
x	Independent variables
β	Regression coefficients
ε	Estimation error (residuals)
n	Number of data points
μ	Mean of the data
W_R	Total uncertainty

References

1. Korean Society of Air-Conditioning and Refrigerating Engineers (SAREK). *Handbook of Air-Conditioning and Refrigeration*, 4th ed.; SAREK: Seoul, Republic of Korea, 2007.
2. Korean Meteorological Administration. *2023 Abnormal Climate Report*; Korea Meteorological Administration: Daejeon, Republic of Korea, 2024.
3. Kim, Y.I. Performance of a Water-Cooled Chiller by Controlling Chilled Water Exit Temperature. In Proceedings of the SAREK 2010 Summer Annual Conference, Pyeongchang, Republic of Korea, 23–25 June 2010; pp. 1136–1141.
4. Lee, C.W.; Seong, N.C.; Choi, W.C. Performance Improvement and Comparative Evaluation of the Chiller Energy Consumption Forecasting Model Using Python. *J. KIAEBS* **2021**, *15*, 252–264.
5. Alsaber, A.; Al-Herz, A.; Pan, J.; Al-Sultan, A.T.; Mishra, D.; KRRD Group. Handling Missing Data in a Rheumatoid Arthritis Registry Using Random Forest Approach. *Int. J. Rheum. Dis.* **2021**, *24*, 1282–1293. [CrossRef] [PubMed]
6. Sun, Y.; Li, J.; Xu, Y.; Zhang, T.; Wang, X. Deep Learning versus Conventional Methods for Missing Data Imputation: A Review and Comparative Study. *Expert Syst. Appl.* **2023**, *227*, 120201. [CrossRef]
7. Burgette, L.F.; Reiter, J.P. Multiple Imputation for Missing Data via Sequential Regression Trees. *Am. J. Epidemiol.* **2010**, *172*, 1070–1076. [CrossRef]
8. Algarni, A.; Ragab, M.; Alamri, W.; Mostafa, S.M. Towards Improving Predictive Statistical Learning Model Accuracy by Enhancing Learning Technique. *Comput. Syst. Sci. Eng.* **2022**, *42*, 304–318. [CrossRef]
9. Li, T.; Hutfless, S.; Scharfstein, D.O.; Daniels, M.J.; Hogan, J.W.; Little, R.J.A.; Roy, J.A.; Law, A.H.; Dickersin, K. Standards Should Be Applied in the Prevention and Handling of Missing Data for Patient-Centered Outcomes Research: A Systematic Review and Expert Consensus. *J. Clin. Epidemiol.* **2014**, *67*, 15–32. [CrossRef]
10. Çengel, Y.A. *Fundamentals of Thermal-Fluid Sciences*, 5th ed.; McGraw Hill Education: New York, NY, USA, 2015; p. 253.
11. Chang, Y.S.; Shin, Y.; Kim, Y.I.; Baik, Y.J. In-Situ Performance Analysis of Centrifugal Chiller According to Varying Conditions of Chilled and Cooling Water. *Trans. Korean Soc. Mech. Eng. B* **2002**, *26*, 482–490.
12. Liu, P.L.; Chuang, B.S.; Lee, W.S.; Yeh, P.L. An Analytical Solution of the Optimal Chillers Operation Problems Based on ASHRAE Guideline 14. *J. Build. Eng.* **2022**, *46*, 103800. [CrossRef]
13. Lee, T.S.; Lu, W.C. An Evaluation of Empirically-Based Models for Predicting Energy Performance of Vapor-Compression Water Chillers. *Appl. Energy* **2010**, *87*, 3486–3493. [CrossRef]
14. Li, J.; Guo, S.; Ma, R.; He, J.; Zhang, X.; Rui, D.; Ding, Y.; Li, Y.; Jian, L.; Cheng, J.; et al. Comparison of the Effects of Imputation Methods for Missing Data in Predictive Modelling of Cohort Study Datasets. *BMC Med. Res. Methodol.* **2024**, *24*, 41. [CrossRef]
15. Anderson, P.; Gupta, S. Identify the Most Appropriate Imputation Method for Handling Missing Values in Clinical Structured Datasets: A Systematic Review. *J. Med. Inform.* **2023**, *29*, 215–230.
16. Van Buuren, S. *Flexible Imputation of Missing Data*; CRC Press: Boca Raton, FL, USA, 2012.
17. Crambes, C.; Daayeb, C.; Gannoun, A.; Henchiri, Y. Multiple imputation in the functional linear model with partially observed covariate and missing values in the response. *Commun. Stat.-Theory Methods* **2025**, *54*, 49–69. [CrossRef]
18. Qu, H.; Zhang, Z. A Time Series Data Augmentation Method Based on SMOTE. In Proceedings of the 2024 36th Chinese Control and Decision Conference (CCDC), Xi'an, China, 25–27 May 2024; pp. 5336–5341.
19. Niako, N.; Melgarejo, J.D.; Maestre, G.E.; Vatcheva, K.P. Effects of Missing Data Imputation Methods on Univariate Blood Pressure Time Series Data Analysis and Forecasting with ARIMA and LSTM. *BMC Med. Res. Methodol.* **2024**, *24*, 320. [CrossRef]

20. Tiwaskar, S.; Rashid, M.; Gokhale, P. Ensemble Technique for Imputing Missing Values in MAR Missingness. In Proceedings of the 2023 6th International Conference on Contemporary Computing and Informatics (IC3I), Gautam Buddha Nagar, India, 14–16 September 2023; IEEE: New York, NY, USA, 2023; pp. 880–883.
21. Rao, A.R.; Reimherr, M. Modern Multiple Imputation with Functional Data. *Stat* **2021**, *10*, e331. [CrossRef]
22. DiazOrdaz, K.; Kenward, M.G.; Gomes, M.; Grieve, R. Multiple Imputation Methods for Bivariate Outcomes in Cluster Randomised Trials. *Stat. Med.* **2016**, *35*, 3482–3496. [CrossRef]
23. Cai, S.; Qin, Y.; Rao, J.N.K.; Winiszewska, M. Empirical Likelihood Confidence Intervals under Imputation for Missing Survey Data from Stratified Simple Random Sampling. *Can. J. Stat.* **2019**, *47*, 281–301. [CrossRef]
24. Wang, C.; Ren, B.; Li, X.; Chen, L. A CNN-BiLSTM and KNN-Based Missing Data Imputation for Wind Power Generation Forecasting. In Proceedings of the 2023 IEEE 6th International Electrical and Energy Conference (CIEEC), Hefei, China, 12–14 May 2023; IEEE: New York, NY, USA, 2023; pp. 4065–4070. [CrossRef]
25. Alabadla, M.; Sidi, F.; Ishak, I.; Ibrahim, H.; Affendey, L.S.; Che Ani, Z.; Jabar, M.A.; Bukar, U.A.; Devaraj, N.K.; Muda, A.S.; et al. Systematic Review of Using Machine Learning in Imputing Missing Values. *IEEE Access* **2022**, *10*, 44483–44502. [CrossRef]
26. Hwang, H.; Min, D. A Transfer Learning for Missing Value Imputation and Its Relationship with Prediction Performance in Time Series Data. *J. Korean Inst. Ind. Eng.* **2023**, *49*, 294–308.
27. Aziz, R.Z.A.; Lestari, S.; Fitria; Arianto, F. Imputation Missing Value to Overcome Sparsity Problems. *Telkomnika* **2024**, *22*, 949–955. [CrossRef]
28. Yoon, Y.R.; Shin, S.H.; Moon, H.J. Analysis of Building Energy Consumption Patterns according to Building Types Using Clustering Methods. *J. Korean Soc. Living Environ. Syst.* **2017**, *24*, 232–237. [CrossRef]
29. Kim, C.; Kang, K.-H. Comparison of Data Reconstruction Methods for Missing Value Imputation. *J. Converg. Cult. Technol.* **2024**, *10*, 603–608.
30. Altukhova, O. Choice of Method Imputation Missing Values for Obstetrics Clinical Data. *Procedia Comput. Sci.* **2020**, *176*, 976–984. [CrossRef]
31. Wang, Y.; Jin, X.; Shi, W.; Wang, J. Online Chiller Loading Strategy Based on the Near-Optimal Performance Map for Energy Conservation. *Appl. Energy* **2019**, *238*, 1444–1451. [CrossRef]
32. Zhang, Y.; Li, H.; Wang, S. Energy Performance Analysis of Multi-Chiller Cooling Systems for Data Centers Concerning Progressive Loading Throughout the Lifecycle Under Typical Climates. *Build. Simul.* **2024**, *17*, 1693–1708. [CrossRef]
33. Kim, S.W.; Kim, Y.I. Performance Prediction of a Water-Cooled Centrifugal Chiller in Standard Temperature Conditions Using In-Situ Measurement Data. *Sustainability* **2025**, *17*, 2196. [CrossRef]
34. Es-sakali, N.; Zoubir, Z.; Kaitouni, S.I.; Mghazli, M.O.; Cherkaoui, M.; Pfafferott, J. Advanced predictive maintenance and fault diagnosis strategy for enhanced HVAC efficiency in buildings. *Appl. Therm. Eng.* **2024**, *254*, 123910. [CrossRef]
35. Dong, W.; Sheng, K.; Huang, B.; Xiong, K.; Liu, K.; Cheng, X. Stretchable self-powered TENG sensor array for human robot interaction based on conductive ionic gels and LSTM neural network. *IEEE Sens. J.* **2024**, *24*, 37962–37969. [CrossRef]
36. Lee, J.W.; Kim, Y.I. Energy saving of a university building using a motion detection sensor and room management system. *Sustainability* **2020**, *12*, 9471. [CrossRef]
37. Xie, L.J.; Jiang, J.C.; Huang, A.C.; Tang, Y.; Liu, Y.C.; Zhou, H.L.; Xing, Z.X. Calorimetric evaluation of thermal stability of organic liquid hydrogen storage materials and metal oxide additives. *Energies* **2022**, *15*, 2236. [CrossRef]

processes

MDPI

Article

Superstructure Optimization Based on Hierarchical Accelerated Branch and Bound Algorithm and Its Application in Feedstock Scheduling

Jian Cao *, Haitao Yang and Yi Ji

College of Automation (College of Artificial Intelligence), Control Science and Engineering, Shahe Campus, Beijing Information Science and Technology University, No. 55 Taihang Road, Changping District, Beijing 102206, China; yanghaitao@bistu.edu.cn (H.Y.); jiyi@bistu.edu.cn (Y.J.)

* Correspondence: caojian@bistu.edu.cn

Abstract

As modern petrochemical systems tend towards larger scales, process structures become more complex and highly integrated. The interaction and overall complexity between the operations of each unit are constantly increasing, and the solution process is becoming increasingly complex. The superstructure generated by P-graph theory can reduce the generation of redundant structures, and the accelerated branch and bound algorithm can efficiently explore a vast solution space. However, this algorithm can only solve single-objective optimization problems. This article mainly focuses on the multi-objective optimization problem of the chemical process flow, with reduced production costs and the mitigation of environmental pollution being the main considerations. A novel algorithm framework based on an improved accelerated branch and bound method and incorporating the core idea of Analytic Hierarchy Process (AHP) was proposed, aiming to improve the high computational cost of traditional methods and solve the problem of high modeling costs in multi-objective optimization. The feedstock scheduling of ethylene is used to verify the effectiveness and superiority of the proposed method, with low-cost and carbon emission solutions being easily selected. This article provides a reference for the study of superstructures in chemical production processes. The experimental results show that with a small increase in energy consumption costs, the optimal solution obtained by our algorithm can reduce carbon emissions by 1.27% compared to the original solution, and reduce 2/3 of the modeling workload.

Keywords: superstructure; P-graph; analytic hierarchy process; accelerated branch and bound; process industry

1. Introduction

As an important component of the global economy, the chemical industry not only provides a large amount of production materials and employment opportunities for society, but also plays an irreplaceable role in promoting technological progress and social development. However, while the chemical industry brings economic benefits, it is also one of the main industrial sources of greenhouse gas emissions [1]. With the increasing global emphasis on environmental protection and sustainable development, the chemical industry is facing enormous pressure to reduce carbon emissions, reduce energy consumption, and improve resource utilization efficiency. Modern chemical production processes are usually very complex, involving numerous unit operations and material flows. The complexity of the process makes optimization of the production process particularly important. In order to achieve low-carbon production, low consumption, and efficient transformation in the chemical industry, the optimization of chemical process flows is necessary [2]. For example, thermal integration is actually designed to achieve energy-saving effects in chemical processes [3,4]. Therefore, modeling, analyzing, and optimizing chemical processes through scientific methods while ensuring process safety and product quality has become a hot and difficult issue in the current field of chemical engineering research.

An important component of process system synthesis is process network synthesis (PNS) [5]. Process network synthesis is very important in dealing with optimal decision-making problems and n-best problem [6]. Ali et al. used a combination of process network synthesis and machine learning as a decision-making tool for urban solid waste management [7]. The process system synthesis methods in process system synthesis include intuitive inference, hierarchical decomposition, mathematical programming, and pinch analysis [8,9]. And in addition to the above methods, Daoutidis et al. explored the integration of machine learning in PNS and focused on applications such as agent modeling, process optimization, and design space exploration [10]. However, the optimization of the hierarchical decomposition factor system sequence in the hierarchical decomposition method may miss the global optimal solution. Intuitive inference and pinch analysis methods cannot achieve trade-offs between subsystems at different levels. Therefore, achieving the global optimum of process system synthesis is very difficult. At the same time, the interpretability of solutions using artificial intelligence and machine learning methods is relatively poor. The commonly used method in process synthesis is mathematical programming, which is mainly based on mathematical optimization models for system synthesis. Compared to the above methods, mathematical models representing multi-level influencing factors in a system can obtain the global optimal design and have a certain degree of interpretability.

However, as process networks become increasingly complex, the complexity of corresponding mathematical models also increases. Subsequently, several problems arise, including an increase in the number of redundant structures in the superstructure, a non-linear increase in the model, and long development cycles for models, etc. Friedler et al. introduced a special directed binary graph called a Progress graph (P-graph) to solve the previous problem, which can achieve better trade-offs between subsystems at different levels [11]. P-graph simplifies the decision-making process, improves the efficiency and effectiveness of strategy formulation, and has become an important research direction in the field of process systems engineering [12–15]. Aviso et al. fuzzified the drought level and applied the fuzzy P-graph method to solve the problem of seeking the optimal operational adjustment solution in wind energy systems under drought conditions [16]. Pimentel et al. set the goal of maximizing the value of the industry chain and studied whether the price of sugarcane was stable by targeting lactic acid as the main product in their research. By simulating the parameters of various operating units, a research model was ultimately integrated using the P-graph theory [17]. P-graph is becoming increasingly popular in addressing sustainability and green development issues [18,19]. At the same time as the final production cost, considerable economic benefits were achieved. Bertok et al. applied the P-graph method to seek the optimal process design network for producing ethanol and acetone from grains [20]. The Branch and Bound (BB) algorithm in the P-graph framework obtains the solution space by analyzing the problem. Then, the feasible solutions in the solution space are partially enumerated, and finally a series of optimal feasible solutions are obtained [21]. The Mixed-Integer Nonlinear Programming (MINLP) model is commonly used to describe strict superstructure models that include both integer and continuous variables. The ABB (Accelerated Branch and Bound) algorithm utilizes the BB algorithm to solve mixed-integer nonlinear programming models, establishing search trees and new custom boundaries at each node to remove uninteresting regions and reduce the decision space [22]. Additionally, the ABB algorithm can simultaneously obtain optimal and suboptimal solutions when solving problems. The current optimization problems in chemical processes face challenges such as multi-objectivity and high complexity. However, traditional methods are not particularly effective in solving these problems. We consider combining AHP and ABB algorithms to solve these problems.

The Analytic Hierarchy Process (AHP) is also widely used in the analysis of the chemical industry. Bakhtari et al. used the hierarchical relationship between the challenges of Industry 4.0 development as a standard, analyzed their interactions and main obstacles, and focused their attention on achieving solutions [23]. Ba et al. conducted a scoring analysis of pipelines using the Analytic Hierarchy Process [24]. In order to solve multi-objective optimization problems, multiple modeling of the superstructure model is required when applying the ABB algorithm [25]. As a multi-criteria decision-making method, AHP can effectively guide and quantify subjective judgments, playing an important role in multi-objective decision-making [26]. This article combines the accelerated branch and bound algorithm with the analytic hierarchy process to solve optimization problems in chemical processes, in order to avoid redundant modeling and simplify the algorithm flow. This combination has to some extent solved the challenges faced by today's industrial systems, such as high integration, strong complexity, and multiple influencing variables. This is especially for the numerous redundant structural challenges and multi-objective optimization problems that exist in the production process.

Based on the above characteristics of the ABB and AHP algorithms, we plan to combine the two algorithms to achieve more efficient and accurate multi-objective optimization results in highly complex chemical processes. The process optimization problem solved under our proposed new algorithm framework not only saves modeling costs, but also fully considers the objectives of multi-objective optimization.

The remainder of this paper is organized as follows. Section 2 describes the proposed methodology. The experiment data, results and analysis are presented in Section 3. Finally, Section 4 summarizes the conclusions of this paper.

2. Materials and Methods

2.1. P-Graph Theory

The basic idea of the superstructure model is to decompose a large system into multiple subsystems or components, and accurately describe the interactions between these components. A component can be a physical component, functional unit, individual, or any other distinguishable entity. Superstructure models typically use graph theory to describe the structure and connectivity relationships of a system. The graphical representation can intuitively understand the structure and behavior of the system, and can also apply methods from graph theory to analyze and interpret the model. P-graph is a special type of directed graph used to represent and solve combinatorial optimization problems in process synthesis [27,28]. P-graph consists of nodes and directed edges, where nodes represent process units such as reactors, separation towers, etc. Among them, directed edges represent the flow of materials, and directed edges are the flow paths of materials. The node pointing to another node represents the flow of materials from one process unit to another. All nodes in the P-graph must be connected to ensure that materials can flow from one process unit to another, and each process unit in the P-graph can only appear once. P-graph theory can represent all possible networks and provide information on their structural feasibility for relevant case studies [29–31]. The P-Graph Studio software (http://p-graph.org, Version: 5.2.5.0, vendor: University of Pannonia, Veszprém, Hungary) for P-graph applications is already a mature and widely used model building platform that enables researchers to apply P-graphs. The superstructure is transformed into a P-graph by constructing a process flow diagram (PFD), as shown in Figure 1.

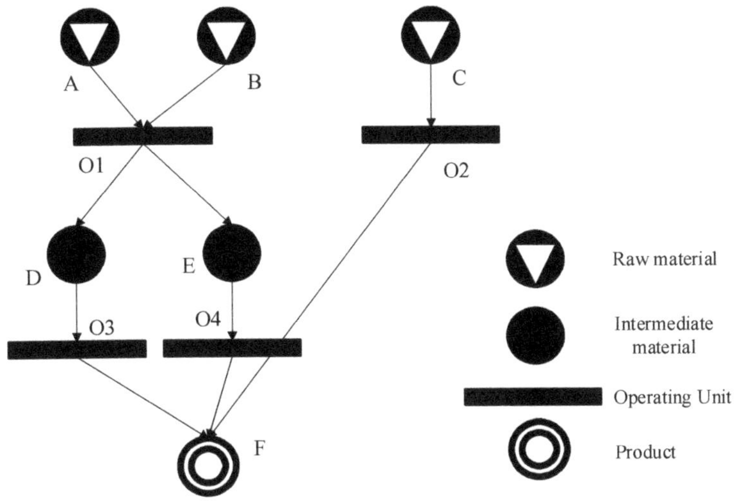

Figure 1. Superstructure modeling by P-graph.

P-graph theory defines four sets, namely P (Product), M (Material), R (Raw material), and O (Operating Unit). P represents the product that needs to be obtained; R represents the raw materials required for the production of the product; M represents the intermediate products obtained in the production process of the product and O represents the operating unit in the production process. The Superstructure in Figure 1 is very intuitive, where A, B, and C are the raw materials or some initial state for producing product F. D and E are intermediate products or intermediate states in the production process. O1, O2, O3, O4 are operating units in the production process, which can represent equipment or heating, cooling, and other operations. In order to solve the problem of structural redundancy in practical engineering, Friedler et al. added 5 axioms on the basis of P-graph representation; the following 5 axioms greatly reduce the solution space of P-graph methods [32]:

(1) The topological network of the structure must include every element in the product set P;
(2) If a material is in the output set of the operating unit, then this material cannot be in the raw material set;
(3) All operating units in the topology structure of the process network need to be explained;
(4) Any element in the operating unit and product set must have at least one path connected;
(5) Material must be interconnected with at least one operating unit.

In addition, P-graph theory also integrates algorithms such as MSG (Maximum Structure Generation), SSG (Solution Structure Generation), and ABB. The solving of the process synthesis problem based on P-graph theory has become more convenient. This paper applies the P-graph theory to the feedstock scheduling of ethylene production. The material content produced is virtually represented as the operating unit of the superstructure model. The framework also provides a reference for other optimization problems in chemical processes, and when solving such problems, virtual data can be added to P-graph modeling.

2.2. Accelerated Branch and Bound

The ABB algorithm is a process optimization method that improves the branch and bound (BB) deterministic algorithm. This algorithm uses the BB algorithm to solve the MINLP problem, without deviating from the maximum structure generated by MSG. The ABB will automatically arrange the optimization process structure. The ABB algorithm has significant differences from traditional branch and bound methods. The general BB algorithm is ineffective in solving mixed-integer programming models for synthesis problems. This is because the BB algorithm generates too many free variables and does not fully utilize the structural characteristics of the process system. For the process synthesis problem of superstructure optimization, with the help of strict superstructures, the ABB algorithm wisely utilizes the structural characteristics of process synthesis and does not contain any redundant process structures. The steps of the ABB algorithm are as follows:

(1) Set the root node as the initial structural space and initialize the current optimal solution to infinity.
(2) Calculate the objective function value for each possible structure and determine the objective function value for each structure based on the objective function.
(3) Compare the objective function values of each structure and update the current optimal solution to the structure with the minimum objective function value.
(4) Perform branching operations on the structural space based on the current optimal solution. Calculate the objective function value for the branched structure and define it accordingly.
(5) If the objective function value of a branch has exceeded the current optimal solution, the branch can be pruned. If a better structure is found, update the current optimal solution.
(6) Repeat the steps of (2–5) until the termination condition is reached.

The key steps of the ABB algorithm are to effectively reduce the search space through branching and boundary operations, thereby accelerating the process of finding the optimal solution. Meanwhile, the continuous updating of the current optimal solution helps to discover better structures in a timely manner, and the efficiency and accuracy of the algorithm are also improved [33]. Researchers have confirmed that the ABB exhibits an accelerated solving effect when searching for the optimal solution or top-*n* optimal solutions, making the solution faster and the results more accurate [34]. This work optimized the ABB algorithm during the pruning process in the P-graph framework, and the AHP score was used as one of the pruning conditions to improve the efficiency of the ABB algorithm.

2.3. Analytic Hierarchy Process

AHP is a method mainly used for solving complex decision-making problems. The establishment of the AHP algorithm aims to decompose problems, construct hierarchical structures, and compare elements at each level pairwise to ultimately determine the priority ranking of decisions. This algorithm is a multi-index comprehensive evaluation algorithm that can be used as a quantitative scheme for indicator weighting selection or to assign weights to indicators. When individuals make decisions, their subjective emphasis on certain factors is inconsistent. The AHP algorithm can assign weights to these indicators without the need to collect data. In order to achieve better results, it is important to pay attention to the key steps when using the AHP algorithm for decision calculations. For example, after the construction of the hierarchical evaluation model, expert scoring is required for the indicators of the criterion layer. The specific requirements for expert scoring are shown in Table 1.

Table 1. Expert scoring criteria.

Numerical Rating	Relative Importance Between Elements *i* and *j*
1	Equal importance
3	*i* is slightly more important than *j*
5	*i* is obviously more important than *j*
7	*i* is extremely more important than *j*
9	*i* is absolutely more important than *j*
2, 4, 6, 8	The median of the adjacent judgments mentioned above
Reciprocal of 1–9	The importance of comparing the exchange order between *i* and *j*

The scoring in the above table adopts the 1–9 scale method proposed by American operations researcher Thomas L. Saaty [35]. Based on the above content, we can construct the criterion layer matrix A, as represented by Equation (1) (where $m = n$); the criterion layer matrix needs to satisfy Equation (2) (where $i \leq m; j \leq n; i$ and j are positive integers):

$$A = \begin{pmatrix} a_{11} & \cdots & a_{1n} \\ \cdots & \cdots & \cdots \\ a_{m1} & \cdots & a_{mn} \end{pmatrix} \tag{1}$$

$$\begin{cases} a_{ij} = Z \\ a_{ij} = \frac{1}{a_{ji}} \\ a_{ii} = 1 \end{cases} \tag{2}$$

After establishing the criterion layer matrix, hierarchical single sorting is required. Hierarchical single sorting refers to comparing all elements in the current layer pairwise against a certain element in the previous layer and performing hierarchical sorting. The calculation can be based on the judgment matrix A, and the eigenvalues and eigenvectors that satisfy $AW = \lambda_{max}W$ need to be guaranteed. The maximum eigenvalue of A is λ_{max}, and the normalized eigenvector corresponding to λ_{max} is W, where W_i is the component of W. The judgment matrix is used to calculate the weights of each factor a_{ij} on the target layer. The calculation steps of the weight vector W and the maximum feature λ_{max} are shown in Equations (3)–(5), where Equation (4) normalizes W_i and the maximum eigenvalue of the judgment matrix is obtained through Equation (5).

$$\overline{W_i} = \sqrt[n]{\prod_{j=1}^{n} a_{ij}} \; i, j = 1, 2, \ldots, n \tag{3}$$

$$W_i = \frac{\overline{W_i}}{\sum\limits_{i=1}^{n} \overline{W_i}} \tag{4}$$

$$\lambda_{max} = \frac{1}{n}\sum_{i=1}^{n} \frac{(AW)_i}{W_i} \tag{5}$$

The maximum eigenvalue and $C.I.$ can be obtained through the following Formulas (6)–(8), where $C.I.$ is the consistency ratio. In the AHP algorithm, $C.I.$ can measure the consistency of each element in the judgment matrix. The median random consistency index $R.I.$ in the following formula is generally taken from a table. $C.R.$ is used to determine whether the obtained hierarchical ranking weights are correct. When $C.R. < 0.1$, the calculated hierarchical ranking weights are considered correct and reasonable.

$$C.I. = \frac{\lambda_{max} - n}{n - 1} \tag{6}$$

$$C.R. = \frac{C.I.}{R.I.} \tag{7}$$

$$BW = \lambda W \tag{8}$$

If B is an n-order judgment matrix, its maximum eigenvalue λ_{max} can be obtained from Equation (8). W is the eigenvector of B, and $C.I.$ can be used to verify the consistency of the judgment matrix A. The random consistency index $R.I.$ value can be obtained from simulating 1000 experiments as shown in Table 2; the matrix order is also the order of the criterion layer.

Table 2. The value of *R.I.*

Matrix Order	1	2	3	4	5	6	7	8	9
R.I.	0	0	0.58	0.90	1.12	1.24	1.32	1.41	1.45

The $C.R.$ can be calculated based on the data in Table 2 and Formulas (7) and (8); then, a judgement can be made. When $C.R. < 0.1$, it indicates that the consistency of the judgment matrix A is considered to be within an acceptable range, and the eigenvectors of A can be used to calculate the weight vector. If $C.R. \geq 0.1$, the judgment matrix A should be adjusted. This work successfully solves the multi-objective optimization problem of ethylene production using the AHP algorithm, and this framework also provides a new approach for multi-objective optimization problems in chemical processes.

3. Methodology and Case Study

3.1. Methodology

The methodology of the hierarchical accelerated branch and bound algorithm is the combination of P-graph and AHP. Roco et al. proposed that the sustainable energy network in the Philippines is generated by the combination of the AHP method and P-graph theory [36]. For a large number of indicators, the superstructure network obtains relative weights through the AHP method and integrates them into a P-graph to evaluate sustainability scores. The AHP method indicates that the most important criterion among the interviewed experts is whether the energy can be self-sufficient. The integrated P-graph and AHP simulation indicate that the optimal and near optimal solution relies on geothermal and hydroelectric energy sources, while retaining other energy sources to meet non-electricity demands. The experiment shows that the optimal energy network generated by combining AHP and P-graph reduces the principal electricity cost expenditure. Estrella et al. cited AHP and P-graph in the cement manufacturing industry [37]. Due to the emission of large amounts of greenhouse gases during the cement manufacturing process, researchers considered limiting carbon emissions by optimizing processes and using alternatives to fossil-derived fuels. The proposed method based on P-graph and Analytic Hierarchy Process (AHP) was constructed to assist in planning the post-CCT retrofitting of cement plants in the total control and trading environment using the Emissions Trading Scheme (ETS). Finally, three alternative post combustion technologies were considered, namely monoethanolamine (MEA) absorption technology, the chilled ammonia process (CAP), and membrane-assisted liquefaction (MEM). The results indicate that under certain conditions, the CAP method is the most effective in helping companies reduce greenhouse gas emissions while saving energy consumption and costs.

The above two cases demonstrate that the combination of the AHP algorithm and P-graph theory can achieve very good results in the field of chemical engineering. This article tightly combines the AHP algorithm with the ABB algorithm under strict superstructure to achieve the goal of solving optimization problems. Figure 2 is the algorithm flowchart used in this study.

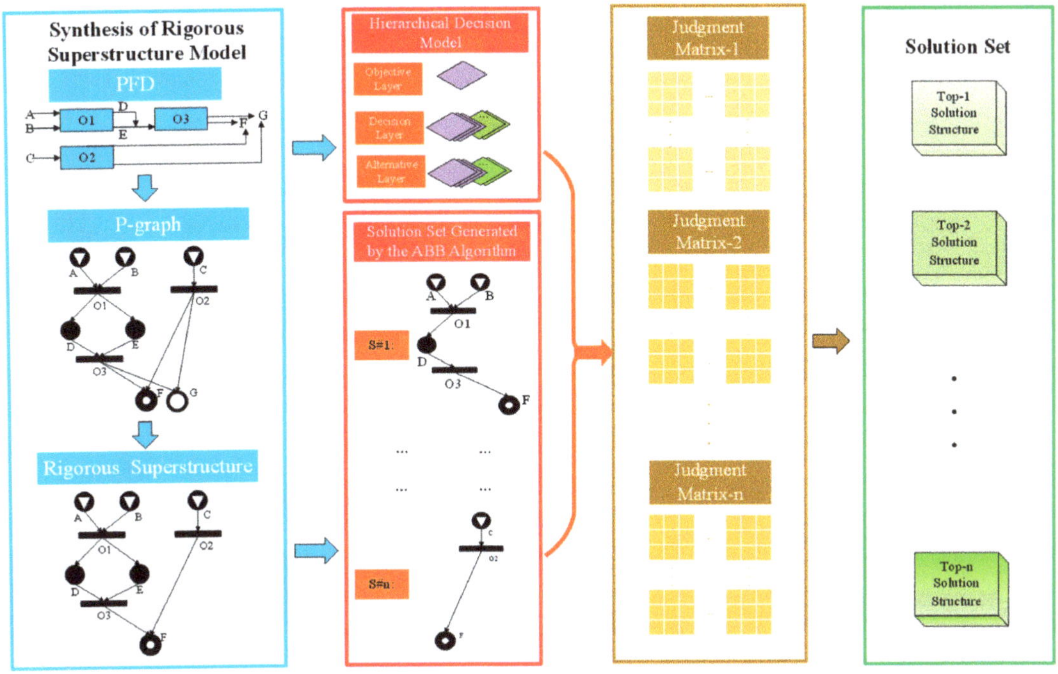

Figure 2. Flowchart of the hierarchical accelerated branch and bound algorithm.

In Figure 2, the key step is to use AHP expert scoring to pre-score feasible solutions, and Top-*K* solutions are selected as the initial boundary to accelerate pruning. Specifically, when pruning using the ABB algorithm, there are two pruning conditions: 1. Pruning when the quantitative target exceeds the cost boundary. 2. The AHP comprehensive score exceeds the tolerance threshold (integrating qualitative objectives), which further improves the efficiency of the ABB algorithm. In Figure 2, it can also be clearly seen that utilizing the characteristics of AHP enables the method proposed in this paper to better solve multi-objective decision-making problems.

When modeling, environmental indicators can be added and then the ABB algorithm can be used to find the optimal and suboptimal solutions. However, the particularity of the ABB algorithm should be taken into account. In the optimization process, only a simple weighting can be applied to the environmental and energy consumption indicators, resulting in solutions that are not as accurate as those obtained by expert scoring. Secondly, environmental indicators are not our main focus, and if the ABB algorithm is used to optimize the solution, secondary modeling is required. At this point, the optimization problem becomes more complex and unnecessary.

However, in actual production, more factors need to be considered. For example, ethylene is an important chemical in the country, with an annual production growth rate of

6.2%. The large amount of carbon dioxide emissions generated from its production seems unreasonable, so carbon emission indicators should be considered when optimizing the scheduling of ethylene production. In the process of optimizing the superstructure model, energy consumption costs, model structure, and carbon emission indicators should be considered. A hierarchical analysis structure can be established as shown in Figure 3. Based on the above discussion, the feedstock scheduling of ethylene production has become ensuring less energy consumption, lower carbon emissions, and a better raw material structure under quantitative ethylene production.

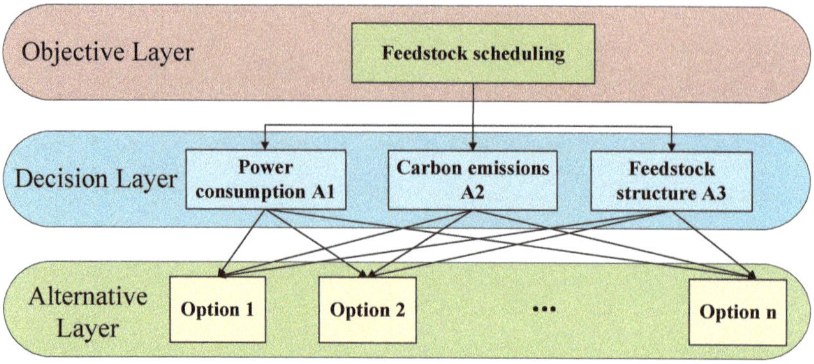

Figure 3. Hierarchical analysis structure.

3.2. Case Study

Ethylene has been widely used in the chemical industry and is a fundamental component of the petrochemical industry [38]. The case study considers the optimal feedstock scheduling problem for ethylene production. There are four main raw materials: C3C4C5 (C5), hydrogenated tail oil (HCR), light diesel oil (light), and naphtha (NAP). Due to the differences in production efficiency and energy consumption among each plant in ethylene production, it is necessary to analyze feedstock scheduling based on actual needs. In order to produce as much ethylene as possible, a reasonable solution method and scientific feedstock scheduling model need to be constructed [39–41]. Assuming that plants A and B have different production processes, in order to reduce energy consumption, the feedstock of plants A and B will also be different. All of these factors need to be considered when establishing a superstructure model. The specific parameters for modeling this case are shown in Table 3.

Table 3. Status involved in production.

Subscript	Content
_A	Material for plant A
_B	Material for plant B
_M	Remaining material after allocating to plant A and B
_0	Raw status
_1	Status after one month
_2	Status after two months
1+	Raw for second month
2+	Raw for the third month
0_1	Status after the first month

From Table 3, it can be seen that we have identified the allocation of raw materials to plants A and B differently. The strict superstructure model generated by the parameters in

Table 3 and P-graph theory is shown in Figure 4. This model consists of four sets, namely R, O, P, and M, with a total of 20 operating units labeled from O1 to O20. The R set is the set of four raw materials C5, HCR, light, and NAP. The M set is the remaining of all materials allocated to plants A and B.

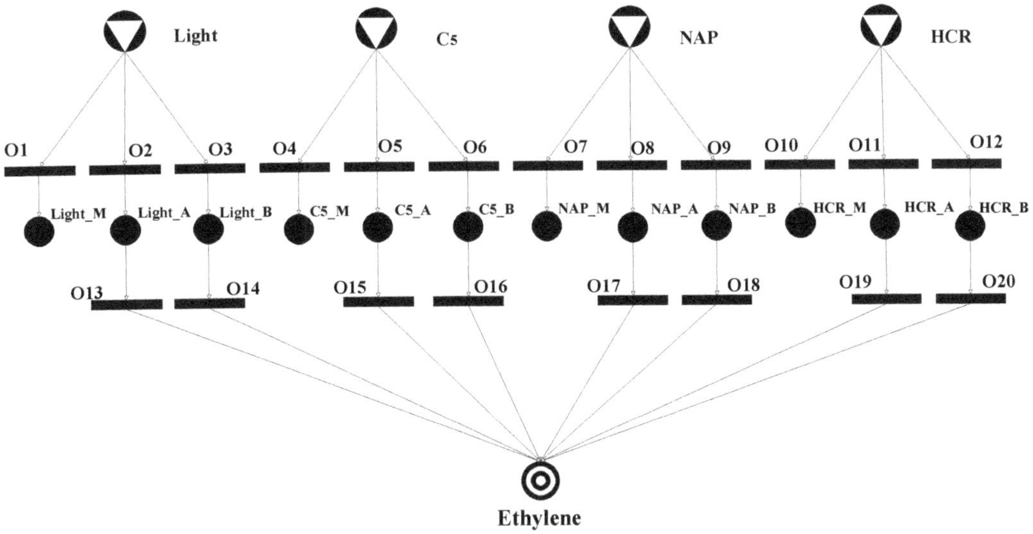

Figure 4. Feedstock Scheduling Superstructure.

According to the basic structure of Figure 4, the cycle of the case is extended to three months, as shown in Figure 5. If there are two operating units that directly obtain the final product from the raw material or intermediate product, then the operating units assigned to plants A and B are assigned to the raw material from left to right. If there are other operating units, then this operating unit is the surplus allocated to plants A and B. Taking raw material light_0 as an example, O4 is the operating unit for allocating materials to plant B; O3 is the operation of allocating materials to plant A; and O2 is the remaining material that has been allocated. The rules for other raw materials are similar. The special operating units O1, O8, O9, and O16 refer to the operating unit of allocating and preparing materials for the second month after the first month's materials have been used up; O21, O28, O29, O36 are the operating units of allocating materials prepared for the third month after the second month's materials have been used up.

In Figure 5, the black marked operating units (such as O2, O7, etc.) represent the remaining portion after raw material allocation, the blue marked operating units (such as O3, O5, etc.) represent raw material allocation to plant A, the green marked operating units (such as O4, O6, etc.) represent raw material allocation to plant B, and the red marked operating units (such as O1, O8, etc.) represent additional raw material addition. After the establishment of the superstructure model, the ABB algorithm ranks the Top-10 solutions based on the cost. Ten solution sets are named S#1, S#2,. . ., S#10; the specific raw material consumption of the Top-10 solutions is shown in Table 4.

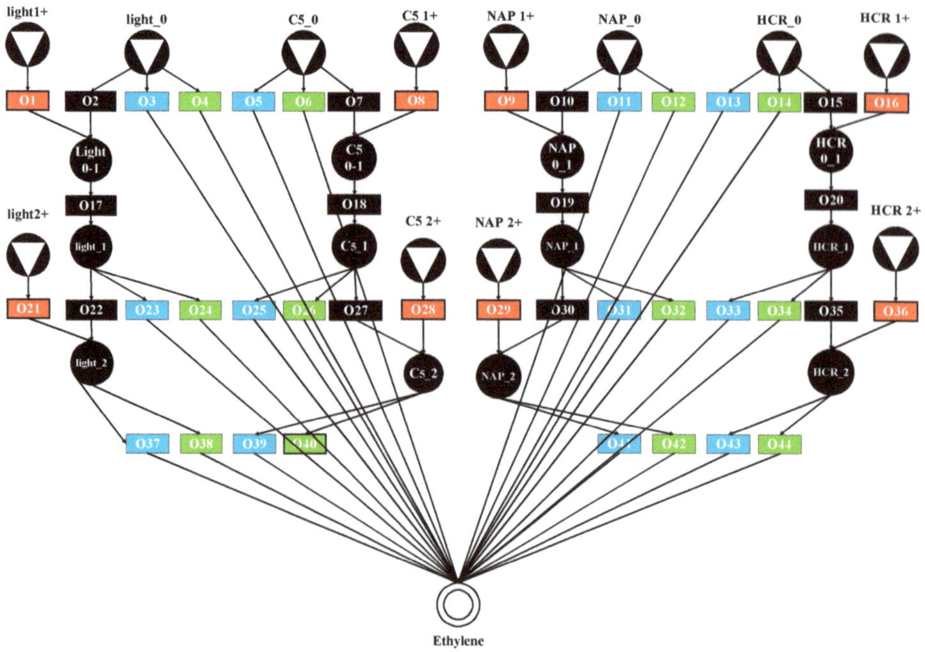

Figure 5. Feedstock Scheduling of four raw materials in plants A and B.

Table 4. Top-10 solutions of ABB.

Solution	Light ($g \cdot y^{-1}$)	C5 ($g \cdot y^{-1}$)	NAP ($g \cdot y^{-1}$)	HCR ($g \cdot y^{-1}$)	Cost ($EUR \cdot y^{-1}$)
S#1	313,728.00	153,817.00	602,552.00	309,120.00	10,188,700.00
S#2	313,728.00	153,817.00	630,553.00	309,120.00	10,189,900.00
S#3	313,728.00	153,817.00	603,380.00	303,960.00	10,190,300.00
S#4	313,728.00	153,817.00	631,381.00	303,960.00	10,191,400.00
S#5	313,728.00	153,817.00	593,357.00	309,120.00	10,192,100.00
S#6	313,728.00	153,817.00	621,358.00	309,120.00	10,193,200.00
S#7	313,728.00	153,817.00	594,186.00	303,960.00	10,193,700.00
S#8	313,728.00	153,817.00	622,187.00	303,960.00	10,194,800.00
S#9	313,728.00	117,836.80	638,079.00	309,120.00	10,197,900.00
S#10	313,728.00	117,836.80	638,907.00	303,960.00	10,199,500.00

represents one of the sets of solutions generated by the ABB.

Because the lowest cost in actual production does not necessarily mean the most optimal scheduling for actual production. Therefore, this case selects the top 10 solutions with the lowest cost consumption for producing the same amount of ethylene in this experiment. Figure 6 can provide a more intuitive comparison of the raw material energy consumption of the 10 solutions.

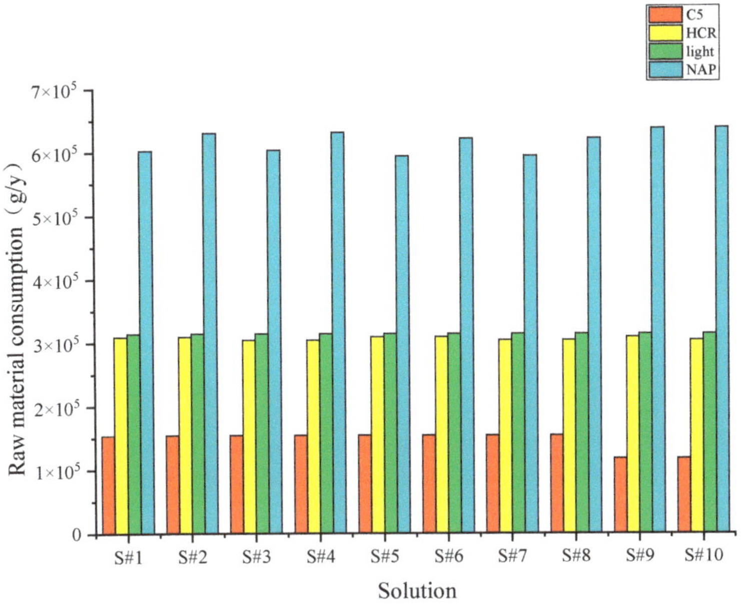

Figure 6. Consumption of raw materials for ten solutions.

In the guidelines for national greenhouse gas inventories prepared by the Intergovernmental Panel on Climate Change (IPCC), the method for calculating the total carbon dioxide emissions from energy consumption is provided [42]. The calculation method for carbon dioxide emissions is as follows: $E = N \times EF$, where E is the carbon dioxide emissions, N is the consumption of natural gas energy, which needs to be converted from the physical units of production and consumption of solid, liquid, and gaseous fuels (such as t or m^3) to ordinary energy units (J), and EF is the carbon dioxide emission factor of natural gas. The carbon dioxide index is shown in Table 5.

Table 5. CO_2 Emission Factors for combustion.

Fuel Type	Carbon Content(kg/GJ)	Carbon Oxidation Factor	Effective CO_2 Emission Factor(kg/TJ)		
			Default Value	95% Confidence Interval	
	A	B	$A \cdot B \cdot 44/12 \times 1000$	lower	upper
Gas/Diesel Oil	20.1	1	74,100	72,600	74,800
Liquefied Petroleum Gases	17.2	1	63,100	61,600	65,600
Ethane	16.8	1	61,600	56,500	68,600
Naphtha	20.0	1	73,300	69,300	76,300
Refinery Gas	15.7	1	57,600	48,200	69,000
Other Petroleum Products	20.0	1	73,300	72,200	74,400
Refinery Feedstocks	20.0	1	73,300	69,300	76,300

According to Table 5 and the carbon dioxide emission calculation formula, the specific carbon emissions of the Top-10 solutions can be calculated, and the results are shown in Table 6.

Table 6. Carbon emissions of Top-10 solutions.

Solution	Carbon Emissions (kg)
S#1	3395.7
S#2	3492.7
S#3	3384.0
S#4	3481.1
S#5	3363.8
S#6	3460.9
S#7	3352.2
S#8	3449.2
S#9	3416.2
S#10	3404.6

represents one of the sets of solutions.

The above has already taken into account the two factors of carbon emissions and energy consumption. As the leader of the petrochemical industry, ethylene's raw materials mainly include naphtha, natural gas, coal, and ethane, accounting for 41%, 17%, 13%, and 12% of global petrochemical raw materials, respectively. Due to resource limitations, the position of naphtha as the main raw material for domestic cracking units will be maintained for a long time. Considering factors such as the main source of raw materials for ethylene production, if the proportion of naphtha in the raw material structure of the substructure is high, the expert rating will also be higher. The comparison of naphtha distribution among the ten solutions obtained is shown in Figure 7.

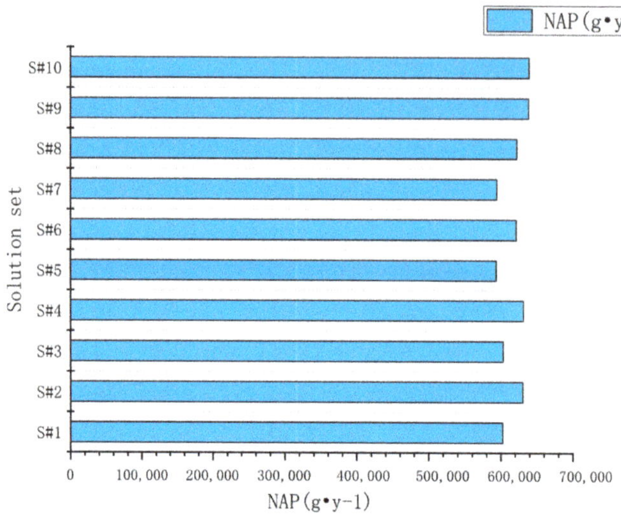

Figure 7. Naphtha consumption of top-10 solutions.

After the case study, Figure 8 shows the differences in three different indicators of the decision-making layer among the ten options. It is evident that there are differences between the decision-level indicators A2 and A3 for the ten options. Solution #1, Solution #5, and Solution #7 have lower naphtha consumption compared to other solutions, and these solutions will receive lower ratings under the A3 index. Solution #6, Solution #2, and Solution #4 have higher CO2 emissions, and these solutions have lower scores under indicator A2.

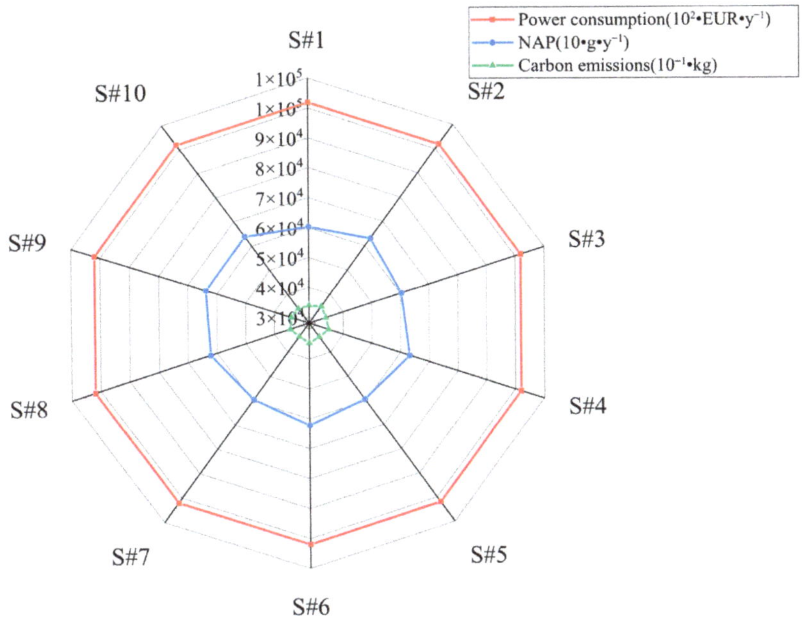

Figure 8. Comparison of values of three indicators.

Experts scored the ten solutions obtained by the ABB algorithm based on the proposed model to obtain a judgment matrix. The elements in the same level need to pass consistency checks with the elements in the previous level, and the weights of criterion layer A1, criterion layer A2, and criterion layer A3 are calculated as 0.64, 0.26, and 0.1. From Figure 9, it can be intuitively seen that the weight proportion of the criterion layer is basically consistent with the initial requirements of this case, with A1 having the highest proportion and A2 following closely.

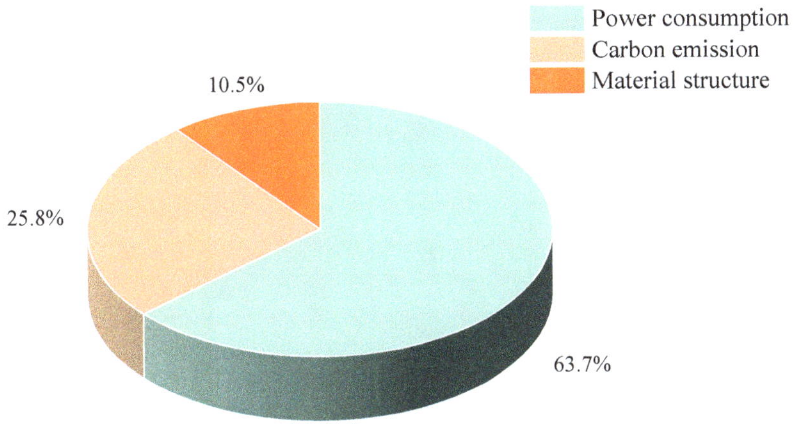

Figure 9. Percentage of weights at the criterions.

From Table 7, it can be seen that the maximum eigenvalue of the criterion layer, $C.R. < 0.1$, passed the consistency test. The expert ratings at the criterion level have all passed the consistency test, proving the validity of the results. The final result is shown in

Figure 10. According to the target layer score, the optimal choice is Solution #1, as shown in Figure 11, while the second-best option has become Solution #3.

Table 7. Results of AHP.

	Maximum Eigenvalue	*C.R.*	Consistency
Rule 1	10.34	0.025	True
Rule 2	10.86	0.064	True
Rule 3	11.27	0.094	True

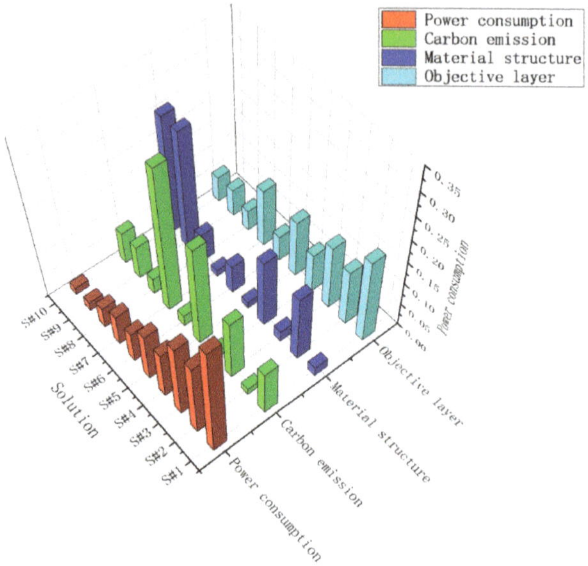

Figure 10. Comparison of AHP scoring by solution.

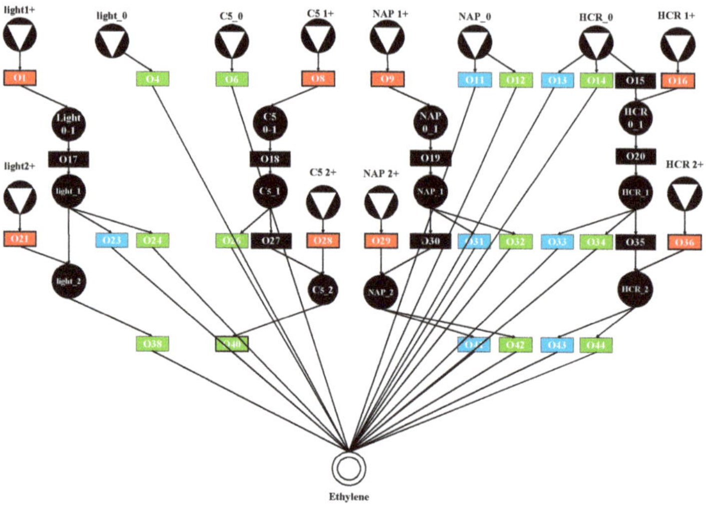

Figure 11. Optimal Solution#1.

Instead of only considering solution 2 obtained by the ABB algorithm in criterion layer 1, after integrating multiple indicators, the AHP algorithm optimized the results of the ABB algorithm. The ranking given by the AHP algorithm combined with the ABB algorithm for the optimization scheduling problem in this case is Solution #1, Solution #3, Solution #2, Solution #5, Solution #7, Solution #4, Solution #10, Solution #6, Solution #9, and Solution #8. The power consumption for each solution is shown in Figure 12.

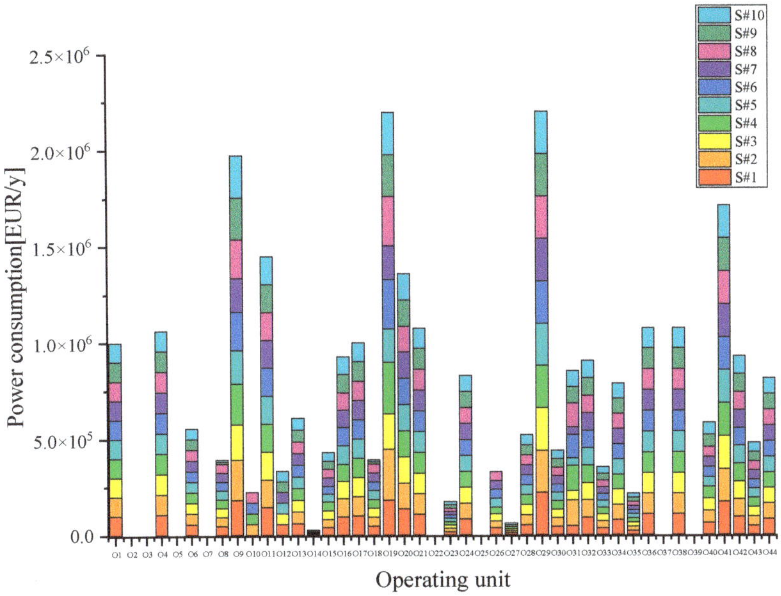

Figure 12. Power consumption of atomic operating units for ten solutions.

4. Conclusions

The strict superstructure in solving the ethylene production problem avoids the interference of redundant structures in the problem, while the ABB algorithm performs pruning operating units during the calculation process. Redundant calculations are avoided, the solution space is further optimized, and the solution set can be calculated faster. Specifically in ethylene production, the ABB algorithm can intelligently search for feedstock scheduling and allocation solutions based on production demand and raw material supply, and the optimal or approximate optimal solution can be quickly obtained. The AHP algorithm has been introduced into the multi-objective optimization problem of ethylene production without the need for remodeling. The experimental results show that the proposed framework greatly reduces the workload of modeling through process optimization, and reduces 2/3 of the modeling workload and model complexity compared to traditional ABB algorithms. During the cooperation with Chinese petrochemical enterprises, the proposed method has achieved good results in optimizing ethylene production. Under the condition of adding a small amount of cost, the optimal solution obtained by this framework can reduce carbon emissions by 1.27% compared to before the improvement, and traditional methods cannot effectively consider multi-objective problems.

The actual production requirements of each plant are different, and the AHP expert scoring used in this article is the result of comprehensive scoring by authoritative personnel such as engineers, managers, and research experts who have been working in the plant for many years. Therefore, we can adjust the judgment matrix of the decision-making layer in the AHP algorithm based on expert ratings, and ultimately obtain a more suitable feedstock scheduling solution for our plant. The integration of the ABB algorithm and AHP algorithm can not only solve the problem of ethylene production, but also provide customized solutions for specific needs in different fields and industries.

In the study of P-graph theory, it was found that the modeling ability for uncertainty, fuzziness, and gradient relationships in the real world is slightly weak. Therefore, it is possible to introduce fuzzy P-graph theory to deal with more problems. It is also possible to integrate nonlinear functions into the model to solve nonlinear problems in complex problems. In order to expand the optimization capability of the model, sustainability indicators such as Carbon Footprint, Water Footprints, and Land Footprints can be incorporated into the model to create a more comprehensive evaluation.

Author Contributions: Conceptualization, J.C.; Methodology, J.C. and H.Y.; Software, J.C. and H.Y.; Validation, J.C. and Y.J.; Formal analysis, H.Y. and J.C.; Writing—original draft, H.Y.; Writing—review and editing, J.C. and Y.J.; Supervision, J.C.; Funding acquisition, J.C. All authors have read and agreed to the published version of the manuscript.

Funding: This research was supported by the Natural Science Foundation of China (No. 52402442) and Beijing Information Science and Technology University research Found (No. 2023XJJ10).

Data Availability Statement: The original contributions presented in this study are included in the article. Further inquiries can be directed to the corresponding author.

Conflicts of Interest: The authors declare no conflict of interest.

Abbreviations

The following abbreviations are used in this manuscript:

PNS	process network synthesis
P-graph	Progress graph
MINLP	The Mixed Integer Nonlinear Programming
ABB	Accelerated Branch and Bound
BB	Branch and Bound
AHP	The Analytic Hierarchy Process
PFD	process flow diagram
MSG	Maximum Structure Generation
SSG	Solution Structure Generation
ETS	Emissions Trading Scheme
MEA	monoethanolamine
CAP	chilled ammonia process
MEM	membrane assisted liquefaction
HCR	hydrogenated tail oil
light	light diesel oil
C5	C3C4C5
NAP	naphtha

References

1. Mallapragada, D.S.; Dvorkin, Y.; Modestino, M.A.; Biddinger, E.; Esposito, D.V.; Smith, W.A.; Hodge, B.M.; Harold, M.P.; Donnelly, V.M.; Nuz, A.; et al. Decarbonization of the chemical industry through electrification: Barriers and opportunities. *Joule* **2023**, *7*, 23–41. [CrossRef]
2. Prawitasari, D.A. Potensi Limbah Hasil Pertanian Padi menjadi Sumber Bioenergi melalui Konsep Biorefineri dengan Pendekatan P-Graph dan Analisis Ekonomi Teknik di Kabupaten Sleman, Yogyakarta. *Asian J. Innov. Entrep.* **2025**, *9*, 16–32. [CrossRef]
3. Velázquez, J.J.H.; Durán, F.M.Z.; Díaz, L.A.C.; Ruiz, J.C.; Avila, J.R.A. Hybrid two-step optimization of internally heat-integrated distillation columns. *J. Taiwan Inst. Chem. Eng.* **2022**, *130*, 103967. [CrossRef]

4. Xu, Y.; Li, J.; Cui, G.; Zhang, G.; Yang, Q. An improved treble-level assisting optimization strategy to enhance algorithm search ability in heat exchanger network design. *J. Taiwan Inst. Chem. Eng.* **2021**, *129*, 162–170. [CrossRef]
5. Friedler, F.; Fan, L.T.; Imreh, B. Process network synthesis: Problem definition. *Networks* **1998**, *31*, 119–124. [CrossRef]
6. Orosz, Á.; How, B.S.; Friedler, F. Multiple-solution heat exchanger network synthesis using P-HENS solver. *J. Taiwan Inst. Chem. Eng.* **2022**, *130*, 103859. [CrossRef]
7. Ali, R.A.; Nik Ibrahim, N.N.L.; Wan Ab Karim Ghani, W.A.; Lam, H.L.; Sani, N.S. Utilization of process network synthesis and machine learning as decision-making tools for municipal solid waste management. *Int. J. Environ. Sci. Technol.* **2022**, *19*, 1985–1996. [CrossRef]
8. Camisa, A.; Notarnicola, I.; Notarstefano, G. Distributed primal decomposition for large-scale MILPs. *IEEE Trans. Autom. Control* **2021**, *67*, 413–420. [CrossRef]
9. Yuan, T.J.; Mao, Y.L. A review of pinch analysis techniques and extended application in power systems. *Renew. Sustain. Energy Rev.* **2024**, *202*, 114684. [CrossRef]
10. Daoutidis, P.; Lee, J.H.; Rangarajan, S.; Chiang, L.; Gopaluni, B.; Schweidtmann, A.M.; Harjunkoski, I.; Mercangöz, M.; Mesbah, A.; Boukouvala, F.; et al. Machine learning in process systems engineering: Challenges and opportunities. *Comput. Chem. Eng.* **2024**, *181*, 108523. [CrossRef]
11. Friedler, F.; Tarjan, K.; Huang, Y.W. Graph-theoretic approach to process synthesis: Polynomial algorithm for maximal structure generation. *Comput. Chem. Eng.* **1993**, *17*, 929–942. [CrossRef]
12. Tan, R.R.; Aviso, K.B.; Klemes, J.J.; Lam, H.L.; Varbanov, P.S.; Friedler, F. Towards generalized process networks: Prospective new research frontiers for the P-graph framework. *Chem. Eng. Trans.* **2018**, *70*, 91–96. [CrossRef]
13. Tick, A.; Molnár, G. Improving Efficiency: P-Graph Modeling for Business Workflows with Dynamic and Fuzzy Extensions. *Acta Polytech. Hung.* **2025**, *22*, 79–100. [CrossRef]
14. Rohman, F.S.; Alwi, S.R.W.; Er, H.A.; Ahmad Termizi, S.N.A.; Abd Manan, Z.; Lim, J.S.; Muhammad, D. Cogeneration system's energy performance improvement by using P-graph and advanced process control. *Chem. Prod. Process Model.* **2025**, *20*, 1–37. [CrossRef]
15. Lim, S.Y.; Chemmangattuvalappil, N.G.; Tapia, J.F.D.; Khoiroh, I.; Chai, P.V.; Ng, L.Y. Optimal chemical reaction pathway for palm process residue recovery using Process Graph (P-graph) framework. *Comput. Chem. Eng.* **2025**, *194*, 109000. [CrossRef]
16. Aviso, K.B.; Marfori III, I.A.V.; Tan, R.R.; Ubando, A.T. Optimizing abnormal operations of off-grid community utility systems with fuzzy P-graph. *Energy* **2020**, *202*, 117725. [CrossRef]
17. Pimentel, J.; Rodríguez, G.; Gil, I. Synthesis of Alternative Cost-Effective Process Flowsheets for Lactic Acid Valorization by Means of the P-graph Methodology. *Ind. Eng. Chem. Res.* **2020**, *59*, 5921–5930. [CrossRef]
18. Vance, L.; Cabezas, H.; Heckl, I.; Bertok, B.; Friedler, F. Synthesis of Sustainable Energy Supply Chain by the P-graph Framework. *Ind. Eng. Chem. Res.* **2013**, *52*, 266–274. [CrossRef]
19. Phuang, Z.X.; Sang, K.; Wang, X.C.; You, F.; Woon, K.S. Sustaining regional infrastructure resiliency with externality for plastic waste circularity via geospatially-enhanced P-graph-life cycle optimization framework. *J. Clean. Prod.* **2025**, *486*, 144413. [CrossRef]
20. Bertok, B.; Barany, M.; Friedler, F. Generating and Analyzing Mathematical Programming Models of Conceptual Process Design by P-graph Software. *Ind. Eng. Chem. Res.* **2013**, *52*, 166–171. [CrossRef]
21. Strouvalis, A.M.; Heckl, I.; Friedler, F.; Kokossis, A.C. An accelerated Branch-and-Bound algorithm for assignment problems of utility systems. *Comput. Chem. Eng.* **2002**, *26*, 617–630. [CrossRef]
22. Belkhiri, H.; Chergui, M.E.A.; Ouaïl, F.Z. Optimizing a linear function over an efficient set. *Oper. Res.* **2022**, *22*, 3183–3201. [CrossRef]
23. Bakhtari, A.R.; Waris, M.M.; Sanin, C.; Szczerbicki, E. Evaluating industry 4.0 implementation challenges using interpretive structural modeling and fuzzy analytic hierarchy process. *Cybern. Syst.* **2021**, *52*, 350–378. [CrossRef]
24. Ba, Z.; Wang, Y.; Fu, J.; Liang, J. Corrosion risk assessment model of gas pipeline based on improved AHP and its engineering application. *Arabian J. Sci. Eng.* **2022**, *47*, 10961–10979. [CrossRef]
25. Huang, M.Y.; Huang, L.Y.; Zhong, Y.X.; Yang, H.W.; Chen, X.M.; Huo, W.; Wang, J.Z.; Zhang, F.; Bai, B.; Shi, L. MILP acceleration: A survey from perspectives of simplex Initialization and learning-based branch and bound. *J. Oper. Res. Soc. China* **2025**, *13*, 1–55. [CrossRef]
26. Tavana, M.; Soltanifar, M.; Santos-Arteaga, F.J. Analytical hierarchy process: Revolution and evolution. *Ann. Oper. Res.* **2023**, *326*, 879–907. [CrossRef]
27. Hemmati, S.; Elnegihi, M.M.; Lee, C.H.; Chong, D.Y.L.; Foo, D.C.Y.; How, B.S.; Yoo, C. Synthesis of Large-Scale Bio-Hydrogen Network Using Waste Gas from Landfill and Anaerobic Digestion: A P-Graph Approach. *Processes* **2020**, *8*, 505. [CrossRef]
28. Sitoh, Y.; Tan, V.W.G.; Tapia, J.F.D.; Tan, R.R.; Foo, D.C.Y. Optimal Pathways for Nitric Acid Synthesis Using P-Graph Attainable Region Technique (PART). *Processes* **2023**, *11*, 2684. [CrossRef]

29. Ji, M.; Zhang, W.; Xu, Y.; Liao, Q.; Klemeš, J.J.; Wang, B. Optimisation of multi-period renewable energy systems with hydrogen and battery energy storage: A P-graph approach. *Energy Convers. Manag.* **2023**, *281*, 116826. [CrossRef]
30. Pimentel, J.; Orosz, Á.; Aviso, K.B.; Tan, R.R.; Friedler, F. Conceptual Design of a Negative Emissions Polygeneration Plant for Multiperiod Operations Using P-Graph. *Processes* **2021**, *9*, 233. [CrossRef]
31. Tóth, Á.; Suta, A.; Pimentel, J.; Argoti, A. A comprehensive, semi-automated systematic literature review (SLR) design: Application to P-graph research with a focus on sustainability. *J. Clean. Prod.* **2023**, *415*, 137741. [CrossRef]
32. Friedler, F.; Varga, J.B.; Fan, L.T. Decision-mapping: A tool for consistent and complete decisions in process synthesis. *Chem. Eng. Sci.* **1995**, *50*, 1755–1768. [CrossRef]
33. Ahn, J.; Kim, H.J. A branch and bound algorithm for scheduling of flexible manufacturing systems. *IEEE Trans. Autom. Sci. Eng.* **2024**, *21*, 4382–4396. [CrossRef]
34. Friedler, F.; Varga, J.B.; Feher, E.; Fan, L.T. Combinatorially accelerated branch-and-bound method for solving the MIP model of process network synthesis. In *State of the Art in Global Optimization*; Springer: Boston, MA, USA, 1996; Volume 7, pp. 609–626. [CrossRef]
35. Saaty, T.L. Deriving the AHP 1–9 scale from first principles. In Proceedings of the 6th ISAHP, Berna, Switzerland, 2–4 August 2001. [CrossRef]
36. Roco, J.B.A.; Promentilla, M.A.B. Philippine Energy Network Design via an Integrated P-graph and Analytic Hierarchy Process (AHP) Framework. In Proceedings of the DLSU Research Congress 2022, Manila, Philippines, 6–8 July 2022. Available online: https://www.dlsu.edu.ph/wp-content/uploads/pdf/conferences/research-congress-proceedings/2022/SEE-05.pdf (accessed on 10 September 2025).
37. Estrella, E.T.; Belaro, C.M.; Promentilla, M.A.B. Retrofitting of Cement Plant with Carbon Capture Technology and Emissions Trading Scheme (ETS): An Integrated P-Graph and AHP Approach. *Chem. Eng. Trans.* **2023**, *103*, 397–402. [CrossRef]
38. Chen, Y.; Kuo, M.J.; Lobo, R.; Ierapetritou, M. Ethylene production: Process design, techno-economic and life-cycle assessments. *Green Chem.* **2024**, *26*, 2903–2911. [CrossRef]
39. Cao, J.; He, Y.L.; Zhu, Q.X. Feedstock scheduling optimization based on novel extensible P-Graph reasoning in ethylene production. *Ind. Eng. Chem. Res.* **2020**, *59*, 18965–18976. [CrossRef]
40. Eisman, E.M.; Navarro, M.; Castro, J.L. A multi-agent conversational system with heterogeneous data sources access. *Expert Syst. Appl.* **2016**, *53*, 172–191. [CrossRef]
41. Xi, J.; Gong, H.; Zhang, Y.; Dai, X.; Chen, L. The evaluation of GHG emissions from Shanghai municipal wastewater treatment plants based on IPCC and operational data integrated methods (ODIM). *Sci. Total Environ.* **2021**, *797*, 148967. [CrossRef]
42. Yu, X.; Tan, C. China's process-related greenhouse gas emission dataset 1990–2020. *Sci. Data* **2023**, *10*, 55. [CrossRef]

sustainability

MDPI

Article

Multi-Objective Sustainable Operational Optimization of Fluid Catalytic Cracking

Shibao Pang [1,2], Yang Lin [1,2], Hongxun Shi [1,2], Rui Yin [1,2], Ran Tao [1,2], Donghong Li [1,2] and Chuankun Li [1,2,*]

[1] State Key Laboratory of Chemical Safety, Qingdao 266000, China
[2] SINOPEC Research Institute of Safety Engineering Co., Ltd., Qingdao 266000, China
[*] Correspondence: lick.qday@sinopec.com

Abstract

Fluid Catalytic Cracking (FCC) constitutes a critical process in petroleum refining, facing increasing pressure to align with sustainable development goals by improving energy efficiency and reducing environmental impact. This study tackles a multi-objective optimization challenge in FCC operations, seeking to simultaneously maximize the gasoline production and minimize the coke yield—the latter being directly linked to CO_2 emissions in FCC. A data-driven optimization model leveraging a dual Long Short-Term Memory architecture is developed to capture complex relationships between operating variables and product yields. To efficiently solve the model, an Improved Multi-Objective Whale Optimization Algorithm (IMOWOA) is proposed, integrating problem-specific adaptive multi-neighborhood search and dynamic restart mechanisms. Extensive experimental evaluations demonstrate that IMOWOA achieves superior convergence characteristics and comprehensive performance compared to established multi-objective algorithms. Relative to the yields before optimization, the proposed methodology increases the gasoline yield by 0.32% on average, coupled with an average reduction of 0.11% in the coke yield. For the studied FCC unit with an annual processing capacity of 2.6 million tons, the coke reduction corresponds to an annual CO_2 emission reduction of approximately 10,277 tons, delivering benefits to sustainable FCC operations.

Keywords: fluid catalytic cracking; sustainable operational optimization; multi-objective optimization; whale optimization algorithm

1. Introduction

The growing global energy challenges and environmental degradation have elevated sustainability to a crucial concern in industrial development [1,2]. Industries worldwide are required to improve energy efficiency and reduce carbon emissions through enhanced operational practices [3–5]. The petroleum refining industries, as significant contributors to energy consumption and greenhouse gas emissions, are under increasing pressure to shift toward more environmentally friendly and energy-efficient operational paradigms [6–8]. This shift is not only an environmental necessity in response to environmental change, but also a strategic move to comply with economic sustainability policies, such as China's Carbon Peaking and Carbon Neutrality Goals [9].

Fluid Catalytic Cracking (FCC) is a central unit in modern petroleum refineries [10,11]. It converts heavy petroleum fractions into high-value products, primarily gasoline, which is an important transportation fuel [12]. The economic performance of a refinery, therefore, is highly dependent on FCC, often incentivizing the maximization of the gasoline yield [13].

https://doi.org/10.3390/su172210045

However, coke is inevitably produced during the FCC gasoline production, which is an undesirable carbonaceous byproduct depositing on the catalyst surface. To restore the activity of the catalyst, the coke-laden catalyst is combusted with air [14]. This process represents a major energy loss and emits substantial amounts of CO_2 within the FCC unit, directly contributing to the refinery's carbon footprint.

Within this context, coke formation represents not merely an operational concern but a critical sustainability metric. The combustion of coke in the regenerator constitutes a major source of greenhouse gas emissions while simultaneously reflecting inefficient carbon utilization. Consequently, suppressing coke yield is essential for advancing FCC sustainability, directly mitigating environmental impact through lower CO_2 emissions and reducing waste through improved carbon utilization (i.e., preventing valuable carbon from being lost as coke). Operational optimization, which systematically adjusts key operating variables to obtain desired product yields, therefore emerges as a crucial approach for improving FCC sustainability. However, the strong nonlinearity and inherent complexity of FCC pose significant challenges to its operational optimization.

Over the past decades, extensive research has been conducted on FCC operational optimization. Early efforts were predominantly dedicated to single-objective optimization [15–17]. John et al. [18] developed a six-lump kinetic model and utilized gPROMS to optimize the catalyst and oil gas mass flow rates, leading to increased propylene yield. Chen et al. [19] applied a random forest algorithm to identify key operating variables affecting gasoline octane number, and established a mapping between these variables and octane number using a Long Short-Term Memory (LSTM) neural network, thereby constructing an optimization model to minimize octane number loss. Chen et al. [20] employed a random weight neural network to develop an optimization model for fossil fuel yields and introduced a fast adaptive differential evolution algorithm to address associated constraints under uncertainty. Li et al. [21] designed a bilevel robust optimization framework that optimizes the feed selection and reaction temperature of an industrial FCC unit to maximize profit under uncertainty of feed properties. Recent years have witnessed growing interest in multi-objective optimization, which more closely aligns with the practical need to simultaneously satisfy multiple performance indicators [22,23]. Jarullah et al. [24] developed a multi-objective optimization model grounded in first principles and solved it using gPROMS. In a separate approach, Liu et al. [25] constructed a multi-objective optimization model by nonlinear multiple regression to simultaneously minimize RON loss and operational risk in FCC gasoline processing. José et al. [26] compared the effectiveness of particle swarm optimization and genetic algorithm in solving the multi-objective optimization problem of the product yield distribution. Despite these advances, most studies have emphasized economic objectives, often overlooking sustainability considerations.

Research on multi-objective sustainable operational optimization for FCC remains relatively limited. Wan et al. [27] developed an optimization framework combining a lumped kinetic model and the SPEA-2 algorithm to reduce pollutant gas emissions while balancing the production of light olefins and other chemicals. Liu et al. [28] established a simulation model for the fractionation and absorption-stabilization units of FCC using Aspen Plus and optimized both gasoline yield and energy consumption with the NSGA-II algorithm. However, these studies rely heavily on the availability of accurate mechanism optimization models, posing significant challenges in practical applications. The mechanism models require detailed knowledge of reaction kinetics, catalyst behavior, and complex multiphase flow dynamics information that is often difficult to acquire in industrial-scale FCC units [29,30]. In addition, minimizing coke yield is a crucial objective in FCC sustainable optimization due to its direct contribution to energy losses and CO_2 emissions. Existing optimization approaches predominantly optimize a certain product yield by adjusting operating variables at a given time point. Yet, FCC product yields are influenced by the cumulative effect of the variables over a preceding period. The current multi-objective optimization models and algorithms do not fully take into account this characteristic, thereby constraining the effectiveness and applicability of proposed optimization strategies in practical FCC operation.

This paper investigates an FCC Multi-Objective Sustainable Operational Optimization (FCC-MOSOO) problem, targeting the dual objectives of maintaining gasoline production while suppressing coke formation. The proposed approach facilitates a reduction in environmental impact while maintaining economic profitability, thereby supporting the greener and more sustainable refining operations. The main contributions of this paper are as follows:

(1) A data-driven modeling approach based on a Dual-LSTM architecture is introduced to capture the complex relationship between operating variables and product yields in the FCC-MOSOO problem, overcoming the computational efficiency limitations of traditional mechanism models.

(2) The time lag and cumulative effects in FCC processes are considered in the optimization by adjusting variable trajectories over a future time window, advancing beyond traditional static setpoint optimization to achieve more realistic and effective operational strategies.

(3) An Improved Multi-Objective Whale Optimization Algorithm (IMOWOA) tailored to the FCC-MOSOO problem is developed, enhanced with Adaptive Multi-Neighborhood Search (AMNS) and Dynamic Restart (DR) mechanisms. The algorithm provides actionable Pareto-optimal solutions, enabling effective trade-off analysis between economic and environmental objectives.

2. Problem Formulation

2.1. Problem Description

FCC is a highly integrated chemical process that primarily consists of a reactor and a regenerator operating in a continuous cycle. In the reactor, the heavy oil gas feedstock contacts the hot powdered catalyst, cracking into valuable lighter products such as gasoline, liquefied petroleum gas, and dry gas. During this process, coke deposits on the catalyst surface, leading to its deactivation. The spent catalyst is subsequently transferred to the regenerator, where the deposit is burned off, thereby restoring catalytic activity and providing the necessary heat required for the endothermic cracking reactions in the reactor. The regenerated catalyst is then recirculated back to the reactor, completing the cycle.

A central challenge in FCC is balancing economic performance and environmental impact. Operating conditions that favor high gasoline production (e.g., higher reaction temperature) often promote higher coke formation, which increases energy consumption and CO_2 emissions, leading to a conflict between profit and sustainability. Further complicating the challenge is the time lag in FCC. As illustrated in Figure 1, the product yield at time $t + \tau$ depends on the cumulative effects of operating variables over a preceding period from $t - q + 1$ to $t + n$. From an optimization perspective, this period is divided into two segments by the current optimization time t: a historical window from $t - q + 1$ to t and a future window from $t + 1$ to $t + n$, where q and n are the lengths of the historical and future windows. In the historical window, non-adjustable observed variables that define the FCC operating state are captured. The future window includes the controlled variables that serve as decision variables in optimization. Mathematically, the observed and controlled variables are denoted as $\boldsymbol{O}_t$ and $\boldsymbol{X}_t$, expressed as Equations (1) and (2). Given a fixed operating state defined by $\boldsymbol{O}_t$, the FCC-MOSOO problem seeks to identify optimal controlled variables $\boldsymbol{X}_t$ that simultaneously maximize gasoline yield and minimize coke yield.

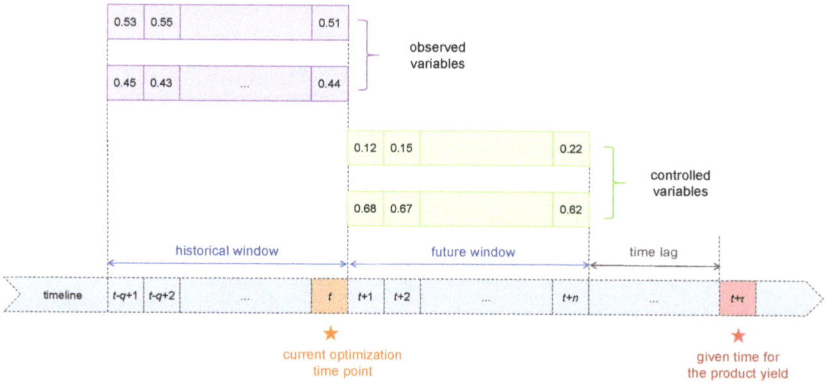

Figure 1. Time lag in the FCC process.

$$O_t = \begin{bmatrix} o_{1,t-q+1} & o_{1,t-q+2} & \cdots & o_{1,t} \\ o_{2,t-q+1} & o_{2,t-q+2} & \cdots & o_{2,t} \\ \vdots & \vdots & \ddots & \vdots \\ o_{p,t-q+1} & o_{p,t-q+2} & \cdots & o_{p,t} \end{bmatrix} \tag{1}$$

$$X_t = \begin{bmatrix} x_{1,t+1} & x_{1,t+2} & \cdots & x_{1,t+n} \\ x_{2,t+1} & x_{2,t+2} & \cdots & x_{2,t+n} \\ \vdots & \vdots & \ddots & \vdots \\ x_{m,t+1} & x_{m,t+2} & \cdots & x_{m,t+n} \end{bmatrix} \tag{2}$$

where p and m are the dimensions of the observed and controlled variables, o_{ij} is the observed variable i sampled at time j, and x_{ij} is the controlled variable i adjusted at time j.

2.2. Optimization Model

Calculating gasoline and coke yields is essential for the FCC-MOSOO problem. Since mechanism methods are often computationally expensive and challenging to calibrate accurately for industrial processes, a data-driven approach based on a Dual-LSTM architecture is employed to calculate the product yields. LSTM is suitable for this application due to its capability to process sequential data and learn long-term dependencies, enabling it to effectively capture the complex temporal relationship between operating variables and the product yield [31–33].

The Dual-LSTM architecture employs two parallel LSTM branches to extract final hidden states h_t^{obs} and h_t^{ctrl} from O_t and X_t, respectively. Each branch consists of a multi-layer stacked LSTM. The resulting states are concatenated into a combined feature $h_t^{\text{concat}} = [h_t^{\text{obs}} | h^{\text{ctrl}} t]$, which is then passed through a dense layer to calculate the product yield $\hat{y}_{t+\tau}$ (i.e., gasoline yield or coke yield) at time $t + \tau$. The complete computational procedure is formalized in Equation (3). The training process utilized historical data collected from an industrial FCC unit at a collaborative refinery. Data acquisition spanned from 8 March 2025 to 14 May 2025, with a sampling interval of 3 min. The raw dataset underwent comprehensive preprocessing, which involved removal of null values, outlier elimination using the 3-σ rule, noise reduction via a Butterworth filter, and final normalization through min–max scaling. This procedure finally yielded a refined dataset containing 15,791 valid samples.

$$\hat{y}_{t+\tau} = f_{\text{Dual-LSTM}}(O_t, X_t) \tag{3}$$

Building upon the Dual-LSTM architecture, the optimization objectives of the FCC-MOSOO problem are formulated as Equations (4) and (5). Based on the reaction mechanism and empirical knowledge of FCC, the selected observed variables and controlled variables are listed in Table 1.

$$\min \quad -\hat{y}_{t+\tau}^{\text{gasoline}} = -f_{\text{Dual-LSTM}}^{\text{gasoline}}(O_t, X_t) \tag{4}$$

$$\min \quad \hat{y}_{t+\tau}^{\text{coke}} = f_{\text{Dual-LSTM}}^{\text{coke}}(O_t, X_t) \tag{5}$$

According to production requirements, the operational optimization problem is required to meet the following constraints:

$$lb_i \leq x_{ij} \leq ub_i \tag{6}$$

$$|x_{ij} - x_{i,j-1}| \leq \Delta_i \tag{7}$$

Constraints (6) ensure that the values of the controlled variables remain within their allowable range at all times, where lb_i and ub_i are the lower and upper bounds of the controlled variable i.

Constraints (7) restrict that the step size of any adjustment must not exceed the allowable limit, where Δ_i is the maximum adjustment limit of the controlled variable i.

Table 1. The observed variables and controlled variables of the FCC-MOSOO problem.

Variable No.	Observed Variable	Controlled Variable
1	Regenerated catalyst temperature	Reaction temperature
2	Spent catalyst temperature	Reaction pressure
3	Oil gas temperature	Regenerator pressure
4	Gasoline yield	Lifting steam flow rate
5	Coke yield	Liquid feedstock temperature

3. Proposed Algorithm

Considering the inherent complexity and strong nonlinearity of the FCC process, swarm intelligence optimization algorithms are particularly suitable for addressing such challenges [34]. Among these, the Whale Optimization Algorithm (WOA) [35] has attracted widespread attention due to its simple parameter adjustment and excellent global search ability. It has been successfully applied in various domains such as traffic scheduling [36], price forecasting [37], and shop scheduling [38]. Based on WOA, the proposed IMOWOA is designed in this section to solve the FCC-MOSOO problem.

3.1. IMOWOA Framework

The basic WOA is inspired by the bubble-net foraging behavior of humpback whales, which comprises three core search operators: spiral updating, shrink encircling, and prey searching, as detailed in the following descriptions. Building upon the WOA search operators, IMOWOA further incorporates AMNS and DR mechanisms that are specifically designed for the FCC-MOSOO problem to improve search performance. The overall framework of IMOWOA is illustrated in Figure 2, with comprehensive details provided in the subsequent subsections.

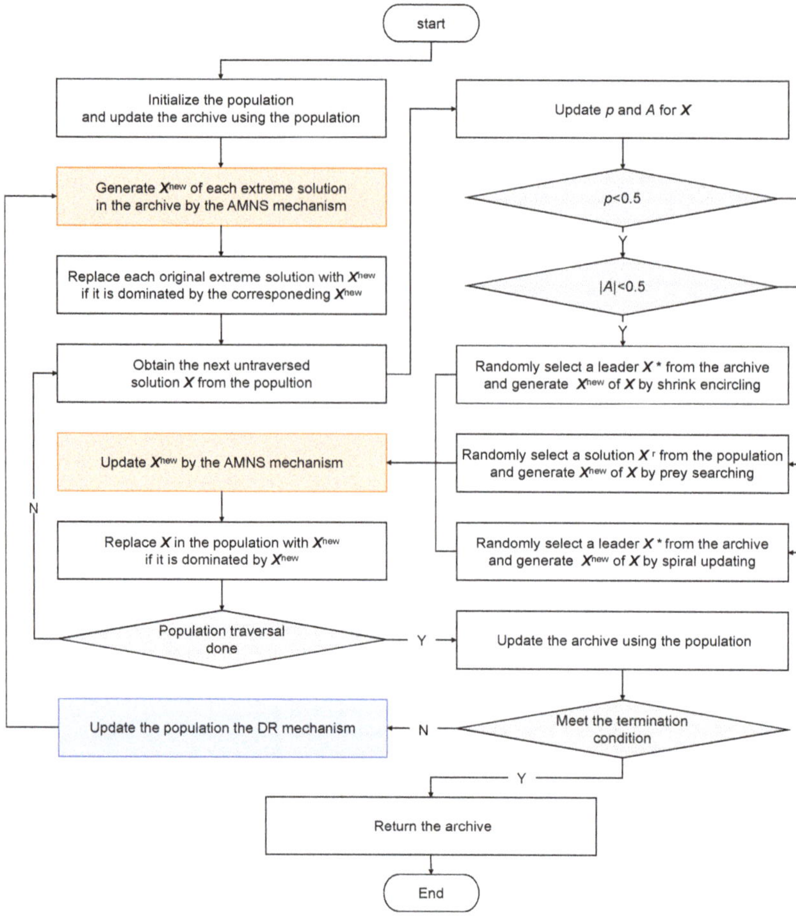

Figure 2. Framework of IMOWOA.

Spiral updating: spiral updating emulates the distinctive spiral-shaped movement of whales to approach the target prey, represented by the following formulation:

$$X^{\text{new}} = |X^* - X| \cdot e^{bl} \cdot \cos(2\pi l) + X^* \tag{8}$$

Shrink encircling: shrink encircling replicates whales encircling the target prey, represented by the following formulation:

$$X^{\text{new}} = X^* - A \cdot |2 \cdot \text{rand}() \cdot X^* - X| \tag{9}$$

Prey searching: this operator is responsible for exploring new prey, represented by the following formulation:

$$X^{\text{new}} = X^r - A \cdot |2 \cdot \text{rand}() \cdot X^r - X| \tag{10}$$

where X is a solution in the population, X^{new} is the new solution generated from X, X^* is the current best solution, X^r is a randomly selected solution, b is a constant used to define the shape of the logarithmic spiral and is recommended to be set as 1, l is a random number between $[-1, 1]$, $A = 2a \cdot \text{rand}() - a$ is a control number, rand() represents a function that returns a random number in the range of [0, 1], and a linearly decreases from 2 to 0.

3.2. Population Initialization

The population is initialized as a set of whale individuals, each representing a candidate solution. In the context of FCC-MOSOO, the controlled variable matrix $\mathbf{X}_t$ is adopted as the solution encoding scheme. To ensure the feasibility of the initial solutions, a constrained random walk strategy is utilized for population initialization. Specifically, each element x_{ij} in $\mathbf{X}_t$ is generated by uniform sampling within dynamic bounds $[lb^*_{ij}, ub^*_{ij}]$. These bounds are determined based on the value of the preceding element $x_{i,j-1}$, thereby introducing temporal dependency into the initial population. The formulations of the dynamic bounds are provided as follows:

$$lb^*_{ij} = \max\{lb_i, x_{i,j-1} - \Delta_i\} \tag{11}$$

$$ub^*_{ij} = \min\{ub_i, x_{i,j-1} + \Delta_i\} \tag{12}$$

3.3. Adaptive Multi-Neighborhood Search Mechanism

To enhance local search capability, the AMNS mechanism is introduced, which incorporates an adaptive execution strategy and three local search operators. The adaptive execution strategy activates the local search using a probability p^{ls}, defined in Equation (13). As optimization progresses, this strategy gradually shifts the focus from global exploration to local refinement, enhancing the quality of the final solutions.

$$p^{\text{ls}} = \frac{1}{2} \cdot \left[1 + \sin\left(\frac{g \cdot \pi}{2 \cdot G_{\max}}\right)\right] \tag{13}$$

where g is the current iteration count and $G_{\max}$ is the maximum number of iterations.

The three local search operators are specifically designed to align with the structural characteristics of the solutions to the FCC-MOSOO problem. Each operator acts either at the element level or the row level of the solution matrix, offering complementary search behaviors to enhance diversity and adaptability during the optimization process. When the local search is activated, one operator is randomly selected and applied to the target solution. The description of each operator is elaborated as follows:

Element-level polynomial search: this operator leverages modified polynomial mutation to generate a new solution. Governed by an adaptive distribution index, it applies controlled perturbations to a randomly selected element. The perturbation magnitude decreases progressively with iterations, enabling fine-grained exploitation of the solution neighborhood while preserving solution stability. The operation on the randomly selected element is formulated as Equations (14)–(16).

$$x^{\text{new}}_{ij} = x_{ij} + \delta_p \cdot (ub^*_{ij} - lb^*_{ij}) \tag{14}$$

$$\delta_p = \begin{cases} 2 \cdot u^{\frac{1}{\eta+1}} - 1 & u \le 0.5 \\ 1 - [2 \cdot (1-u)]^{\frac{1}{\eta+1}} & u > 0.5 \end{cases} \tag{15}$$

$$\eta = 20 \cdot \frac{g}{G_{\max}} \tag{16}$$

where x_{ij} is a randomly selected element; x^{new}_{ij} is the new allele element; δ_p is the perturbation magnitude; u is a random number in [0, 1]; η is the adaptive distribution index.

Row-level Gaussian random walk search: this operator implements the Gaussian random walk on a randomly selected row. Each walk step follows a Gaussian distribution with the previous element as its mean and a decaying variance. This process generates smooth modifications along the row, avoiding disruptive jumps. The operation on each element in the randomly selected row is formulated as Equations (17) and (18).

$$x_{ij}^{\text{new}} = \text{Gaussian}(x_{i,j-1}^{\text{new}}, \sigma^2) \tag{17}$$

$$\sigma = 0.1 \cdot \left(1 - \frac{g}{G_{\max}}\right) \tag{18}$$

where $\text{Gaussian}(x_{i,j-1}^{\text{new}}, \sigma^2)$ return a Gaussian random number with mean $x_{i,j-1}^{\text{new}}$ and variance σ^2.

Row-level Cauchy search: this operator utilizes the long-tailed Cauchy distribution to perturb a row randomly selected from the original solution. The long tail facilitates both occasional large-magnitude perturbations and frequent small-step adjustments, making the operator effective in escaping shallow local optima and exploring broader regions surrounding the original solution. The operation on each element in the randomly selected row is formulated as Equations (19) and (20).

$$x_{ij}^{\text{new}} = x_{ij} + \text{Cauchy}(0,1) \cdot \delta_c \tag{19}$$

$$\delta_c = 0.1 \cdot e^{-\frac{g}{G_{\max}}} \tag{20}$$

where δ_c is perturbation magnitude, and $\text{Cauchy}(0,1)$ returns a random number following the standard Cauchy distribution.

The pseudocode of the AMNS mechanism is shown in Algorithm 1.

Algorithm 1 AMNS mechanism

Require: the solution $\boldsymbol{X}$,
the current iteration count g,
the maximum number of iterations $G_{\max}$

Update p^{ls} by Equation (13)
$\boldsymbol{X}^{\text{new}} \leftarrow \boldsymbol{X}$
if rand() $< p^{\text{ls}}$ **then**
    $p \leftarrow$ rand()
    if $p < 1/3$ **then**
        Generate $\boldsymbol{X}^{\text{new}}$ by performing element-level polynomial serach on $\boldsymbol{X}$
    else if $p < 2/3$ **then**
        Generate $\boldsymbol{X}^{\text{new}}$ by performing row-level Gaussian random walk search on $\boldsymbol{X}$
    else
        Generate $\boldsymbol{X}^{\text{new}}$ by performing row-level Cauchy search on $\boldsymbol{X}$
    end if
end if
return $\boldsymbol{X}^{\text{new}}$

3.4. Dynamic Restart Mechanism

The DR mechanism is embedded in the IMOWOA to inject diversity when search stagnation is detected. This disrupts entrenched solution clusters and facilitates the exploration of undiscovered regions in the objective space. To detect stagnation, the improvement in both objectives is monitored across iterations. For each iteration, stagnation is detected if either objective shows no improvement for g_{th} consecutive iterations, where g_{th} is the stagnation threshold decreasing according to Equation (21).

$$g_{\text{th}} = \frac{G_{\max}}{e \cdot g + 1} + 2 \tag{21}$$

The DR mechanism includes two operators: the complete random walk and the partial random walk. The complete random walk is identical to the random walk in the population initialization. The partial random walk selects a pivot column randomly and then performs an independent random walk for each row starting from the pivot column element. These two designed operators correspond to wide restarts and narrow restarts, respectively.

Upon detecting stagnation, 10% solutions in the population are randomly selected. For each selected solution, one of the restart operators is applied according to an increased probability $p^{\mathrm{rt}} = g/G_{\max}$: the partial random walk is applied if a random number is less than p^{rt}; otherwise, the complete random walk is applied. This selection method ensures the early dominance of the complete random walk (wide restarts), which facilitates radical exploration to escape deep local optima. Meanwhile, late-stage preference for the partial random walk (narrow restarts) injects diversity without disrupting near-converged solutions.

The pseudocode of the dynamic mechanism is shown in Algorithm 2.

Algorithm 2 DR mechanism

Require: the population *pop*,
the current iteration count g,
the maximum number of iterations $G_{\max}$,
the stagnation iteration count for objective 1 c_1,
the stagnation iteration count for objective 2 c_2,

Update the g_{th} by Equation (21)
if $c1 < g_{\mathrm{th}}$ or $c2 < g_{\mathrm{th}}$ **then**
    $p^{\mathrm{rt}} \leftarrow g/G_{\max}$
    Randomly select 10% solutions from *pop*
    for each solution $\boldsymbol{X}$ in the selected solutions **do**
        if rand()< p^{rt} **then**
            Randomly select a column from $\boldsymbol{X}$
            Generate $\boldsymbol{X}^{\mathrm{new}}$ by performing the partial random walk on $\boldsymbol{X}$
            starting from the selected column
        else
            Generate $\boldsymbol{X}^{\mathrm{new}}$ by performing the complete random walk on $\boldsymbol{X}$
        end if
        Replace $\boldsymbol{X}$ in the population with $\boldsymbol{X}^{\mathrm{new}}$
    end for
end if
return the updated *pop*

3.5. Archive Maintenance

The proposed IMOWOA utilizes an external archive to retain elite non-dominated solutions identified throughout the optimization process. This archive is dynamically updated according to Pareto dominance relations and crowding distance to ensure both convergence and diversity. The update process operates as follows: (1) a new solution is incorporated only if it is not dominated by any solution currently in the archive; (2) existing solutions that are dominated by the new solution are immediately removed; and (3) when the archive reaches its capacity limit, solutions with larger crowding distances are prioritized for retention to maintain a well-distributed and diverse solution set.

4. Experiment Results

Extensive experiments were conducted to evaluate the performance of the proposed IMOWOA in solving the FCC-MOSOO problem. All programs of the experiments were implemented with Python 3.11.9 and executed on a laptop manufactured by Lenovo in

Beijing, China, with 16 GB RAM and a 2.3 GHz Intel Core i7-12700H under a Windows 11 (64) system.

4.1. Evaluation Metric and Test Instance

In the experiments, generational distance (GD), inverted generational distance (IGD), and hypervolume (HV) were employed as evaluation metrics. These metrics provide insights into the convergence and overall quality of the non-dominated solution set obtained by the algorithm.

GD is primarily a performance indicator of the convergence ability, as formulated in Equation (22). A smaller GD value signifies a better performance.

$$GD = \frac{1}{|U|}\sqrt{\sum_{u \in U} \min_{p \in P^*}(d_{p,u})} \tag{22}$$

where U is the non-dominated solution set, P^* is the true Pareto front, and $d_{p,u}$ is the Euclidean distance between solutions p and u in the objective space.

IGD is a comprehensive performance indicator that evaluates convergence and diversity, as formulated in Equation (23). A solution set with a smaller IGD value is preferable.

$$IGD = \frac{1}{|P^*|}\sqrt{\sum_{p \in P^*} \min_{u \in U}(d_{p,u})} \tag{23}$$

HV is calculated as the volume in the objective space dominated by the non-dominated solution set and bounded by a predefined reference point, as formulated in Equation (24). A larger HV value indicates a better performance.

$$HV = \text{volume}\left(\bigcup_{u \in U} c_u\right) \tag{24}$$

where c_u is the hypercube defined by the solution u and the reference point in the objective space.

Since the true Pareto front is a priori unknown, a reference set was constructed for each test instance to calculate GD and IGD. This reference set consists of non-dominated solutions from the combined results of all compared algorithms in the following subsections. For HV, the reference point was defined per instance using the maximum value of each objective.

Test instances were generated from historical operating data of the FCC unit to ensure that the evaluation was grounded in real-world processes. Specifically, fifteen distinct time points were randomly selected from the data. For each time point t, a corresponding test instance was constructed using the observed variables $\mathbf{O}_t$. Each case represents a unique and realistic initial state from which the optimization algorithm is tasked to find the optimal operating conditions.

4.2. Performance Comparison

To verify the effectiveness of the IMOWOA, comprehensive comparisons were conducted with both classical and emerging multi-objective algorithms, including NSGA-II [39], SPEA-2 [40], and MOGWO [41]. For a fair comparison, all algorithms were configured with a population size of 50 and a maximum iteration of 100. The crossover and mutation probabilities in NSGA-II and SPEA-2 were set to 0.8 and 0.2, respectively, based on parameter tuning by the Taguchi experimental design method.

Each algorithm was independently run 20 times to ensure statistical reliability. Table 2 presents the means and standard deviations of each metric for all algorithms, with the best values highlighted in bold. The last column provides Welch's t-test results between

IMOWOA and each compared algorithm. The symbols "+" and "=" denote IMOWOA is statistically superior and similar to the compared algorithm.

Table 2. Means, standard deviations, and Welch's *t*-test results of GD, IGD, and HV for NSGA-II, SPEA-2, MOGWO, and IMOWOA on all test instances.

Test Instance	NSGA-II (Mean/Std)	SPEA-2 (Mean/Std)	MOGWO (Mean/Std)	IMOWOA (Mean/Std)
	GD IGD HV	GD IGD HV	GD IGD HV	GD IGD HV
1	$5.74 \times 10^{-3}/8.00 \times 10^{-4}$ $1.50 \times 10^{-2}/1.57 \times 10^{-3}$ $1.23 \times 10^{-3}/5.81 \times 10^{-5}$	$5.85 \times 10^{-3}/1.30 \times 10^{-3}$ $1.50 \times 10^{-2}/1.40 \times 10^{-3}$ $1.23 \times 10^{-3}/5.26 \times 10^{-5}$	$6.61 \times 10^{-3}/2.26 \times 10^{-3}$ $1.52 \times 10^{-2}/1.70 \times 10^{-3}$ $1.22 \times 10^{-3}/8.25 \times 10^{-5}$	$\mathbf{1.83 \times 10^{-3}/9.62 \times 10^{-4}}$ (+,+,+) $\mathbf{3.83 \times 10^{-3}/2.10 \times 10^{-3}}$ (+,+,+) $\mathbf{1.81 \times 10^{-3}/1.31 \times 10^{-4}}$ (+,+,+)
2	$7.81 \times 10^{-3}/1.54 \times 10^{-3}$ $1.72 \times 10^{-2}/1.62 \times 10^{-3}$ $1.11 \times 10^{-3}/5.28 \times 10^{-5}$	$7.66 \times 10^{-3}/2.07 \times 10^{-3}$ $1.71 \times 10^{-2}/1.66 \times 10^{-3}$ $1.12 \times 10^{-3}/5.88 \times 10^{-5}$	$7.50 \times 10^{-3}/3.28 \times 10^{-3}$ $1.67 \times 10^{-2}/2.71 \times 10^{-3}$ $1.10 \times 10^{-3}/9.72 \times 10^{-5}$	$\mathbf{2.44 \times 10^{-3}/1.23 \times 10^{-3}}$ (+,+,+) $\mathbf{4.51 \times 10^{-3}/2.24 \times 10^{-3}}$ (+,+,+) $\mathbf{1.67 \times 10^{-3}/1.20 \times 10^{-4}}$ (+,+,+)
3	$5.78 \times 10^{-3}/4.89 \times 10^{-4}$ $1.11 \times 10^{-2}/1.79 \times 10^{-3}$ $3.78 \times 10^{-4}/3.37 \times 10^{-5}$	$6.17 \times 10^{-3}/6.40 \times 10^{-4}$ $1.14 \times 10^{-2}/1.01 \times 10^{-3}$ $3.57 \times 10^{-4}/3.89 \times 10^{-5}$	$7.64 \times 10^{-3}/6.92 \times 10^{-4}$ $1.36 \times 10^{-2}/1.29 \times 10^{-3}$ $2.95 \times 10^{-4}/3.16 \times 10^{-5}$	$\mathbf{7.03 \times 10^{-4}/6.20 \times 10^{-4}}$ (+,+,+) $\mathbf{1.45 \times 10^{-3}/5.19 \times 10^{-4}}$ (+,+,+) $\mathbf{8.17 \times 10^{-4}/3.11 \times 10^{-5}}$ (+,+,+)
4	$6.87 \times 10^{-3}/5.68 \times 10^{-4}$ $1.45 \times 10^{-2}/1.43 \times 10^{-3}$ $3.01 \times 10^{-4}/3.35 \times 10^{-5}$	$6.81 \times 10^{-3}/5.50 \times 10^{-4}$ $1.48 \times 10^{-2}/9.11 \times 10^{-4}$ $3.01 \times 10^{-4}/2.62 \times 10^{-5}$	$7.61 \times 10^{-3}/7.15 \times 10^{-4}$ $1.56 \times 10^{-2}/1.23 \times 10^{-3}$ $2.61 \times 10^{-4}/2.85 \times 10^{-5}$	$\mathbf{5.32 \times 10^{-4}/3.13 \times 10^{-4}}$ (+,+,+) $\mathbf{1.68 \times 10^{-3}/5.87 \times 10^{-4}}$ (+,+,+) $\mathbf{7.91 \times 10^{-4}/2.09 \times 10^{-5}}$ (+,+,+)
5	$6.32 \times 10^{-3}/5.02 \times 10^{-4}$ $1.39 \times 10^{-2}/1.58 \times 10^{-3}$ $3.78 \times 10^{-4}/3.28 \times 10^{-5}$	$6.26 \times 10^{-3}/6.39 \times 10^{-4}$ $1.36 \times 10^{-2}/1.08 \times 10^{-3}$ $3.79 \times 10^{-4}/3.76 \times 10^{-5}$	$7.71 \times 10^{-3}/8.53 \times 10^{-4}$ $1.57 \times 10^{-2}/1.03 \times 10^{-3}$ $3.13 \times 10^{-4}/3.18 \times 10^{-5}$	$\mathbf{5.26 \times 10^{-4}/2.82 \times 10^{-4}}$ (+,+,+) $\mathbf{2.63 \times 10^{-3}/8.03 \times 10^{-4}}$ (+,+,+) $\mathbf{8.65 \times 10^{-4}/2.38 \times 10^{-5}}$ (+,+,+)
6	$6.96 \times 10^{-3}/5.39 \times 10^{-4}$ $1.37 \times 10^{-2}/1.48 \times 10^{-3}$ $1.88 \times 10^{-4}/2.58 \times 10^{-5}$	$7.06 \times 10^{-3}/5.20 \times 10^{-4}$ $1.34 \times 10^{-2}/1.24 \times 10^{-3}$ $1.90 \times 10^{-4}/2.26 \times 10^{-5}$	$7.90 \times 10^{-3}/5.39 \times 10^{-4}$ $1.50 \times 10^{-2}/1.13 \times 10^{-3}$ $1.51 \times 10^{-4}/2.17 \times 10^{-5}$	$\mathbf{6.59 \times 10^{-4}/4.81 \times 10^{-4}}$ (+,+,+) $\mathbf{8.86 \times 10^{-4}/1.46 \times 10^{-4}}$ (+,+,+) $\mathbf{6.55 \times 10^{-4}/1.47 \times 10^{-5}}$ (+,+,+)
7	$7.11 \times 10^{-3}/4.62 \times 10^{-4}$ $1.62 \times 10^{-2}/1.16 \times 10^{-3}$ $2.05 \times 10^{-4}/2.18 \times 10^{-5}$	$7.02 \times 10^{-3}/4.98 \times 10^{-4}$ $1.53 \times 10^{-2}/1.46 \times 10^{-3}$ $2.16 \times 10^{-4}/2.34 \times 10^{-5}$	$7.94 \times 10^{-3}/6.21 \times 10^{-4}$ $1.73 \times 10^{-2}/1.16 \times 10^{-3}$ $1.70 \times 10^{-4}/2.27 \times 10^{-5}$	$\mathbf{6.59 \times 10^{-4}/4.40 \times 10^{-4}}$ (+,+,+) $\mathbf{2.02 \times 10^{-3}/6.50 \times 10^{-4}}$ (+,+,+) $\mathbf{6.95 \times 10^{-4}/3.04 \times 10^{-5}}$ (+,+,+)
8	$6.55 \times 10^{-3}/5.57 \times 10^{-4}$ $1.35 \times 10^{-2}/1.30 \times 10^{-3}$ $7.65 \times 10^{-4}/4.20 \times 10^{-5}$	$6.57 \times 10^{-3}/4.18 \times 10^{-4}$ $1.34 \times 10^{-2}/1.25 \times 10^{-3}$ $7.69 \times 10^{-4}/4.10 \times 10^{-5}$	$7.79 \times 10^{-3}/7.17 \times 10^{-4}$ $1.47 \times 10^{-2}/7.01 \times 10^{-4}$ $6.97 \times 10^{-4}/2.49 \times 10^{-5}$	$\mathbf{4.63 \times 10^{-4}/2.64 \times 10^{-4}}$ (+,+,+) $\mathbf{1.54 \times 10^{-3}/4.87 \times 10^{-4}}$ (+,+,+) $\mathbf{1.44 \times 10^{-3}/3.20 \times 10^{-5}}$ (+,+,+)
9	$5.77 \times 10^{-3}/5.93 \times 10^{-4}$ $1.34 \times 10^{-2}/1.45 \times 10^{-3}$ $5.35 \times 10^{-4}/4.37 \times 10^{-5}$	$5.73 \times 10^{-3}/7.33 \times 10^{-4}$ $1.34 \times 10^{-2}/1.09 \times 10^{-3}$ $5.40 \times 10^{-4}/4.41 \times 10^{-5}$	$7.05 \times 10^{-3}/7.04 \times 10^{-4}$ $1.52 \times 10^{-2}/1.04 \times 10^{-3}$ $4.58 \times 10^{-4}/3.19 \times 10^{-5}$	$\mathbf{6.55 \times 10^{-4}/3.28 \times 10^{-4}}$ (+,+,+) $\mathbf{3.01 \times 10^{-3}/7.88 \times 10^{-4}}$ (+,+,+) $\mathbf{1.04 \times 10^{-3}/2.85 \times 10^{-5}}$ (+,+,+)
10	$6.08 \times 10^{-3}/6.07 \times 10^{-4}$ $1.26 \times 10^{-2}/1.05 \times 10^{-3}$ $1.56 \times 10^{-4}/2.52 \times 10^{-5}$	$6.13 \times 10^{-3}/5.35 \times 10^{-4}$ $1.25 \times 10^{-2}/9.36 \times 10^{-4}$ $1.58 \times 10^{-4}/2.06 \times 10^{-5}$	$6.90 \times 10^{-3}/4.92 \times 10^{-4}$ $1.37 \times 10^{-2}/8.90 \times 10^{-4}$ $1.19 \times 10^{-4}/1.87 \times 10^{-5}$	$\mathbf{6.08 \times 10^{-4}/3.11 \times 10^{-4}}$ (+,+,+) $\mathbf{1.91 \times 10^{-3}/5.44 \times 10^{-4}}$ (+,+,+) $\mathbf{5.39 \times 10^{-4}/2.79 \times 10^{-5}}$ (+,+,+)
11	$6.60 \times 10^{-3}/6.82 \times 10^{-4}$ $1.41 \times 10^{-2}/1.28 \times 10^{-3}$ $8.41 \times 10^{-4}/4.54 \times 10^{-5}$	$6.42 \times 10^{-3}/9.44 \times 10^{-4}$ $1.36 \times 10^{-2}/1.56 \times 10^{-3}$ $8.63 \times 10^{-4}/6.68 \times 10^{-5}$	$7.88 \times 10^{-3}/1.58 \times 10^{-3}$ $1.61 \times 10^{-2}/1.58 \times 10^{-3}$ $7.74 \times 10^{-4}/5.81 \times 10^{-5}$	$\mathbf{6.33 \times 10^{-4}/2.88 \times 10^{-4}}$ (+,+,+) $\mathbf{2.82 \times 10^{-3}/8.92 \times 10^{-4}}$ (+,+,+) $\mathbf{1.45 \times 10^{-3}/2.92 \times 10^{-5}}$ (+,+,+)
12	$5.97 \times 10^{-3}/8.94 \times 10^{-4}$ $1.34 \times 10^{-2}/1.18 \times 10^{-3}$ $3.29 \times 10^{-4}/3.69 \times 10^{-5}$	$5.62 \times 10^{-3}/5.73 \times 10^{-4}$ $1.21 \times 10^{-2}/1.38 \times 10^{-3}$ $3.50 \times 10^{-4}/3.43 \times 10^{-5}$	$6.83 \times 10^{-3}/8.03 \times 10^{-4}$ $1.44 \times 10^{-2}/1.37 \times 10^{-3}$ $2.70 \times 10^{-4}/3.22 \times 10^{-5}$	$\mathbf{8.45 \times 10^{-4}/3.62 \times 10^{-4}}$ (+,+,+) $\mathbf{2.24 \times 10^{-3}/5.65 \times 10^{-4}}$ (+,+,+) $\mathbf{7.41 \times 10^{-4}/3.58 \times 10^{-5}}$ (+,+,+)
13	$5.81 \times 10^{-3}/6.54 \times 10^{-4}$ $1.20 \times 10^{-2}/1.46 \times 10^{-3}$ $3.07 \times 10^{-4}/4.06 \times 10^{-5}$	$5.70 \times 10^{-3}/5.23 \times 10^{-4}$ $1.20 \times 10^{-2}/1.22 \times 10^{-3}$ $3.16 \times 10^{-4}/2.38 \times 10^{-5}$	$7.09 \times 10^{-3}/8.31 \times 10^{-4}$ $1.39 \times 10^{-2}/1.31 \times 10^{-3}$ $2.40 \times 10^{-4}/2.65 \times 10^{-5}$	$\mathbf{4.91 \times 10^{-4}/2.12 \times 10^{-4}}$ (+,+,+) $\mathbf{1.51 \times 10^{-3}/5.21 \times 10^{-4}}$ (+,+,+) $\mathbf{7.30 \times 10^{-4}/2.79 \times 10^{-5}}$ (+,+,+)
14	$6.13 \times 10^{-3}/5.97 \times 10^{-4}$ $1.38 \times 10^{-2}/1.23 \times 10^{-3}$ $1.52 \times 10^{-4}/2.87 \times 10^{-5}$	$6.22 \times 10^{-3}/9.34 \times 10^{-4}$ $1.29 \times 10^{-2}/1.04 \times 10^{-3}$ $1.70 \times 10^{-4}/3.08 \times 10^{-5}$	$7.00 \times 10^{-3}/4.76 \times 10^{-4}$ $1.46 \times 10^{-2}/9.68 \times 10^{-4}$ $1.11 \times 10^{-4}/2.02 \times 10^{-5}$	$\mathbf{7.20 \times 10^{-4}/3.38 \times 10^{-4}}$ (+,+,+) $\mathbf{1.14 \times 10^{-3}/3.95 \times 10^{-4}}$ (+,+,+) $\mathbf{5.24 \times 10^{-4}/2.72 \times 10^{-5}}$ (+,+,+)
15	$5.95 \times 10^{-3}/5.97 \times 10^{-4}$ $1.26 \times 10^{-2}/1.25 \times 10^{-3}$ $2.83 \times 10^{-4}/2.63 \times 10^{-5}$	$5.90 \times 10^{-3}/5.90 \times 10^{-4}$ $1.21 \times 10^{-2}/9.10 \times 10^{-4}$ $2.91 \times 10^{-4}/2.92 \times 10^{-5}$	$7.00 \times 10^{-3}/5.69 \times 10^{-4}$ $1.36 \times 10^{-2}/8.37 \times 10^{-4}$ $2.33 \times 10^{-4}/2.03 \times 10^{-5}$	$\mathbf{5.91 \times 10^{-4}/3.52 \times 10^{-4}}$ (+,+,+) $\mathbf{1.92 \times 10^{-3}/5.86 \times 10^{-4}}$ (+,+,+) $\mathbf{6.99 \times 10^{-4}/2.97 \times 10^{-5}}$ (+,+,+)

Notes: Best values are highlighted in bold. "+" and "=" denote that IMOWOA is statistically superior and similar to the compared algorithm, respectively.

As illustrated in Table 2, the proposed IMOWOA obtained the best results of each metric on all test instances, demonstrating its strong convergence capability and well-balanced comprehensive performance in FCC-MOSSO. Among the remaining compared algorithms, NSGA-II and SPEA-2 performed comparably, yet both surpassed MOGWO.

For a more intuitive comparison, Figure 3 visualizes the final non-dominated solution sets obtained from 20 runs on four representative test instances (2, 6, 10, and 14). As depicted, MOGWO converged to narrow regions, indicating limited exploration capability and a tendency toward premature convergence. NSGA-II exhibited better distribution uniformity than SPEA-2, though with slightly inferior exploration capability. In contrast, the proposed IMOWOA achieved significantly broader coverage in the objective space. Its solution sets not only spanned a larger area but also dominated most of the solutions derived by the other algorithms.

Overall, the thorough results suggest that IMOWOA exhibits superior search capabilities in the FCC-MOSOO problem, outperforming all compared algorithms in terms of both convergence behavior and comprehensive performance.

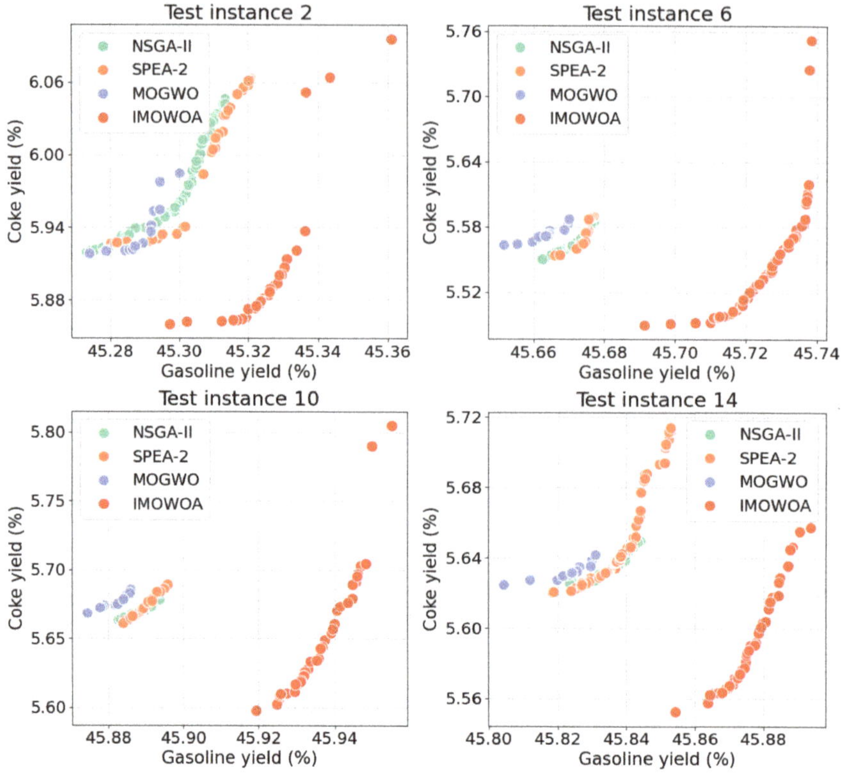

Figure 3. Non-dominated solutions of NSGA-II, SPEA-2, MOGWO, and IMOWOA on test instances 2, 6, 10, and 14. (IMOWOA exhibits wider Pareto front coverage, indicating superior exploration ability).

4.3. Ablation Study

To evaluate the effectiveness of AMNS and DR mechanisms, an ablation study was carried out comparing IMOWOA against three variants: the basic multi-objective WOA without either mechanism (MOWOA), MOWOA with only DR(MOWOA-D), and MOWOA with only AMNS (MOWOA-A).

The comparison results of GD, IGD, and HV for the above algorithms are summarized in Table 3. For a more detailed analysis, Figure 4 plots the metrics of 20 independent runs of each algorithm on test instances 3, 8, and 13 (each gray dot represents a metric) along with the box plots reflecting statistical distributions. In terms of GD, IMOWOA and MOWOA-A performed similarly, and the best results were produced between them, indicating the contribution of the AMNS mechanism to convergence. Although MOWOA-D performed worse than IMOWOA and MOWOA-A, it still surpassed MOWOA. As illustrated in Figure 4, GD of MOWOA-D exhibited a wider distribution, attributed to the influence of the DR mechanism. In contrast, both IMOWOA and MOWOA-A demonstrated more concentrated and better distributions.

Table 3. Means, standard deviations, and Welch's *t*-test results of GD, IGD, and HV for MOWOA, IMOWOA-D, IMOWOA-A, and IMOWOA on all test instances.

Test Instance	MOWOA (Mean/Std)	MOWOA-D (Mean/Std)	MOWOA-A (Mean/Std)	IMOWOA (Mean/Std)
	GD IGD HV	GD IGD HV	GD IGD HV	GD IGD HV
1	$3.24 \times 10^{-3}/1.87 \times 10^{-3}$ $8.98 \times 10^{-3}/2.95 \times 10^{-3}$ $1.56 \times 10^{-3}/1.81 \times 10^{-4}$	$2.87 \times 10^{-3}/9.39 \times 10^{-4}$ $5.87 \times 10^{-3}/2.15 \times 10^{-3}$ $1.66 \times 10^{-3}/1.18 \times 10^{-4}$	**$1.78 \times 10^{-3}/9.80 \times 10^{-4}$** $7.24 \times 10^{-3}/1.60 \times 10^{-3}$ $1.75 \times 10^{-3}/1.28 \times 10^{-4}$	$1.83 \times 10^{-3}/9.62 \times 10^{-4}$ (+,+,=) **$3.83 \times 10^{-3}/2.10 \times 10^{-3}$** (+,+,+) **$1.81 \times 10^{-3}/1.31 \times 10^{-4}$** (+,+,=)
2	$4.97 \times 10^{-3}/4.75 \times 10^{-3}$ $1.23 \times 10^{-2}/5.33 \times 10^{-3}$ $1.33 \times 10^{-3}/2.41 \times 10^{-4}$	$3.01 \times 10^{-3}/1.26 \times 10^{-3}$ $6.42 \times 10^{-3}/2.57 \times 10^{-3}$ $1.57 \times 10^{-3}/1.38 \times 10^{-4}$	**$1.77 \times 10^{-3}/8.29 \times 10^{-4}$** $6.77 \times 10^{-3}/1.86 \times 10^{-3}$ $1.62 \times 10^{-3}/1.07 \times 10^{-4}$	$2.44 \times 10^{-3}/1.23 \times 10^{-3}$ (+,=,=) **$4.51 \times 10^{-3}/2.24 \times 10^{-3}$** (+,+,+) **$1.67 \times 10^{-3}/1.20 \times 10^{-4}$** (+,+,=)
3	$2.24 \times 10^{-3}/1.09 \times 10^{-3}$ $4.74 \times 10^{-3}/2.06 \times 10^{-3}$ $6.31 \times 10^{-4}/8.05 \times 10^{-5}$	$2.16 \times 10^{-3}/1.21 \times 10^{-3}$ $3.09 \times 10^{-3}/1.45 \times 10^{-3}$ $7.08 \times 10^{-4}/5.73 \times 10^{-5}$	**$6.28 \times 10^{-4}/3.04 \times 10^{-4}$** $3.43 \times 10^{-3}/1.25 \times 10^{-3}$ $7.56 \times 10^{-4}/3.43 \times 10^{-5}$	$7.03 \times 10^{-4}/6.20 \times 10^{-4}$ (+,+,=) **$1.45 \times 10^{-3}/5.19 \times 10^{-4}$** (+,+,+) **$8.17 \times 10^{-4}/3.11 \times 10^{-5}$** (+,+,+)
4	$2.01 \times 10^{-3}/6.99 \times 10^{-4}$ $5.49 \times 10^{-3}/1.57 \times 10^{-3}$ $6.16 \times 10^{-4}/6.32 \times 10^{-5}$	$1.87 \times 10^{-3}/1.07 \times 10^{-3}$ $3.62 \times 10^{-3}/1.08 \times 10^{-3}$ $6.93 \times 10^{-4}/4.66 \times 10^{-5}$	$6.28 \times 10^{-4}/2.62 \times 10^{-4}$ $3.67 \times 10^{-3}/1.03 \times 10^{-3}$ $7.27 \times 10^{-4}/3.03 \times 10^{-5}$	**$5.32 \times 10^{-4}/3.13 \times 10^{-4}$** (+,+,=) **$1.68 \times 10^{-3}/5.87 \times 10^{-4}$** (+,+,+) **$7.91 \times 10^{-4}/2.09 \times 10^{-5}$** (+,+,+)
5	$2.25 \times 10^{-3}/7.31 \times 10^{-4}$ $6.76 \times 10^{-3}/1.63 \times 10^{-3}$ $6.63 \times 10^{-4}/5.60 \times 10^{-5}$	$1.72 \times 10^{-3}/8.36 \times 10^{-4}$ $4.07 \times 10^{-3}/1.45 \times 10^{-3}$ $7.62 \times 10^{-4}/6.23 \times 10^{-5}$	$6.04 \times 10^{-4}/2.98 \times 10^{-4}$ $4.83 \times 10^{-3}/1.43 \times 10^{-3}$ $8.08 \times 10^{-4}/3.74 \times 10^{-5}$	**$5.26 \times 10^{-4}/2.82 \times 10^{-4}$** (+,+,=) **$2.63 \times 10^{-3}/8.03 \times 10^{-4}$** (+,+,+) **$8.65 \times 10^{-4}/2.38 \times 10^{-5}$** (+,+,+)
6	$1.93 \times 10^{-3}/6.74 \times 10^{-4}$ $4.45 \times 10^{-3}/1.54 \times 10^{-3}$ $4.88 \times 10^{-4}/4.92 \times 10^{-5}$	$1.80 \times 10^{-3}/1.07 \times 10^{-3}$ $3.13 \times 10^{-3}/1.33 \times 10^{-3}$ $5.59 \times 10^{-4}/4.90 \times 10^{-5}$	**$4.46 \times 10^{-4}/2.47 \times 10^{-4}$** $2.03 \times 10^{-3}/8.28 \times 10^{-4}$ $6.06 \times 10^{-4}/2.83 \times 10^{-5}$	$6.59 \times 10^{-4}/4.81 \times 10^{-4}$ (+,+,=) **$8.86 \times 10^{-4}/1.46 \times 10^{-4}$** (+,+,+) **$6.55 \times 10^{-4}/1.47 \times 10^{-5}$** (+,+,+)
7	$1.95 \times 10^{-3}/1.08 \times 10^{-3}$ $5.78 \times 10^{-3}/2.24 \times 10^{-3}$ $5.34 \times 10^{-4}/7.85 \times 10^{-5}$	$2.00 \times 10^{-3}/9.47 \times 10^{-4}$ $4.07 \times 10^{-3}/1.31 \times 10^{-3}$ $5.95 \times 10^{-4}/5.36 \times 10^{-5}$	**$5.05 \times 10^{-4}/3.04 \times 10^{-4}$** $3.42 \times 10^{-3}/8.75 \times 10^{-4}$ $6.50 \times 10^{-4}/3.48 \times 10^{-5}$	$6.59 \times 10^{-4/4.40} \times 10^{-4}$ (+,+,=) **$2.02 \times 10^{-3}/6.50 \times 10^{-4}$** (+,+,+) **$6.95 \times 10^{-4}/3.04 \times 10^{-5}$** (+,+,+)
8	$2.08 \times 10^{-3}/7.38 \times 10^{-4}$ $4.81 \times 10^{-3}/1.47 \times 10^{-3}$ $1.19 \times 10^{-3}/8.42 \times 10^{-5}$	$1.94 \times 10^{-3}/7.47 \times 10^{-4}$ $3.56 \times 10^{-3}/1.42 \times 10^{-3}$ $1.24 \times 10^{-3}/7.32 \times 10^{-5}$	$5.50 \times 10^{-4}/2.57 \times 10^{-4}$ $3.87 \times 10^{-3}/1.74 \times 10^{-3}$ $1.37 \times 10^{-3}/3.21 \times 10^{-5}$	**$4.63 \times 10^{-4}/2.64 \times 10^{-4}$** (+,+,=) **$1.54 \times 10^{-3}/4.87 \times 10^{-4}$** (+,+,+) **$1.44 \times 10^{-3}/3.20 \times 10^{-5}$** (+,+,+)
9	$1.98 \times 10^{-3}/7.52 \times 10^{-4}$ $6.30 \times 10^{-3}/1.91 \times 10^{-3}$ $8.51 \times 10^{-4}/7.98 \times 10^{-5}$	$1.82 \times 10^{-3}/6.27 \times 10^{-4}$ $3.74 \times 10^{-3}/7.39 \times 10^{-4}$ $9.20 \times 10^{-4}/4.10 \times 10^{-5}$	$6.70 \times 10^{-4}/3.25 \times 10^{-4}$ $5.33 \times 10^{-3}/1.48 \times 10^{-3}$ $9.80 \times 10^{-4}/3.34 \times 10^{-5}$	**$6.55 \times 10^{-4}/3.28 \times 10^{-4}$** (+,+,=) **$3.01 \times 10^{-3}/7.88 \times 10^{-4}$** (+,+,+) **$1.04 \times 10^{-3}/2.85 \times 10^{-5}$** (+,+,+)
10	$2.00 \times 10^{-3}/6.70 \times 10^{-4}$ $5.10 \times 10^{-3}/1.20 \times 10^{-3}$ $3.89 \times 10^{-4}/4.51 \times 10^{-5}$	$1.95 \times 10^{-3}/6.70 \times 10^{-4}$ $3.64 \times 10^{-3}/8.90 \times 10^{-4}$ $4.41 \times 10^{-4}/4.44 \times 10^{-5}$	$8.65 \times 10^{-4}/2.58 \times 10^{-4}$ $4.27 \times 10^{-3}/1.15 \times 10^{-3}$ $4.75 \times 10^{-4}/2.60 \times 10^{-5}$	**$6.08 \times 10^{-4}/3.11 \times 10^{-4}$** (+,+,+) **$1.91 \times 10^{-3}/5.44 \times 10^{-4}$** (+,+,+) **$5.39 \times 10^{-4}/2.79 \times 10^{-5}$** (+,+,+)
11	$2.68 \times 10^{-3}/1.17 \times 10^{-3}$ $7.40 \times 10^{-3}/2.07 \times 10^{-3}$ $1.19 \times 10^{-3}/1.06 \times 10^{-4}$	$2.04 \times 10^{-3}/1.06 \times 10^{-3}$ $4.83 \times 10^{-3}/1.70 \times 10^{-3}$ $1.29 \times 10^{-3}/9.87 \times 10^{-5}$	$6.62 \times 10^{-4}/3.03 \times 10^{-4}$ $5.03 \times 10^{-3}/1.05 \times 10^{-3}$ $1.39 \times 10^{-3}/3.71 \times 10^{-5}$	**$6.33 \times 10^{-4}/2.88 \times 10^{-4}$** (+,+,=) **$2.82 \times 10^{-3}/8.92 \times 10^{-4}$** (+,+,+) **$1.45 \times 10^{-3}/2.92 \times 10^{-5}$** (+,+,+)
12	$1.83 \times 10^{-3}/7.61 \times 10^{-4}$ $4.66 \times 10^{-3}/1.73 \times 10^{-3}$ $5.99 \times 10^{-4}/5.42 \times 10^{-5}$	$2.46 \times 10^{-3}/9.87 \times 10^{-4}$ $3.67 \times 10^{-3}/1.51 \times 10^{-3}$ $6.35 \times 10^{-4}/5.83 \times 10^{-5}$	**$7.61 \times 10^{-4}/3.24 \times 10^{-4}$** $4.00 \times 10^{-3}/9.15 \times 10^{-4}$ $6.79 \times 10^{-4}/3.67 \times 10^{-5}$	$8.45 \times 10^{-4}/3.62 \times 10^{-4}$ (+,+,=) **$2.24 \times 10^{-3}/5.65 \times 10^{-4}$** (+,+,+) **$7.41 \times 10^{-4}/3.58 \times 10^{-5}$** (+,+,+)
13	$2.17 \times 10^{-3}/8.44 \times 10^{-4}$ $4.57 \times 10^{-3}/1.68 \times 10^{-3}$ $5.52 \times 10^{-4}/6.19 \times 10^{-5}$	$1.98 \times 10^{-3}/8.95 \times 10^{-4}$ $2.92 \times 10^{-3}/1.03 \times 10^{-3}$ $6.27 \times 10^{-4}/5.14 \times 10^{-5}$	$6.46 \times 10^{-4}/3.84 \times 10^{-4}$ $3.52 \times 10^{-3}/1.38 \times 10^{-3}$ $6.63 \times 10^{-4}/3.09 \times 10^{-5}$	**$4.91 \times 10^{-4}/2.12 \times 10^{-4}$** (+,+,=) **$1.51 \times 10^{-3}/5.21 \times 10^{-4}$** (+,+,+) **$7.30 \times 10^{-4}/2.79 \times 10^{-5}$** (+,+,+)
14	$1.85 \times 10^{-3}/1.11 \times 10^{-3}$ $4.45 \times 10^{-3}/2.22 \times 10^{-3}$ $3.89 \times 10^{-4}/6.91 \times 10^{-5}$	$2.77 \times 10^{-3}/2.02 \times 10^{-3}$ $2.83 \times 10^{-3}/7.88 \times 10^{-4}$ $4.39 \times 10^{-4}/3.73 \times 10^{-5}$	**$5.68 \times 10^{-4}/2.63 \times 10^{-4}$** $1.95 \times 10^{-3}/6.96 \times 10^{-4}$ $4.84 \times 10^{-4}/2.28 \times 10^{-5}$	$7.20 \times 10^{-4}/3.38 \times 10^{-4}$ (+,+,=) **$1.14 \times 10^{-3}/3.95 \times 10^{-4}$** (+,+,+) **$5.24 \times 10^{-4}/2.72 \times 10^{-5}$** (+,+,+)
15	$2.10 \times 10^{-3}/9.01 \times 10^{-4}$ $5.36 \times 10^{-3}/1.55 \times 10^{-3}$ $5.32 \times 10^{-4}/6.39 \times 10^{-5}$	$1.64 \times 10^{-3}/5.87 \times 10^{-4}$ $3.51 \times 10^{-3}/1.02 \times 10^{-3}$ $5.96 \times 10^{-4}/4.34 \times 10^{-5}$	$6.74 \times 10^{-4}/2.75 \times 10^{-4}$ $3.73 \times 10^{-3}/1.23 \times 10^{-3}$ $6.49 \times 10^{-4}/2.20 \times 10^{-5}$	**$5.91 \times 10^{-4}/3.52 \times 10^{-4}$** (+,+,=) **$1.92 \times 10^{-3}/5.86 \times 10^{-4}$** (+,+,+) **$6.99 \times 10^{-4}/2.97 \times 10^{-5}$** (+,+,+)

Note: Best values are highlighted in bold. "+" denote that IMOWOA is statistically superior to the compared algorithm.

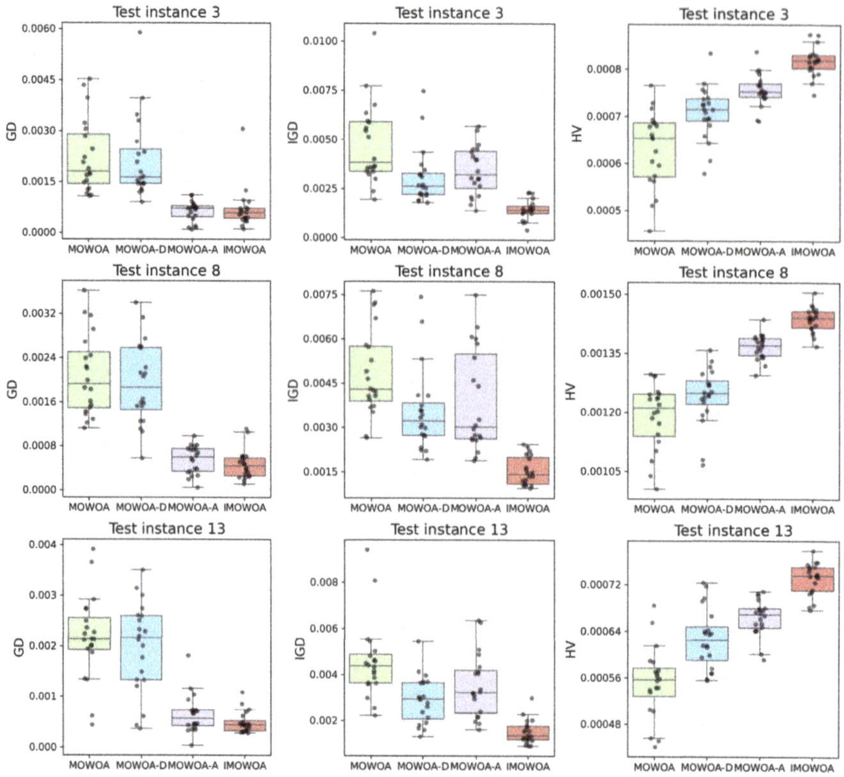

Figure 4. GD, IGD, and HV along with box plots of 20 independent runs of MOWOA, IMOWOA-D, IMOWOA-A, and IMOWOA on test instances 3, 8, and 13. (IMOWOA achieves the best mean values and the most concentrated distribution across all metrics, demonstrating enhanced search capability and stability).

IMOWOA achieved optimal IGD and HV values in all test instances, indicating better comprehensive performance. Both MOWOA-A and MOWOA-D consistently exceeded the performance of MOWOA. Statistically, IMOWOA significantly outperformed all other algorithms in IGD on each test instance. Regarding HV, it surpassed MOWOA-A in 14 instances and dominated MOWOA-D and MOWOA in all instances. Figure 4 further reveals a trade-off between MOWOA-D and MOWOA-A: the former achieved lower IGD values, whereas the latter excelled in HV. IMOWOA, however, displayed the most concentrated distribution in both metrics, reflecting greater robustness and stability.

Based on the above analysis, both proposed mechanisms effectively enhance the search capability of the algorithm. AMNS plays a dominant role in improving convergence and also contributes to the comprehensive performance. Meanwhile, the DR mechanism injects diversity into the search process, further boosting the comprehensive performance.

4.4. Result Evaluation

This subsection presents a comparative analysis between the optimized results and the original actual product yields on each test instance to evaluate the practical industrial value of the proposed optimization method. Since the proposed IMOWOA generates a Pareto non-dominated solution set, the Technique for Order Preference by Similarity to an Ideal Solution (TOPSIS) was employed to select the most suitable operational solution from the available alternatives for each instance. TOPSIS is a multi-criteria decision-making method that ranks alternatives based on their relative distance from the ideal and negative-ideal solutions. In this implementation, gasoline yield was defined as a benefit criterion to be maximized, while coke yield was treated as a cost criterion to be minimized. Both criteria were assigned equal weights to enable the selection of an optimal compromise solution.

The comparison results of gasoline and coke yields of the selected optimized solution with their respective original yields on each test instance are summarized in Figure 5. It is clear that the optimized operational solution consistently enhanced gasoline yield while simultaneously reducing coke yield. Specifically, the gasoline yield showed an average increase of 0.32%, while the coke yield exhibited an average reduction of 0.11%. Taking the studied FCC unit as a representative case, a further sustainability impact analysis was conducted. With an annual processing capacity of 2.6 million tons and a coke-to-CO_2 conversion rate of approximately 98%, the unit can achieve an annual CO_2 emission reduction of about 10,277 tons through the coke yield reduction.

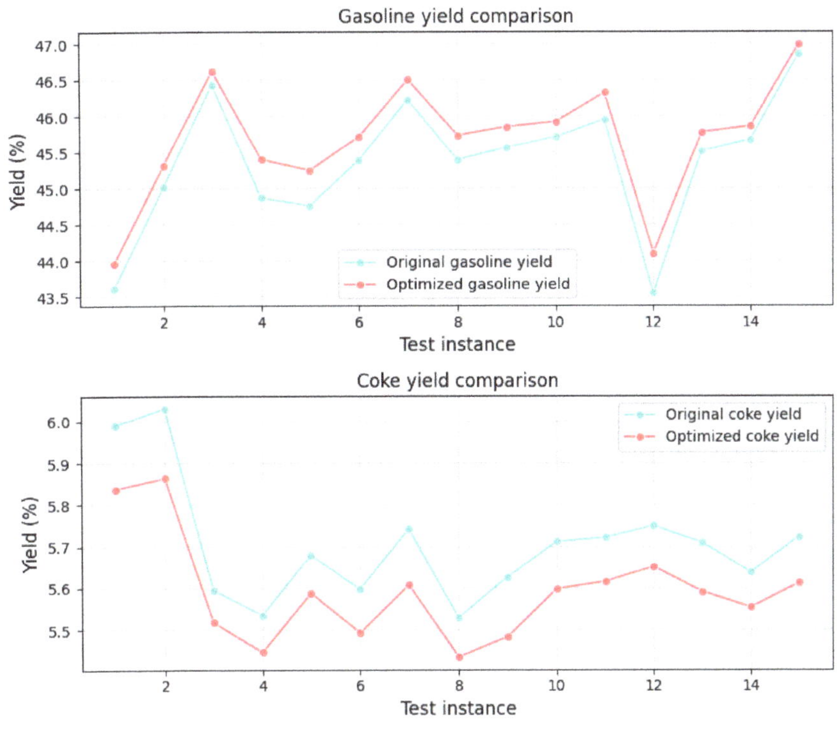

Figure 5. Comparison results of gasoline and coke yields of the selected optimized solution with their respective original yields on 15 test instances. (Optimized gasoline and coke yields show improvement and reduction, respectively, confirming the effectiveness of the proposed approach).

The demonstrated reduction in carbon emissions, along with the simultaneous improvement in gasoline production, validates the effectiveness of the Dual-LSTM modeling approach and the proposed IMOWOA. The proposed approach also exhibits good industrial implementation feasibility, as the data-driven model provides rapid predictions suitable for dynamic optimization, while the optimized operational adjustments can be integrated into existing refinery Distributed Control Systems. These characteristics highlight the significant potential of the proposed approach in supporting refinery sustainability goals while maintaining economic viability.

5. Conclusions

This paper has addressed an FCC-MOSOO problem, targeting the enhancement of gasoline production and the suppression of coke formation to achieve concurrent economic and environmental benefits. A data-driven multi-objective optimization model founded on a Dual-LSTM architecture is established to accurately capture variable-yield relationships with high computational efficiency, thus avoiding the high computational cost and limited applicability of mechanism models. The model optimizes the gasoline and coke yields through adjusting the trajectory of controlled variables over a future window, effectively handling the significant time lag characteristics of the FCC process.

To efficiently solve this model, an IMOWOA enhanced with AMNS and DR mechanisms is designed. Empirical evaluations confirm the superiority of IMOWOA, which outperforms established algorithms, including NSGA-II, SPEA-2, and MOGWO in convergence efficiency and comprehensive performance. Furthermore, the ablation validates the critical role of the proposed AMNS and DR, as IMOWOA consistently exceeds the performance of its variants without these mechanisms. Compared with the yields before optimization, the proposed approach can increase the gasoline yield by an average of 0.32% and reduce the coke yield by an average of 0.11%, demonstrating its capability to improve both economic and environmental performance in FCC operations.

This work provides a practical and intelligent decision-support tool for FCC operators seeking to align with sustainable development goals. Future research will focus on extending the proposed method to incorporate additional sustainability indicators, such as energy consumption, and testing its applicability under wider operational scenarios.

Author Contributions: Conceptualization, S.P. and Y.L.; Methodology, S.P.; Software, S.P.; Resources, H.S.; Validation, R.Y.; Data curation, R.T. and D.L.; Writing—original draft preparation, S.P.; Writing—review and editing, R.Y.; Supervision, H.S. and C.L.; Project administration, Y.L.; Funding acquisition, C.L. All authors have read and agreed to the published version of the manuscript.

Funding: This research was funded by "Research and Demonstration Application of Key Technologies for Low Carbon and Efficient Operation of Typical Chemical Processes Based on Digital Twin Technology" OF the National Key Research and Development Program of China, grant number 2024YFE0212400.

Institutional Review Board Statement: Not applicable.

Informed Consent Statement: Not applicable.

Data Availability Statement: The data presented in this study are available on request from the corresponding author due to privacy concerns.

Acknowledgments: The authors are thankful to the anonymous reviewers and editors for their valuable comments and suggestions.

Conflicts of Interest: Shibao Pang, Yang Lin, Hongxun Shi, Rui Yin, Ran Tao, Donghong Li and Chuankun Li were employed by SINOPEC Research Institute of Safety Engineering Co. Ltd. The remaining authors declare that the research was conducted in the absence of any commercial or financial relationships that could be construed as a potential conflict of interest.

References

1. Liu, Y.; Lu, S.; Yan, X.; Gao, S.; Cui, X.; Cui, Z. Life cycle assessment of petroleum refining process: A case study in China. *J. Clean. Prod.* **2020**, *256*, 120422. [CrossRef]
2. Choy, Y.K.; Onuma, A.; Lee, K.E. The Nexus of Industrial–Urban Sustainability, the Circular Economy, and Climate–Ecosystem Resilience: A Synthesis. *Sustainability* **2025**, *17*, 2620. [CrossRef]
3. Sundaramoorthy, S.; Kamath, D.; Nimbalkar, S.; Price, C.; Wenning, T.; Cresko, J. Energy Efficiency as a Foundational Technology Pillar for Industrial Decarbonization. *Sustainability* **2023**, *15*, 9487. [CrossRef]

4. Chen, J.; Zhu, J.; Qin, X.; Xie, W. Reducing Octane Number Loss in Gasoline Refining Process by Using the Improved Sparrow Search Algorithm. *Sustainability* **2023**, *15*, 6571. [CrossRef]
5. Bechara, C.A.; Alnouri, S.Y. Energy assessment strategies in carbon-constrained industrial clusters. *Energy Convers. Manag.* **2022**, *254*, 115204. [CrossRef]
6. Johnson, E.; Vadenbo, C. Modelling Variation in Petroleum Products' Refining Footprints. *Sustainability* **2020**, *12*, 9316. [CrossRef]
7. Simayi, M.; Hao, Y.; Li, J.; Shi, Y.; Ren, J.; Xi, Z.; Xie, S. Historical volatile organic compounds emission performance and reduction potentials in China's petroleum refining industry. *J. Clean. Prod.* **2021**, *292*, 125810. [CrossRef]
8. Talaei, A.; Oni, A.O.; Ahiduzzaman, M.; Roychaudhuri, P.S.; Rutherford, J.; Kumar, A. Assessment of the impacts of process-level energy efficiency improvement on greenhouse gas mitigation potential in the petroleum refining sector. *Energy* **2020**, *191*, 116243. [CrossRef]
9. Zhao, S.; Song, Q.; Liu, L.; Li, J.; Zhao, D. Uncovering the lifecycle carbon emissions and its reduction pathways: A case study of petroleum refining enterprise. *Energy Convers. Manag.* **2024**, *301*, 118048. [CrossRef]
10. Ferella, F.; D'Adamo, I.; Leone, S.; Innocenzi, V.; De Michelis, I.; Vegliò, F. Spent FCC E-Cat: Towards a Circular Approach in the Oil Refining Industry. *Sustainability* **2019**, *11*, 113. [CrossRef]
11. Bai, P.; Etim, U.J.; Yan, Z.; Mintova, S.; Zhang, Z.; Zhong, Z.; Gao, X. Fluid catalytic cracking technology: Current status and recent discoveries on catalyst contamination. *Catal. Rev.* **2019**, *61*, 333–405. [CrossRef]
12. Khaldi, M.K.; Al-Dhaifallah, M.; Taha, O. Artificial intelligence perspectives: A systematic literature review on modeling, control, and optimization of fluid catalytic cracking. *Alex. Eng. J.* **2023**, *80*, 294–314. [CrossRef]
13. Wang, L.; Long, J.; Li, X.; Peng, H.; Ye, Z. Industrial units modeling using self-attention network based on feature selection and pattern classification. *Chem. Eng. Res. Des.* **2023**, *200*, 176–185. [CrossRef]
14. Luan, H.; Wu, C.; Xiu, G.; Ju, F.; Ling, H.; Pan, H. Study on emission factors of FCC flue gas pollutants in petroleum refineries. *Environ. Sci. Pollut. Res.* **2023**, *29*, 33400–33410. [CrossRef] [PubMed]
15. Huang, M.; Zheng, Y.; Li, S.; Xu, S. Enhancing Transient Event Trigger Real-Time Optimization for Fluid Catalytic Cracking Unit Operation with Varying Feedstock. *Ind. Eng. Chem. Res.* **2019**, *58*, 20340–20356. [CrossRef]
16. Otten-Weinschenker, J.; Mönnigmann, M. Robust optimization of stiff delayed systems: Application to a fluid catalytic cracking unit. *Optim. Eng.* **2022**, *23*, 2025–2050. [CrossRef]
17. Santander, O.; Kuppuraj, V.; Harrison, C.A.; Baldea, M. Integrated Production Planning and Model Predictive Control of a Fluidized Bed Catalytic Cracking-Fractionator Unit. *Ind. Eng. Chem. Res.* **2023**, *62*, 2752–2767. [CrossRef]
18. John, Y.M.; Patel, R.; Mujtaba, I.M. Maximization of propylene in an industrial FCC unit. *Appl. Petrochem. Res.* **2018**, *8*, 79–95. [CrossRef]
19. Chen, C.; Lu, N.; Wang, L.; Xing, Y. Intelligent selection and optimization method of feature variables in fluid catalytic cracking gasoline refining process. *Comput. Chem. Eng.* **2021**, *150*, 107336. [CrossRef]
20. Chen, Q.; Ding, J.; Chai, T.; Pan, Q. Evolutionary Optimization Under Uncertainty: The Strategies to Handle Varied Constraints for Fluid Catalytic Cracking Operation. *IEEE Trans. Cybern.* **2022**, *52*, 2249–2262. [CrossRef]
21. Li, T.; Long, J.; Zhao, L.; Du, W.; Qian, F. A bilevel data-driven framework for robust optimization under uncertainty – applied to fluid catalytic cracking unit. *Comput. Chem. Eng.* **2022**, *166*, 107989. [CrossRef]
22. Wang, P.; Yang, C.; Tian, X.; Cao, Y. Multi-objective optimization of two-stage-riser fluidized catalytic pyrolysis process for maximizing propylene. In Proceedings of the 2016 35th Chinese Control Conference (CCC), Chengdu, China, 27–29 July 2016; pp. 9471–9474. [CrossRef]
23. Tai, X.Y.; Ocone, R.; Christie, S.D.; Xuan, J. Multi-objective optimisation with hybrid machine learning strategy for complex catalytic processes. *Energy AI* **2022**, *7*, 100134. [CrossRef]
24. Jarullah, A.T.; Awad, N.A.; Mujtaba, I.M. Optimal design and operation of an industrial fluidized catalytic cracking reactor. *Fuel* **2017**, *206*, 657–674. [CrossRef]
25. Liu, X.; Liu, Y.; He, X.; Xiao, M.; Jiang, T. Multi-Objective Nonlinear Programming Model for Reducing Octane Number Loss in Gasoline Refining Process Based on Data Mining Technology. *Processes* **2021**, *9*, 721. [CrossRef]
26. Bohorquez, J.F.C.; Tovar, L.P.; Maciel, M.R.W.; Melo, D.C.; Filho, R.M. Surrogate-model-based, particle swarm optimization, and genetic algorithm techniques applied to the multiobjective operational problem of the fluid catalytic cracking process. *Chem. Eng. Commun.* **2020**, *207*, 612–631. [CrossRef]
27. Wan, L.; Deng, K.; Li, X.; Zhao, L.; Long, J. Multi-objective optimization strategy for industrial catalytic cracking units: Kinetic model and enhanced SPEA-2 algorithm with economic, CO_2, and SO_2 emission considerations. *Chem. Eng. Sci.* **2023**, *282*, 119331. [CrossRef]
28. Liu, Y.; Chu, M.; Ye, Q.; Li, J.; Han, D. Multi-objective optimization of FCC separation system based on NSGA-II. *Chem. Eng. Sci.* **2025**, *302*, 120829. [CrossRef]
29. Khaldi, M.K.; Al-Dhaifallah, M.; Taha, O.; Mahmood, T.; Alharbi, A. Computational modeling of a Fluid Catalytic Cracking Unit. *Ain Shams Eng. J.* **2025**, *16*, 103661. [CrossRef]

30. Wang, Y.; Chu, J.; Lu, K. Neural Network Prediction of Gasoline and Diesel Production at Catalytic Cracking Main Fractionator. In Proceedings of the 2019 International Conference on Machine Learning, Big Data and Business Intelligence (MLBDBI), Taiyuan, China, 8–10 November 2019; pp. 233–236. [CrossRef]
31. Bengio, Y.; Simard, P.; Frasconi, P. Learning long-term dependencies with gradient descent is difficult. *IEEE Trans. Neural Netw.* **1994**, *5*, 157–166. [CrossRef]
32. Almuhaini, S.H.; Sultana, N. Bayesian-Optimization-Based Long Short-Term Memory (LSTM) Super Learner Approach for Modeling Long-Term Electricity Consumption. *Sustainability* **2023**, *15*, 13409. [CrossRef]
33. Zhang, X.; Zou, Y.; Li, S.; Xu, S. Product Yields Forecasting for FCCU via Deep Bi-directional LSTM Network. In Proceedings of the 2018 37th Chinese Control Conference (CCC), Wuhan, China, 25–27 July 2018; pp. 8013–8018. [CrossRef]
34. Bannoud, M.A.; da Silva, C.A.M.; Martins, T.D. Applications of metaheuristic optimization algorithms in model predictive control for chemical engineering processes: A systematic review. *Annu. Rev. Control* **2024**, *58*, 100973. [CrossRef]
35. Mirjalili, S.; Lewis, A. The Whale Optimization Algorithm. *Adv. Eng. Softw.* **2016**, *95*, 51–67. [CrossRef]
36. Zhang, P.; Liu, X.; Yi, Z.; He, Q. Improved Multi-Objective Beluga Whale Optimization Algorithm for Truck Scheduling in Open-Pit Mines. *Sustainability* **2024**, *16*, 6939. [CrossRef]
37. Sapnken, F.E.; Khalili Tazehkandgheshlagh, A.; Salomon Diboma, B.; Hamaidi, M.; Gopdjim Noumo, P.; Wang, Y.; Gaston Tamba, J. A whale optimization algorithm-based multivariate exponential smoothing grey-holt model for electricity price forecasting. *Expert Syst. Appl.* **2024**, *255*, 124663. [CrossRef]
38. Pang, S.; Guo, S.; Wang, L.; Li, Y.; Li, X.; Liu, Z. Mass personalization-oriented integrated optimization of production task splitting and scheduling in a multi-stage flexible assembly shop. *Comput. Ind. Eng.* **2021**, *162*, 107736. [CrossRef]
39. Deb, K.; Pratap, A.; Agarwal, S.; Meyarivan, T. A fast and elitist multiobjective genetic algorithm: NSGA-II. *IEEE Trans. Evol. Comput.* **2002**, *6*, 182–197. [CrossRef]
40. Zitzler, E.; Laumanns, M.; Thiele, L. SPEA2: Improving the strength Pareto evolutionary algorithm. *Tik Rep.* **2001**, *103*.
41. Mirjalili, S.; Saremi, S.; Mirjalili, S.M.; dos S. Coelho, L. Multi-objective grey wolf optimizer: A novel algorithm for multi-criterion optimization. *Expert Syst. Appl.* **2016**, *47*, 106–119. [CrossRef]

smart cities

MDPI

Article

Cloud-Enabled Hybrid, Accurate and Robust Short-Term Electric Load Forecasting Framework for Smart Residential Buildings: Evaluation of Aggregate vs. Appliance-Level Forecasting

Kamran Hassanpouri Baesmat *, Emma E. Regentova and Yahia Baghzouz

Department of Electrical and Computer Engineering, University of Nevada, Las Vegas, NV 89154, USA; emma.regentova@unlv.edu (E.E.R.); yahia.baghzouz@unlv.edu (Y.B.)
* Correspondence: kamran@unlv.edu

Highlights:

What are the main findings?

- A cloud-enabled hybrid framework combining Seasonal ARIMAX, Random For-est, and LSTM achieves high accuracy for a short-term residential load forecasting.
- Appliance-level (AL) and aggregate forecasts (AF) are compared, showing that AF outperforms AL in reliability and accuracy.

What are the implications of the main findings?

- Multi-cloud parallel calculations with a two-out-of-three voting scheme enhances forecasting robustness by mitigating single-cloud failures.
- The proposed scalable solution supports smart-city energy management, demand response, and grid reliability.

Abstract

Accurate short-term load forecasting is vital for smart-city energy management, enabling real-time grid stability and sustainable demand response. This study introduces a cloud-enabled hybrid forecasting framework that integrates Seasonal Autoregressive Integrated Moving Average with Exogenous variables (SARIMAX), Random Forest (RF), and Long Short-Term Memory (LSTM) models, unified through a residual-correction mechanism to capture both linear seasonal and nonlinear temporal dynamics. The framework performs fine-grained 5 min forecasting at both appliance and aggregate levels, revealing that the aggregate forecast achieves higher stability and accuracy than the sum of appliance-level predictions. To ensure operational resilience, three independent hybrid models are deployed across distinct cloud platforms with a two-out-of-three voting scheme, that guarantees continuity if a single-cloud interruption occurs. Using a real residential dataset from a house in Summerlin, Las Vegas (2022), the proposed system achieved a Root Mean Squared Logarithmic Error (RMSLE) of 0.0431 for aggregated load prediction representing a 35% improvement over the next-best model (Random Forest) and maintained consistent prediction accuracy during simulated cloud outages. These results demonstrate that the proposed framework provides a scalable, fault-tolerant, and accurate energy forecasting.

Keywords: energy consumption; smart-city energy management; electrical load forecasting; machine learning; statistical analysis; cloud-based computations

1. Introduction

In the modern era, energy management is shaping the world: nearly every sector, whether directly or indirectly, depends on efficient energy utilization, from AI-driven data centers and the protection of the ozone layer to the preservation of ecosystems, advancements in construction, industry, medicine, and tourism.

Energy management in smart cities ensures a balance between production and demand, prevents resource waste, improves air quality, and safeguards the planet.

"Smart" turns appliances into measuring, labeled, and controllable assets, and that radically improves AL data collection and forecasting accuracy/efficiency.

"Smart" appliances offer multiple advantages for accuracy of data, and hence for forecasting. Smart appliances can natively stream instantaneous power, voltage, current, and cumulative energy, not only periodic kWh. Also, they secure machine-readable semantics, not raw points. Devices can expose states (e.g., "wash cycle", "defrost", "setpoint") with consistent naming/units. That makes feature engineering automatic and portable across vendors/homes. Forecasting upgrades to "state-aware", that is one can move from generic time-series to state-space/HSMM or sequence models (RNN/Transformer) that understand cycles and dwell times (e.g., washer, fridge, heat pump). Papers on bottom-up/AL STLF show markedly better fidelity when appliance characteristics and states are included [1].

If devices are truly smart, one can read the appliance directly. This means less dependencies on NILM: Non-Intrusive Load Monitoring (a.k.a. energy disaggregation): Authors of [2] show sampling, transient features, and deep learning, including federated learning push accuracy, however under limits of overlapping loads.

Accurate load forecasting, enables companies to optimally scheduling their activities and determining the appropriate amount of energy to generate in order to maintain stability.

1.1. Load Forecasting Categories and Challenges

Electrical load consumption (ELC) forecasting is typically categorized into: short-term load forecasting (STLF), mid-term load forecasting (MTLF), and long-term load forecasting (LTLF). This study focuses on developing an efficient framework for STLF in residential buildings and addresses several critical research questions (this STLF framework directly supports smart-city energy operations by using smart-meter and weather streams to forecast demand at fine granularity, enabling demand response, EV-charging coordination, and optimal dispatch of solar-plus-storage across neighborhoods):

- Are statistical methods (SM) more effective for load prediction in residential buildings compared to machine learning (ML) techniques?
- Does a hybrid approach that combines SM and ML provide superior results?
- Is it more accurate to predict the overall consumption of a residential building (RB), or to forecast the consumption of individual appliances and then aggregate the results?

Furthermore, this paper examines the impact of feature-selection techniques, methods for combining statistical and machine learning (ML) models, residual (remainder) prediction (RP), and cloud-based computing.

Tables 1 and 2 summarizes challenges for statistical modeling and ML modeling.

Table 1. Challenges of statistical load forecasting.

Parametric Statistical Methods Gap
Handling high variability and peaks
Non-stationarity of load data
Difficulty handling missing data

Table 1. *Cont.*

Parametric Statistical Methods Gap
Inaccuracy at different time horizons
Difficulty incorporating external variables
Inefficiency in real-time forecasting
Computational constraints with large datasets
Autocorrelation and multicollinearity
Risk of overfitting in complex models
Capturing seasonality and periodicity

Table 2. Challenges of ML load forecasting.

Machine Learning Techniques Gap
Generalization challenges across different regions or time periods
Need for large, high-quality datasets
Hyperparameter tuning complexity
Overfitting with complex models
Limited effectiveness in incorporating domain knowledge
Difficulty interpreting model decisions (black-box nature)
Model retraining requirements for real-time data
Difficulty handling rare events or outliers
High computational requirements
Sensitivity to feature selection and engineering

1.2. Feature Selection

Feature selection techniques have long been employed in scientific research, and their effectiveness has been widely studied [3–6]. Each method has its distinct advantages. For instance, the ANOVA F-test assesses whether the mean of a continuous (numeric) feature differs significantly across the categories of the target, by comparing between-class to within-class variance [7], mutual information measures nonlinear dependencies between features and the target variable [8], information gain and other entropy-based scores are commonly applied in classification tasks [9], LASSO suppresses irrelevant features and thus performing implicit feature selection [10], and genetic algorithms offer evolutionary optimization for selecting feature subsets [11]. Research in [12] investigates the role of correlation matrices in prediction and showed that features with high correlation can enhance dataset performance. In [13], the authors evaluated the Chi-Square statistical test as a feature selection method, demonstrating its ability to assess whether features have a meaningful relationship with the target. Furthermore, the authors of [14] highlighted the strength of the Maximum Relevance Minimum Redundancy (MRMR) method in identifying features that are both relevant and non-redundant, thereby mitigating overfitting caused by redundant information.

This research proposes a novel framework that integrates a voting-based system for feature selection. By combining statistical methods with a hybrid approach, the framework leverages correlation analysis, feature independence, and redundancy avoidance to maximize predictive performance in residential load forecasting.

1.3. Statistical Methods (SM)

Statistical methods [15,16], machine learning (ML) techniques [17,18], and hybrid models [19,20] have continually evolved over the past decades, leading to significant advances in prediction accuracy.

For example, research in [21] demonstrated the effectiveness of the autoregressive integrated moving average (ARIMA) model for time-series forecasting, particularly in scenarios driven by temporal dependencies. However, ARIMA struggles to capture seasonality, a limitation that has been addressed by seasonal ARIMA (SARIMA) [22] and SARIMAX (seasonal ARIMA with exogenous regressors) [23]. Despite these improvements, statistical methods generally lack the scalability and flexibility of ML techniques when dealing with large datasets and complex nonlinear relationships.

1.4. Machine Learning (ML)

Machine learning methods, on the other hand, have proven highly powerful and are now widely applied across nearly all domains of science. In [24], the authors highlighted the accuracy of support vector machines (SVMs), which are supervised learning models used in both classification and regression tasks. Tree-based and sequence-based models have also emerged as dominant approaches in forecasting. Decision trees (DTs) [25] produce hierarchical decision rules by recursively splitting on feature values, offering interpretability but often suffering from overfitting. Random Forests (RFs) [26] overcome this drawback by aggregating multiple randomized trees, thereby reducing variance and improving robustness. Extreme Gradient Boosting (XGBoost) [27] further enhances ensemble learning by constructing trees sequentially to correct residual errors, incorporating gradient optimization, regularization, and efficient computation for superior predictive performance. In parallel, Long Short-Term Memory networks (LSTMs) [28] are designed to capture long-range temporal dependencies using gated memory cells, making them especially effective for sequential and time-series forecasting.

Nevertheless, a limitation of many ML-based approaches is their tendency to focus on identifying patterns without explicitly incorporating seasonality or linear temporal dynamics, areas where statistical models such as ARIMA, SARIMA, and SARIMAX excel. This contrast motivates development of hybrid frameworks, which combine the nonlinear modeling power of ML with trend- and seasonality-capturing strengths of statistical approaches. By integrating these complementary capabilities, hybrid systems achieve higher accuracy, reliability, and generalizability.

Authors in [29] present a short-term residential load forecasting framework that integrates evolutionary algorithms for neural architecture search with transfer learning to adapt models across multiple households. The method demonstrates superior performance over benchmark techniques, with an average accuracy gain of 3.17 percentage points compared to the state-of-the-art LSTM one-shot approach. Researchers in [30] introduced a hybrid forecasting framework for urban power load that identifies and quantifies abnormal variation periods via residual decomposition, augments scarce abnormal samples using WGAN-GP, and employs TimesNet to capture complex nonlinear patterns. The method boosts forecasting accuracy for abnormal load periods by about 38.9%. Although it incurs only a minor drop (2.65%) across all periods, it outperforms baseline deep-learning models. The authors of [31] proposed a hybrid deep-learning framework (GNN-ViGNet) enhanced by the use of IoT sensor data and novel preprocessing, feature engineering, and imbalance correction methods, coupled with a Coati–Northern Goshawk Optimization algorithm for real-time route optimization in EV fleets. Their model achieves 98.9% accuracy in charging load forecasting and reduces route distance to 511 km, significantly outperforming baseline techniques such as PSO and Firefly in both convergence speed and prediction error metrics. The authors of [32] introduced radian scaling, a new normalization method that transforms differences of consecutive electricity load values into radians to maintain normalized distributions under seasonal concept drift without retraining machine learning models. Their experiments using lightweight ANN models show radian scaling reduces RMSE dramatically from 158.63 to 43.38 and cuts training epochs by about 67%, statistically outperforming traditional normalization methods. The work in [33] develops a learning framework for forecasting electrical load in renewable energy communities with dynamic portfolios, synthesizing non-stationary, discontinuous, and NON-IID time series, adapting preprocessing and model input (past, present, future features) to achieve domain-invariant feature learning.

Their approach achieves forecasting accuracy comparable to that of centrally trained models, while outperforming single-source models by more than 18% in terms of MAE, and demonstrating strong transferability to unseen data. This research proposes a hybrid forecasting framework that integrates RF, LSTM, and SARIMAX models to address these challenges and enhance short-term load forecasting performance in residential buildings. As hybrid models have demonstrated high performance, this paper also investigates this approach and extends it by incorporating additional techniques to further enhance forecasting accuracy and robustness. Specifically, the study integrates efficient feature-selection methods to improve predictive performance and employs concurrent cloud computing together with a voting-based strategy to increase the model's resilience and scalability.

1.5. The Contributions of the Presented Work Can Be Summarized as Follows

- Integration across paradigms: A unified forecasting architecture that synergistically combines Seasonal ARIMAX, Random Forest, and LSTM models through a residual-correction layer, effectively merging statistical trend modeling with nonlinear and sequential learning.
- Feature–model customization: Each sub-model employs an optimized feature selection strategy (Correlation Matrix, Chi-Square, or MRMR), enabling the framework to capture diverse predictive relationships with minimal redundancy.
- Multi-cloud redundancy and voting: A two-out-of-three cloud voting ensures continuous operation under individual cloud or model failures, an aspect rarely addressed in prior forecasting frameworks.
- Fine-grained temporal resolution: The framework is validated on 5 min interval residential load data, demonstrating that aggregate forecasts outperform the summation of appliance-level predictions (sum of predictions per appliances) in both stability and accuracy.
- Demonstrated accuracy and robustness: The system achieves a 35% reduction in RMSLE relative to the best single-model benchmark and maintains reliable performance under simulated cloud outages.

A detailed comparison between aggregate prediction and appliance-level forecasting in terms of accuracy and reliability. Altogether builds a scalable, fault-tolerant, and statistically grounded hybrid methodology for short-term load forecasting in smart residential environments.

The rest of the paper is organized as follows: Section 2: Data description and preparation, Section 3: Feature selection, Section 4: Methodology, Section 5: Discussion and comparison of the results, Section 6: Conclusions.

2. Methodology

This research proposes a novel method that integrates both statistical approaches and machine learning techniques, termed SSRLR (Sparse voting feature-selection technique, Seasonal Autoregressive Integrated Moving Average with Exogenous variables model, Random Forest, Long Short-Term Memory, and Remainder Prediction). This integration significantly enhances the overall predictive accuracy of the proposed framework. The machine learning component is responsible for capturing complex, non-linear patterns and intricate relationships within the data, while the statistical component simultaneously models seasonality effects and time-dependent correlations. By operating in parallel, these two components complement and reinforce each other: the statistical models address temporal regularities that machine learning alone might overlook, and the machine learning algorithms uncover high-dimensional interactions that are beyond the capacity of purely statistical methods. This mutual coverage enables the framework to extract the maximum amount of useful information from the data while mitigating the risk of overfitting. Furthermore, the synergy between the two paradigms ensures that weaknesses in one component are compensated for by the strengths of the other, thereby improving robustness and generalization performance across diverse scenarios.

Figure 1 illustrates the flow of the embedded procedures. After collecting data on electricity consumption and relevant weather variables, preprocessing is performed by applying the Holiday/Weekend factor and modeling the seasonal effects. Then, the process proceeds to the feature selection phase using three feature selection techniques: Chi-Square test, MRMR, and CM. These methods are applied to identify and retain the most informative features. By systematically selecting only the most relevant variables, this step extracts an optimal set of features, thereby improving generalization and reducing the risk of overfitting. Consequently, the feature selection phase plays a pivotal role in enhancing the robustness and accuracy of the forecasting framework.

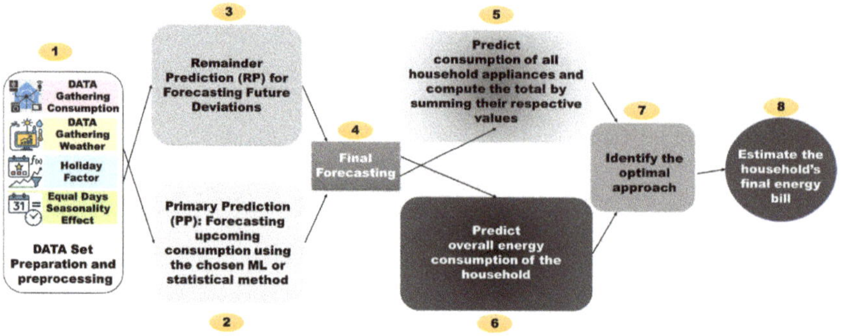

Figure 1. Architecture and procedures of proposed method.

Next, the process of primary prediction (PP) employs both machine learning and statistical methods, specifically, the Long Short-Term Memory (LSTM), the Seasonal Autoregressive Integrated Moving Average with Exogenous variables (SARIMAX) model, and the Random Forest (RF) algorithm. The SARIMAX model is utilized to capture and exploit the most informative seasonal effects and other time-dependent features, thereby maximizing the extraction of temporal patterns from the data. The RF method enables the exploration of diverse forecasting scenarios by leveraging its ensemble-based decision-making capability, which enhances robustness against overfitting. In parallel, the LSTM component contributes to the modeling of both short-term and long-term temporal dependencies, thus complementing the strengths of SARIMAX and RF. This hybrid integration substantially enhances the learning potential of the proposed method, enabling it to adapt effectively to complex and dynamic patterns present in the dataset.

A primary trend is first established to capture the general direction of the target variable. This trend enables the derivation of an additional and highly informative feature the Remainder Prediction (RP) in which deviations from the trend are referred to as remainders. Once the primary trend is obtained, the deviations of the actual values from the underlying forecasted trajectory are calculated. These deviations are then treated as a separate time series, which can be independently modeled to predict the future remainders. The final forecast, i.e., overall consumption, all individual appliance-level forecasting, and also their total consumption. that is subsequently obtained by summing the predicted primary consumption values and the predicted remainders. This decomposition-based strategy reduces forecasting error because it allows the model to focus on different components of data separately: the trend model specializes on capturing long-term patterns, while the remainder model focuses on short-term fluctuations and irregularities. By isolating these components, the approach mitigates the risk of error propagation that can occur when a single model attempts to fit both large-scale and fine-grained variations simultaneously.

This research further enhances the reliability, processing speed, and real-time computing capability of the proposed framework through the use of cloud computing. In this setup, three independent cloud instances operate in parallel. The final forecast is obtained through an averaging mechanism, while a majority-voting mechanism is employed to ensure that a correct prediction is selected. Specifically, if at least two of the three instances exhibit the same level of error (the difference is below a threshold), their output is accepted as correct. If the third instance shows a significantly larger error, it is excluded from the forecasting process and flagged for expert review and retuning. This redundancy significantly improves the accuracy and robustness, allowing the system to operate effectively even in the event of a single instance failure, and it minimizes the risk of erroneous outputs caused by transient faults or network disruptions. The adoption of this dual-stage forecasting methodology trend plus Remainder Prediction within a fault-tolerant cloud infrastructure offers several benefits. From a modeling perspective, separating long-term and short-term patterns allows for more specialized learning and reduces cumulative forecasting errors. From a system architecture perspective, the two-out-of-three cloud framework ensures high availability, operational continuity, and consistent performance under varying computational loads. Leveraging cloud computing brings additional advantages. The elastic scalability of cloud resources allows the system to handle sudden increases in data volume without sacrificing processing speed. Geographic distribution of cloud servers reduces latency, enabling near real-time decision-making. Furthermore, cloud platforms facilitate seamless integration of advanced analytics tools, automated backups, and security protocols, ensuring that forecasting models remain both robust and secure while being accessible from anywhere. This combination of advanced forecasting methodology and resilient cloud-based deployment provides a high-performance, dependable solution for real-world operational environments.

2.1. *Long Short-Term Memory (LSTM) Networks*

Long Short-Term Memory (LSTM) [34] networks, sketched in Figure 2, are a form of Recurrent Neural Network (RNN) specifically built to capture dependencies that span long intervals in sequential data. Unlike standard RNNs, which struggle to learn from distant inputs due to vanishing gradients during backpropagation, LSTMs introduce a dedicated memory cell that can preserve and carry information for many time steps. This design makes them especially powerful for applications like forecasting time series, understanding speech, and processing text.

At each time step, the memory cell's content is carefully managed by three gates forget, input, and output. The forget gate decides which past information should be removed, the input gate controls what new information enters the cell, and the output gate determines what part of the memory contributes to the next hidden state. By adjusting these gates dynamically, an LSTM can flexibly balance learning short-term fluctuations and long-term trends within the data.

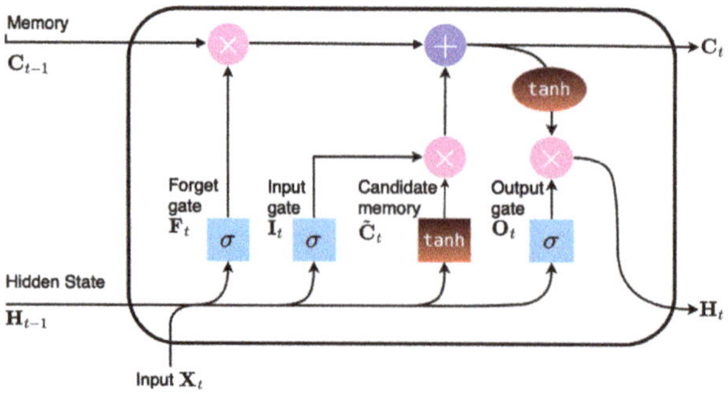

Figure 2. Standard Long Short-Term Memory layer, where ⊗ indicates element-wise multiplication and ⊕ indicates element-wise addition; σ denotes the sigmoid activation function, and tanh generates the candidate cell state. The forget gate decides which information from previous time steps should be removed, the input gate incorporates new, relevant data into the cell state, and the output gate regulates which information is propagated to the next time step.

LSTMs have demonstrated strong performance in numerous areas. In natural language processing, they learn the flow of sentences to generate or translate text. In speech recognition, they interpret audio sequences to convert spoken words into text. They are also effective in spotting unusual patterns in time-series data for anomaly detection. Despite their benefits, LSTMs require substantial computational resources and their inherently sequential updates limit the ability to parallelize training compared to architectures like Convolutional Neural Networks (CNNs). Nevertheless, their capacity to handle both isolated observations and entire sequences makes them a cornerstone model in deep learning research involving temporal or ordered inputs.

2.2. Random Forest (RF)

The Random Forest [35] algorithm (see Figure 3) is an ensemble-based technique that leverages the simplicity of individual decision trees while dramatically improving their reliability. Instead of relying on a single tree, it grows dozens or even hundreds of them, each trained on a different "bootstrap" sample of the data (i.e., random draws with replacement). At each split in a tree, only a randomly chosen subset of the available features is considered, which encourages diversity among trees and helps prevent overfitting. When making a prediction, the forest either tallies a vote among all trees (for classification) or computes the mean of their outputs (for regression), yielding a consensus result that is typically more accurate and stable than any single tree.

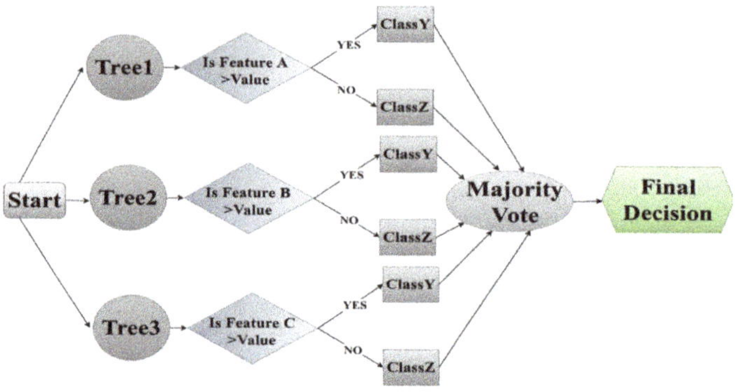

Figure 3. Random Forest: general architecture.

One of Random Forest's greatest assets is its ability to estimate the relative importance of each feature: by measuring how much the model's error increases when a given variable's values are permuted, insight is gained into which inputs truly drive predictions. This makes RF a valuable tool both for high-stakes applications such as detecting fraudulent transactions, forecasting electrical loads, or diagnosing diseases and for exploratory analysis, where understanding the data is as important as prediction itself. On the downside, training and storing a large collection of trees can demand considerable computation and memory, and making real-time predictions may be slower if the forest is very deep. Moreover, while each decision tree is easy to follow on its own, the combined forest becomes something of a "black box", making it harder to explain individual decisions than with simpler, single-tree models. Despite these trade-offs, Random Forest remains a go to choose whenever robustness to noise, missing values, and outliers is required together with strong predictive power.

2.3. Seasonal Autoregressive Integrated Moving Average with Exogenous Variables (SARIMAX)

The SARIMAX [36] model extends the familiar ARIMA approach by explicitly handling seasonal fluctuations and allowing for outside influences. It is ideal when the data repeat patterns at regular intervals such as monthly sales cycles or annual weather variations while also being affected by additional variables like temperature or economic indicators. In essence, SARIMAX merges three pieces:

1. Non-seasonal ARIMA component:

Autoregressive (AR)p)):

This component models the current value of the series by referencing its own previous observations. The order parameter p designates how many past points are used to predict the present.

Integrated (I)d)):

To stabilize the series, the data are differenced d times. This process removes trends and helps ensure that statistical characteristics such as the mean and variance remain roughly constant throughout the sequence.

Moving Average (MA)q)):

Here, the forecast is formed by combining current and past forecast errors. The parameter q indicates how many of these lagged error terms are included in the smoothing equation.

Mathematically, this is written as:

$$\phi(\mathrm{B})\ (1-\mathrm{B})^{d} y_t = \theta(\mathrm{B})\ \epsilon_t \tag{1}$$

where

- $\phi(\mathrm{B})$: is the non-seasonal AR polynomial
- $\theta(\mathrm{B})$: is the non-seasonal MA polynomial
- B: is the backshift operator
- y_t: is the series at time t, and
- ϵ_t: is a white-noise error term

2. Seasonal Component (s)

To capture repeating patterns every "s" period, SARIMAX uses seasonal AR and MA polynomials of orders P and Q, together with seasonal differencing D. This portion is given by:

$$\Phi(\mathrm{B}^s)(1-\mathrm{B}^s)^{D}\ y_t = \Theta(\mathrm{B}^s)\ \epsilon_t \tag{2}$$

where

- $\Phi(B^s)$: Seasonal autoregressive polynomial of order P.
- $\Theta(B^s)$: Seasonal moving average polynomial of order Q.
- S: Seasonality period
- D: Number of seasonal differences.

3. Exogenous regressors (X)

$$y_t = \beta X_t + \epsilon_t \tag{3}$$

where

- X_t: External (exogenous) regressors at time t.
- β: Coefficient vector for the exogenous variables.

Combining all components, the SARIMAX model can be represented as:

$$\phi(B)\ (1-B)^{\ d}\ \Phi(B^s)(1-B^s)^D\ y_{t=\theta}(B)\ \Theta(B^s)\epsilon_t +\ \beta X_t \tag{4}$$

By combining these three elements non-seasonal dynamics, periodic behavior, and outside influences SARIMAX delivers a flexible framework for forecasting. It is widely used in examples like projecting retail sales with marketing spend as an exogenous input, estimating electricity demand driven by temperature and humidity, or any scenario where patterns repeat and external drivers matter.

2.4. SSRLR

The pseudocode of the proposed SSRLR is illustrated below (Algorithm 1):

SSRLR, a novel hybrid method that integrates statistical and machine learning techniques including a Sparse voting feature selection technique, Seasonal Autoregressive Integrated Moving Average Exogenous model, Random Forest, Long Short-Term Memory, and a Remainder Prediction step to leverage both linear trends and complex nonlinear relationships.

Algorithm 1. Pseudocode of proposed algorithm

```
Input: Historical load data, Weather features, Forecast horizon = 12
Step 1: Data Preprocessing
    Normalize data using MinMaxScaler to range [0,1]
Step 2: Feature Selection
    Apply CM
    Select relevant features → CM_Output
    Use this for LSTM
    Apply Chi-Square
    Select relevant features → Chi-Square_Output
    Use this for SARIMAX
    Apply MRMR
    Select relevant features → MRMR_Output
    Use this for RF
Step 3: Model Training and Prediction
    Train SARIMAX model with seasonal and non-seasonal components
    Predict next 12 points → SARIMAX_Output
    Train Random Forest model using weather and historical data
    Predict next 12 points → RF_Output
```

Algorithm 1. *Cont.*

Train LSTM model for time-series forecasting
Predict next 12 points → LSTM_Output
Step 4: Remainder Prediction (RP)
Perform additional adjustments if needed:
Use SARIMAX for LSTM on cloud 1
Use LSTM for RF on cloud 2
Use RF for SARIMAX on cloud 3
Step 5: Ensemble Averaging
SARIMAX Final Output (PP + RP) → cloud 1_Output
RF Final Output (PP + RP) → cloud 2_Output
LSTM Final Output (PP + RP) → cloud 3_Output
Step 6: Ensemble Averaging
Ensemble_Output = (cloud 2_Output + cloud 2_Output + cloud 3_Output)/3
Step 7: Output Final Prediction
Return Ensemble_Output

The dataset is divided into 80% for training and 20% for testing, after which the following procedures are applied:

- The process begins with feature selection using the CM method, followed by primary forecasting with LSTM. A trend is then identified for the remainders, and SARIMAX is applied to forecast the subsequent remainders. The final output for this stage is obtained by summing the LSTM predictions with the forecasted remainders.
- Feature selection is performed using the MRMR method, followed by primary forecasting with Random Forest (RF). A trend is then identified for the remainders, and LSTM is applied to forecast the subsequent remainders. The final output for this stage is obtained by summing the RF predictions with the LSTM-forecasted remainders.
- Feature selection is performed using the Chi-Square method, followed by primary forecasting with SARIMAX. A trend is then identified for the remainders, and Random Forest (RF) is applied to forecast the subsequent remainders. The final output for this stage is obtained by summing the SARIMAX predictions with the RF-forecasted remainders.

Figure 4 presents the detailed schematic of the proposed method.

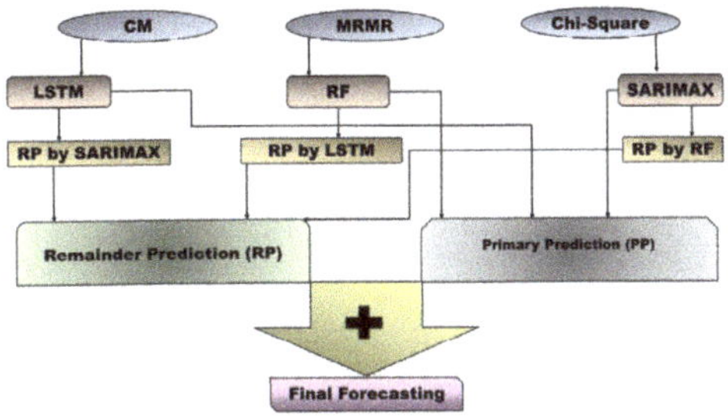

Figure 4. Method in detail.

2.5. *Two-Out-of-Three Multi-Cloud Voting Architecture*

Single cloud has been used in load forecasting in previous research [37]. To enhance the robustness, accuracy, and real-time performance of the forecasting system, the three pipelines described in Section 2.4 are deployed across three independent cloud instances. As illustrated in Figure 5, each cloud instance executes one of the proposed hybrids forecasting strategies:

- Cloud 1: Implements the CM → LSTM → SARIMAX pipeline.
- Cloud 2: Implements the MRMR → RF → LSTM pipeline.
- Cloud 3: Implements the Chi-Square → SARIMAX → RF pipeline.

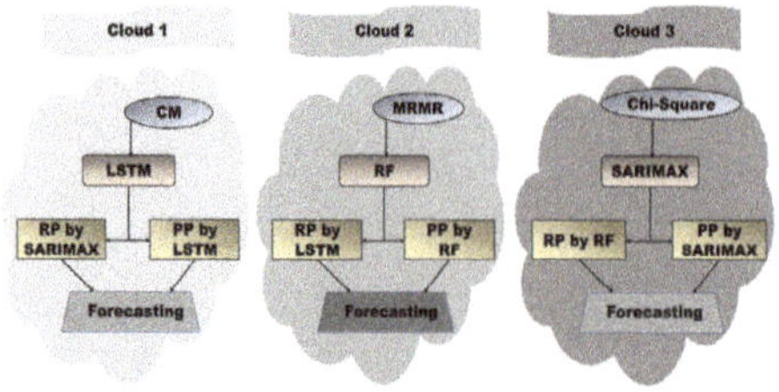

Figure 5. Proposed two-out of-three cloud-based system.

Each cloud processes the same input dataset in parallel, but with different feature-selection methods, primary forecasting models, and remainder forecasting techniques. This methodological diversity significantly reduces the risk of correlated model errors and increases the resilience of the overall system.

Preliminary Averaging: The forecasts from all three clouds are averaged to produce an initial consensus prediction.

Error-Based Majority Voting: The errors from each cloud are evaluated against the actual values. If at least two clouds yield similar and low error magnitudes, their results are accepted as correct. The third cloud, if it shows significantly higher error, is excluded from the current forecasting decision and flagged for expert review and model retuning.

This redundancy ensures that the system maintains robustness remaining operational and accurate even in the presence of network disruptions, computational errors, or degraded model performance in one instance. By leveraging agreement between at least two independent pipelines, the architecture minimizes the likelihood of passing erroneous forecasts to the end user or decision-making process. Figure 5 shows this cloud-based architecture.

The deployment of the two-out of-three architecture in a cloud environment provides several operational and strategic benefits:

- High Availability and Robustness: The redundancy of multiple clouds ensures continuous operation even if one instance experiences downtime or degraded performance.
- Elastic Scalability: Cloud resources can be dynamically allocated to handle varying computational demands, enabling real-time or near real-time processing for large-scale datasets.
- Parallel Computation: Hosting clouds in different regions improving forecast delivery speed and responsiveness.
- Security and Data Integrity: Cloud platforms provide encryption, controlled access, and automated backup systems, ensuring the safety and reliability of both input data and forecasting models.
- Cost Efficiency: Pay-as-you-go billing models allow for optimal use of resources without the overhead of maintaining extensive on-premises infrastructure.
- Seamless Integration and Maintenance: Cloud ecosystems support easy deployment of updates, integration of analytics tools, and automated monitoring of model performance with cheaper price compare to on- premise systems.
- Collaboration and Accessibility: Authorized users can securely access the forecasting system from anywhere, enabling collaborative work across geographically distributed teams.

In the suggested two-out-of-three cloud-based redundancy model, reliability is expressed in terms of R, the probability of each cloud (e.g., AWS) functioning correctly:

- R^3 denotes the case where all three systems perform without error.
- $3R^2\,(1 - R)$ corresponds to the case where two systems deliver correct outputs and one produces an error, which is automatically overruled by the majority decision.

The system failure occurs only if two or more clouds malfunction at the same time, a situation not covered by this formula.

By combining diverse forecasting models with a fault-tolerant two-out-of-three multi-cloud deployment, the proposed system achieves a high degree of accuracy, resilience, and computational efficiency. This design ensures that forecasts remain consistent, secure, and responsive under real-world operational conditions, even in the presence of hardware failures, software errors, or sudden spikes in data volume.

3. Data Description and Preparation

The dataset consists of power usage readings for a household in Summerlin, Las Vegas, Nevada, USA, sampled at five-minute intervals for 365 days of 2022. This dataset was officially collected from a residential customer of a local electric utility as part of its demand response program. Each measurement is given in kilowatts (kW) and reflects the consumption of the following household systems and appliances:

- Forced-air heating unit on the lower level
- Air conditioner on the lower level
- Refrigerator
- Air conditioner on the upper level of the house
- General (miscellaneous) household load
- Laundry machines (washer and dryer)
- Dishwasher and garbage disposal
- Microwave oven
- Forced-air heating unit on the upper level of the house
- Kitchen circuit (GFI) including the toaster

The dataset offers a full coverage with no gaps or missing entries, which makes it a solid foundation for any analysis. Electricity usage was logged every five minutes throughout 2022, giving a detailed picture that picks up even the smallest variations in consumption. Before feeding the data into the forecasting models, preprocessing is performed and relevant features are carefully selected. This step both smooths out irregularities and reduces the risk of overfitting. Figure 6 illustrates the data profile.

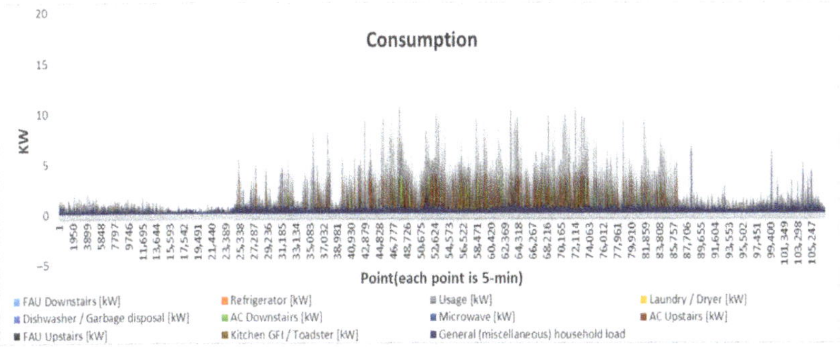

Figure 6. Household one-year power consumption profile—Summerlin, Las Vegas.

Estimating the Holiday/Weekend Adjustment Factor and Modeling the Seasonal Effect

Before feeding the data into the forecasting model, the dataset is preprocessed to adjust for unusual consumption patterns that occur during holidays and weekends, when usage often departs from typical weekday behavior. Social traditions, leisure activities, and economic regulations all play a role: some people travel or gather with friends, while others stay home using appliances or hosting events. Many people in Las Vegas work during the weekends, holidays, or change shifts frequently. In addition, manufacturing schedules may shift factories in some areas shut down for holidays, whereas others continue operating further altering the load profile. To quantify this effect, a Holiday Adjustment Factor is computed by first tagging each timestamp as either a "holiday/weekend" or a "weekday." The data are then divided into two groups one representing holidays and weekends and the other representing regular weekdays and the mean energy consumption is calculated for each group. The adjustment factor is obtained by dividing the holiday/weekend average (1.89 kW) by the weekday average (1.94 kW), yielding a value of approximately 0.98. This value shows that, on average, consumption during holidays and weekends is about 98% of typical weekday use, and incorporating this factor into our model helps dampen the extra variability introduced by these irregular days, resulting in more accurate forecasts. (Holiday Coefficient = (Average Consumption on Holidays-Weekends)/(Average Consumption on Regular Days)).

Holidays included in 2022: 1 January (New Year's Day, Saturday), 17 January (Martin Luther King Jr. Day, Monday), 21 February (Presidents Day, Monday), 30 May (Memorial Day, Monday), 4 July (Independence Day, Monday), 5 September (Labor Day, Monday), 10 October (Columbus Day, Monday), 11 November (Veterans Day, Friday), 24 November (Thanksgiving Day, Thursday), 25 December (Christmas Day, Sunday).

Next, each calendar month is standardized to a fixed length of 31 days. When a month contains fewer than 31 days, the readings from its final day are carried forward to fill the gap. This step creates a uniform time grid, enabling the model to learn genuine seasonal patterns without distortion from varying month lengths.

Seasonal effects are regular, repeating variations in energy consumption that occur over time hourly, daily, weekly, or monthly. By assigning each month a uniform 31-day span, artificial discontinuities in the data are eliminated, allowing the algorithm to detect genuine periodic patterns.

- Applying this procedure to our 2022 dataset produces:
- A total of 372 "normalized" days (12 months × 31 days)
- 372 days × 24 h = 8928 h
- At 5 min resolution: 8928 h × 12 readings per hour = 107,136 data points

To ensure temporal alignment across months, we standardized all monthly sequences to 31 days. This choice is consistent with prior studies in residential and urban load forecasting where month-length normalization is required. Importantly, the 31-day format preserves the complete set of real observations: months with fewer than 31 days are expanded using a small number of interpolated entries, while months with 31 days remain unchanged. In contrast, a 30-day standardization would require deleting one full day of real measurements from every 31-day month, which removes genuine load patterns and weakens the statistical richness of the dataset. Additionally, the 31-day format requires fewer temporal adjustments (7 days per year) compared to the 30-day approach (9 days per year), reducing distortion and maintaining closer fidelity to the original temporal structure. For these reasons, the 31-day standardization provides a more reliable and less intrusive alignment strategy for short-term load forecasting.

This expanded and smoothened dataset preserves the natural ebb and flow of consumption across hours, days, and months, enabling the forecasting model to capture seasonal trends with finer precision.

4. Applying Feature Selection

Selecting appropriate predictors is essential for improving model accuracy and avoiding redundant information. Common approaches include constructing a Correlation Matrix, applying Minimum Redundancy Maximum Relevance (MRMR), performing Analysis of Variance (ANOVA), and employing the Chi-Square test. For this analysis, seventeen atmospheric indicators were obtained from the Visual Crossing service, as illustrated in Figure 7. Several reputable weather data providers including the National Weather Service, the Weather Channel, AccuWeather, Weather Underground, and Visual Crossing were examined and compared. Based on a comprehensive assessment of data completeness, spatial coverage, and reliability, Visual Crossing was selected as the primary source of meteorological information, offering the most consistent dataset with 5 min resolution for the case study location.

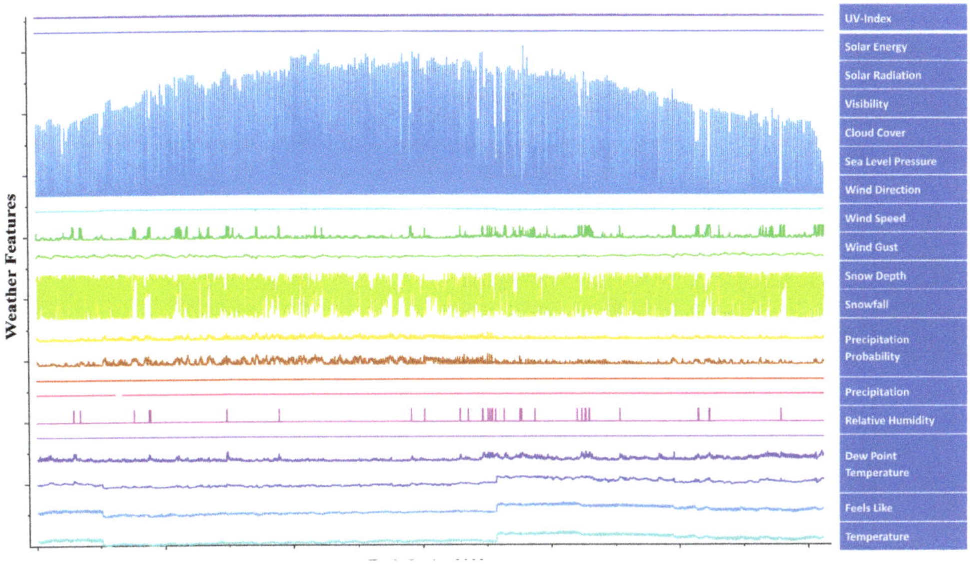

Figure 7. Weather data for location of case study household.

These variables are: Temperature, the ambient air temperature reported in °C or °F; Apparent Temperature (Feels-Like), a composite measure that accounts for temperature, humidity, and wind chill; Dew Point, the temperature at which air becomes saturated and which indicates moisture content; Relative Humidity, the percentage of water vapor present relative to the maximum possible at that temperature; Precipitation, any form of water liquid or frozen falling from the sky; Precipitation Probability, the likelihood (as a percentage) of precipitation occurring; Snowfall Amount, the depth of snow accumulation; Snow Depth, the total height of snow on the ground; Wind Gust, short bursts of high wind speed exceeding the average; Wind Speed, the sustained wind velocity typically measured in km/h or mph; Wind Direction, the compass bearing (in degrees from true north) from which the wind originates; Sea Level Pressure, atmospheric pressure corrected to sea level; Cloud Cover, the fraction of the sky obscured by clouds expressed as a percentage; Visibility, the maximum distance at which objects or lights can be clearly seen;

Solar Radiation, the instantaneous power of sunlight reaching the surface in W/m^2; Solar Energy, the cumulative solar irradiance over a specified period in kWh/m^2; and UV Index, a standardized scale indicating the strength of ultraviolet radiation.

Because our dataset is both large and subject to irregular fluctuations, it is difficult to discern the interdependencies among variables by simple inspection. Traditional statistical analyses often fall short when faced with the non-linear and multi-factor relationships present in meteorological data. To address this, machine learning algorithms were employed to detect subtle patterns, uncover complex associations, and highlight trends that would otherwise remain hidden.

4.1. Chi-Square Test

The Chi-Square Test [38] is a nonparametric technique used to determine whether two categorical variables are related. In load-forecasting applications, continuous weather measurements are first converted into discrete bins (for example, temperature grouped into "low", "medium", and "high"), and energy usage is similarly categorized into consumption ranges. The test then computes a Chi-Square statistic by comparing the observed frequencies in each combination of weather-bin and load-bin with the expected frequencies under the assumption of independence. A larger Chi-Square value indicates a stronger association between a given weather feature and energy consumption.

Procedure:

1. Split weather variables (temperature, UV index, etc.) into a set of categories.
2. Bin the target: Organize historical load values into energy consumption intervals.
3. Compute frequencies: Tally how often each weather-category/load-category pair occurs.
4. Calculate statistic: For each feature, sum over all bins using the term:

$$\frac{Observed - Expected}{Expected} \tag{5}$$

5. Rank features: Order variables by their Chi-Square scores higher scores indicate greater predictive value.

When applied to our dataset, the highest scores belong to temperature, feels-like temperature, UV index, sea level pressure, solar energy, and solar radiation (see Table 3). These results imply that these six weather factors most strongly influence when and how much energy is consumed for heating, cooling, and solar-powered operations. Dew point and humidity produced moderate Chi-Square values, whereas visibility and cloud cover scored relatively low, showing only weak associations with load. All examined variables yielded p-values below the usual significance threshold, confirming that none of these relationships arose by chance.

Table 3. Results of Chi-Square feature selection.

Feature	Temperature	Feels-Like	UV Index	Sea Level Pressure	Solar Energy	Solar Radiation	Visibility
Score	7162.332	5907.64	4932.46	4775.704	4176.774	4151.809	55.890
Feature	Temperature	Feels-Like	UV Index	Sea Level Pressure	Solar Energy	Solar Radiation	
Score	3776.295	2270.17	867.17	676.616	600.80606	284.650	

By selecting top-scoring features such as temperature and solar related variables leaner and more accurate forecasting models can be constructed, while safely omitting the least influential factors.

4.2. Minimum Redundancy Maximum Relevance (MRMR)

The MRMR [39] approach seeks a set of inputs that are both highly informative about the target (energy consumption) and as mutually non-redundant as possible. It does so by ranking features based on mutual information scores, which quantify how much knowing one variable reduces uncertainty about another.

Implementation steps:

1. Standardize each numeric feature:

$$StdV = \frac{V - \mu(V)}{\sigma(V)} \tag{6}$$

where V is the original value, μ(V) its mean, and σ(V) its standard deviation. This puts all features on the same scale, preventing any single unit-scale variable from dominating the MI calculation.

2. Compute mutual information between each feature and the energy-consumption target.
3. Rank features by high relevance (large MI) and low redundancy (little overlap with already-selected features).

Table 4 shows the ranked results. In summary:

- Temperature and feels-like temperature emerge as the strongest predictors.
- Dew point, sea level pressure, and humidity have moderate importance.
- Solar energy and visibility register relatively low relevance scores.

Table 4. Results of MRMR feature selection.

Feature	Temperature	Feels-Like	UV Index	Sea Level Pressure	Solar Energy	Solar Radiation	Visibility
Score	0.386	0.381	0.052	0.203	0.073	0.153	0.040
Feature	Temperature	Feels-Like	UV Index	Sea Level Pressure	Solar Energy	Solar Radiation	
Score	0.242	0.159	0.139	0.106	0.114	0.085	

Because building heating and cooling demand is sensitive to temperature-related inputs, those variables drive most of the predictive power. Less-relevant features can be dropped to simplify the model without sacrificing accuracy.

4.3. Correlation Matrix

A correlation matrix (CM) [40] in Figure 8 shows pairwise linear relationships between numeric features and the energy-consumption. Each entry is the Pearson correlation coefficient, ranging from −1 (perfect inverse) through 0 (no linear relationship) to +1 (perfect direct).

The analysis yields the following results:

- Sea level pressure has the strongest negative correlation (−0.34), implying that higher pressures generally coincide with lower energy use (likely due to milder weather).
- Temperature, feels-like temperature, and dew point have strong positive correlations (approximately +0.32, +0.31, and +0.22), reflecting increased cooling demand under warmer or more humid conditions.
- Solar variables (UV index, solar radiation, solar energy) exhibit modest positive correlations (around +0.17).
- Cloud cover and visibility feature near-zero correlations, indicating negligible direct effects on consumption.

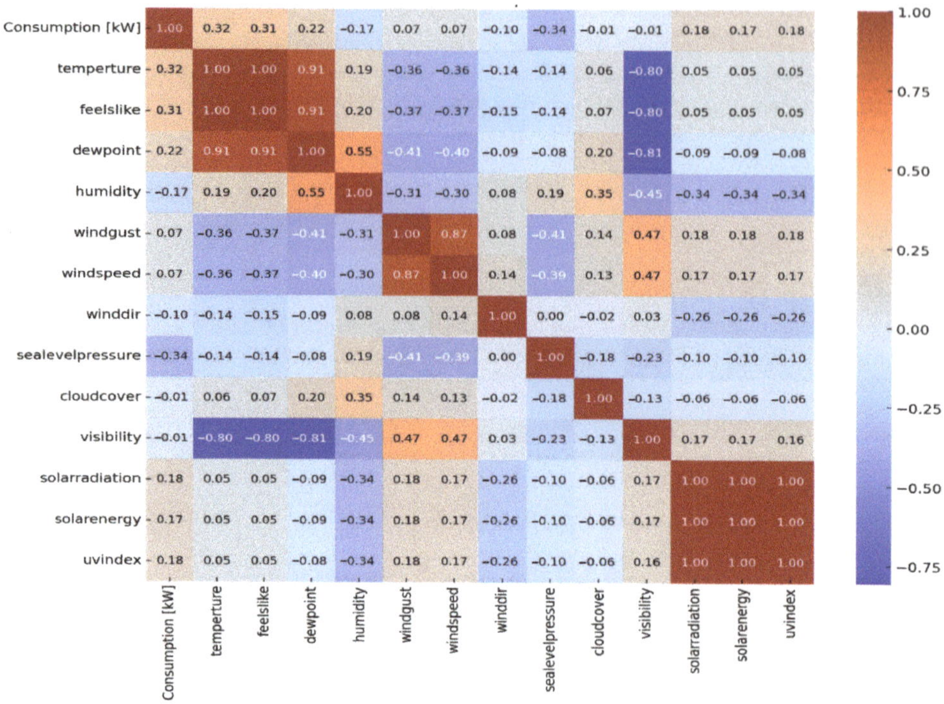

Figure 8. Features selected based on CM.

This correlation analysis reinforces that temperature and related humidity measures are dominant drivers of energy demand, while factors such as cloud cover and visibility play only minor roles.

The top-ranked features for each selection method are: CM: sea level pressure, temperature, feels-like temperature, and dew point. MRMR: temperature, feels-like temperature, dew point, and sea level pressure. Chi-Square: temperature, feels-like temperature, UV index, and sea level pressure.

Each method serves a different role in the proposed approach and is paired with a different machine learning technique or statistical method.

The proposed framework is not dependent on any specific set of weather variables. All meteorological features used in this study (temperature, relative humidity, sea-level pressure, dew point, UV index, etc.) are publicly available through standard weather services and do not require specialized sensors. The pipeline can operate with smaller or alternative feature sets, depending on what is available in a given region. The feature selection component is included to demonstrate how the model behaves when richer predictor sets exist, but the framework remains fully functional even when only the commonly available variables (e.g., temperature, pressure, humidity, wind, precipitation) are present.

5. Discussion and Comparison of the Results: Overall Consumption Prediction vs. Aggregated Appliance Forecasts

This paper introduced a new cloud-enabled hybrid machine learning–statistical framework for short-term load forecasting in smart buildings. The proposed approach integrates the strengths of both machine learning algorithms and statistical models to maximize forecasting accuracy while simultaneously leveraging the reliability, high processing speed, and cost efficiency of cloud computing. By employing a parallel multi-cloud architecture with a two-out-of-three voting mechanism, the framework achieves even greater speed and reliability compared to conventional single-cloud deployments. This architecture also provides a scalable and fault-tolerant foundation that enables researchers to extend the methodology beyond load forecasting into other critical areas of the power grid, such as fault detection, protection, and predictive maintenance monitoring.

In this study, the next hour of electrical consumption was predicted, corresponding to 12 time steps, with each step representing a 5 min interval. The performance of the proposed method is compared against several well-known forecasting techniques including: Linear Regression (LR), Random Forest (RF), Decision Tree (DT), Support Vector Machine (SVM), SARIMAX, Extreme Gradient Boosting (XGBoost), Gated Recurrent Unit (GRU), Long Short-Term Memory (LSTM), and Deep Belief Network (DBN). These comparative results are presented in Tables A1–A11 (in the Appendix A), highlighting the improvements achieved through the dual-stage trend–remainder modeling process in combination with the robust cloud-based deployment.

Table A1 and Figure 9: Overall consumption, Table A2 and Figure 10: FAU (Forced Air Unit) Downstairs consumption, Table A3 and Figure 11: AC (Air Conditioning) Downstairs consumption, Table A4 and Figure 12: Refrigerator consumption, Table A5 and Figure 13: AC Upstairs consumption, Table A6 and Figure 14: Laundry/Dryer consumption, Table A7 and Figure 15: Dishwasher/Garbage Disposal consumption, Table A8 and Figure 16: Microwave consumption, Table A9 and Figure 17: FAU Upstairs consumption, Table A10 and Figure 18: Kitchen GFI (Ground Fault Circuit Interrupter)/Toaster consumption, Table A11 and Figure 19: General (miscellaneous) load consumption and Table A12 and Figure 20: Sum of results of Tables A2–A11.

It is important to note that occasional negative consumption values may appear in the dataset. These values do not indicate that appliances such as air conditioners are generating energy; rather, they are a result of the building's integration with rooftop solar panels. During periods of high solar production particularly when household energy demand is low excess electricity is exported back to the grid.

Five well-established evaluation metrics (Table 5) were employed to compare the results. Since all models were evaluated on the same single-building dataset, normalization measures (e.g., cvRMSE, NMBE) were not applied, as they become relevant only in multi-building comparisons, in accordance with ASHRAE Guideline 14 and IPMVP.

Although the model is trained and evaluated at a 5 min resolution, the output for each horizon is a 12-point vector covering the next hour. For deployment, building EMS/DR (EMS = Energy Management System, DR = Demand Response) controllers can map these to their operating cadence by forming block averages (10/15/30/60 min) or energy-integrated setpoints. This hierarchical use of the same forecasts attenuates short-term noise and matches actuator latencies, while retaining sub-hour detail for rapid interventions (e.g., staggered EV charging). Integrating the proposed forecasting framework with demand response (DR) controllers or optimization models could therefore enhance load flexibility, grid stability, and energy-cost savings an aspect identified as a promising direction for future work.

Aggregate vs. Sum of Appliances:

On the aggregate meter, SSRLR is the best model on every metric (RMSLE = 0.04311, RMSE = 0.07807, MAE = 0.05850). The next best RMSLE is RF = 0.06610, so SSRLR improves AF, RMSLE by ~34.8%.

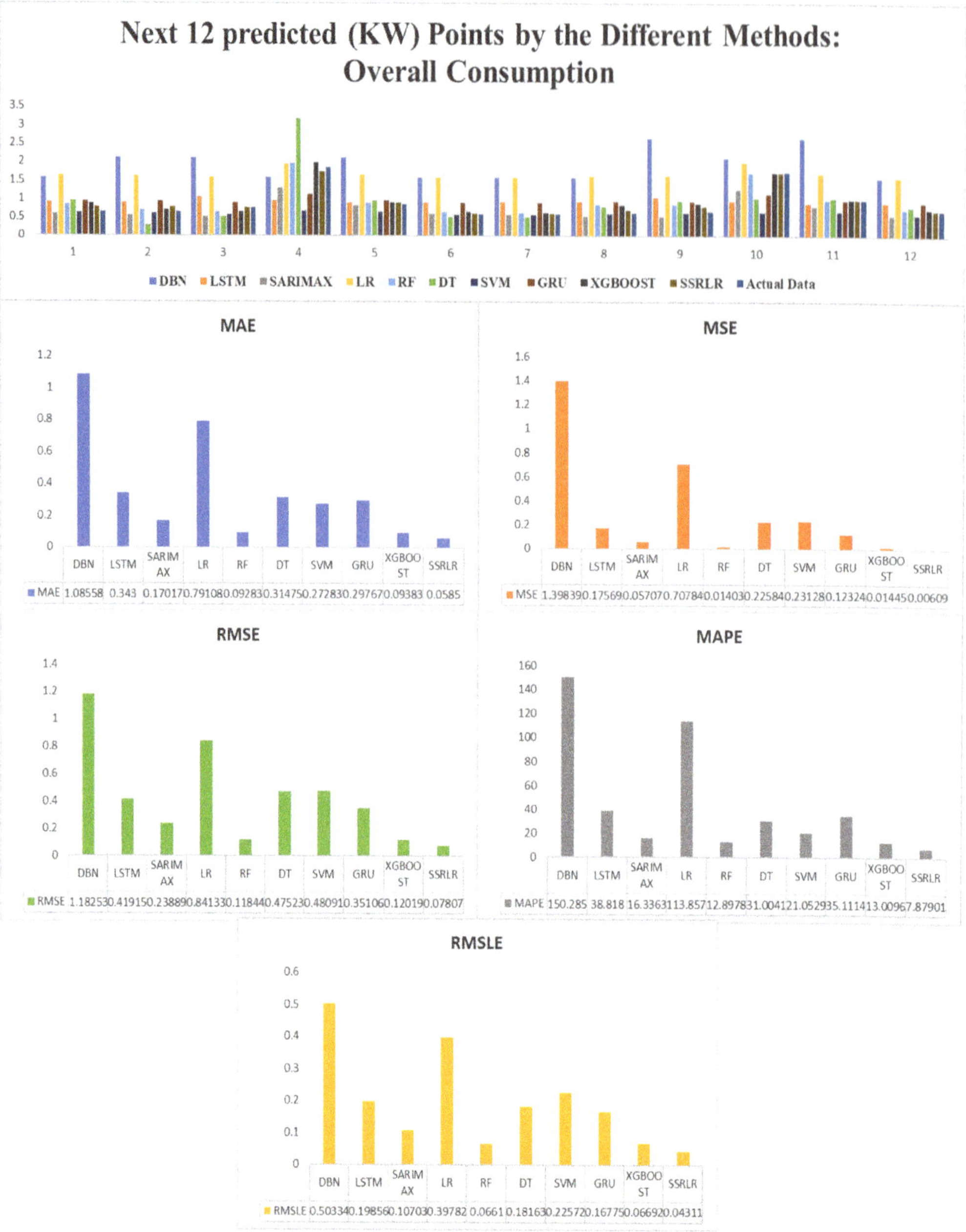

Figure 9. Comparison of models for overall consumption (12-step ahead).

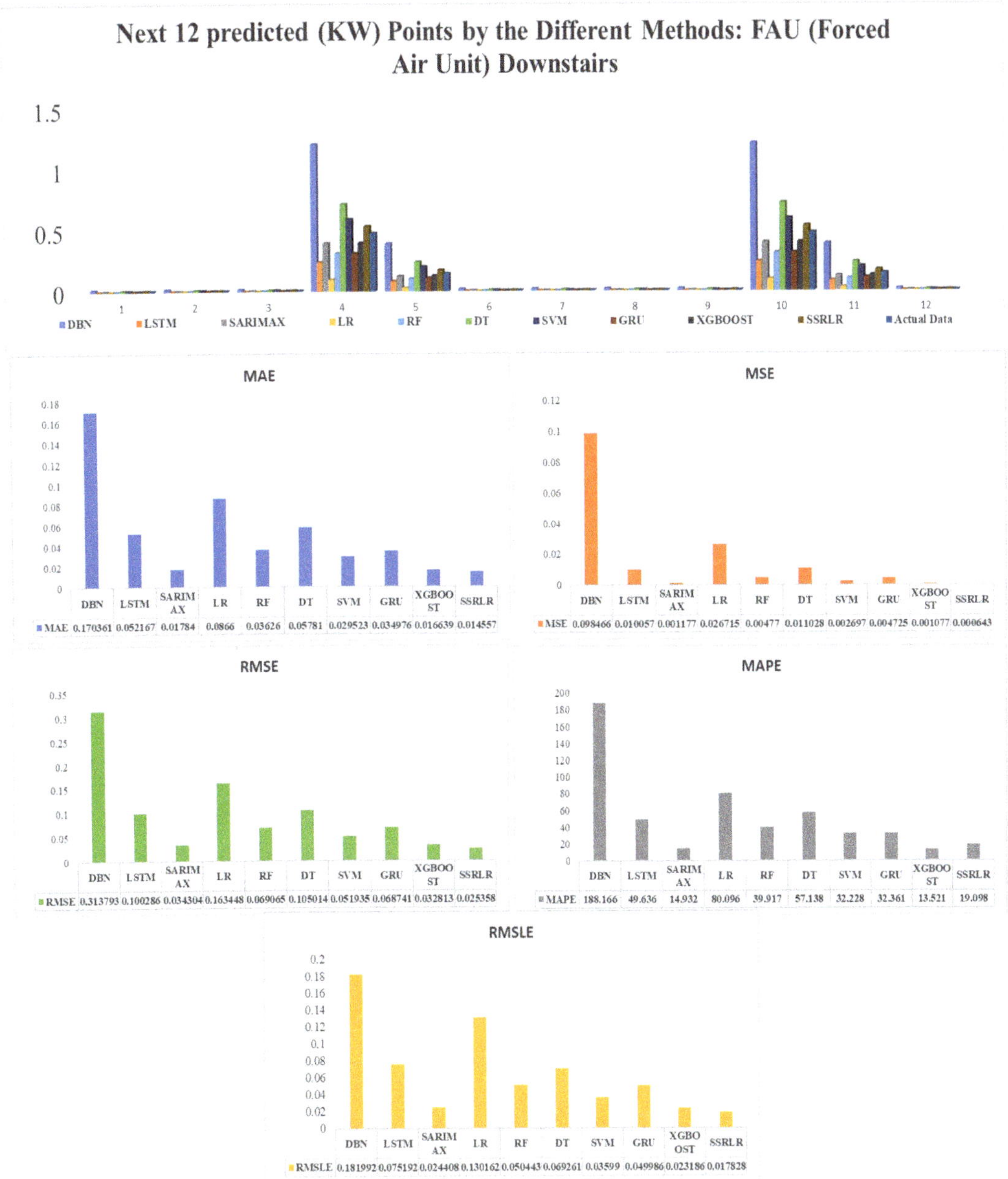

Figure 10. Comparison of models on FAU downstairs consumption (12-step ahead).

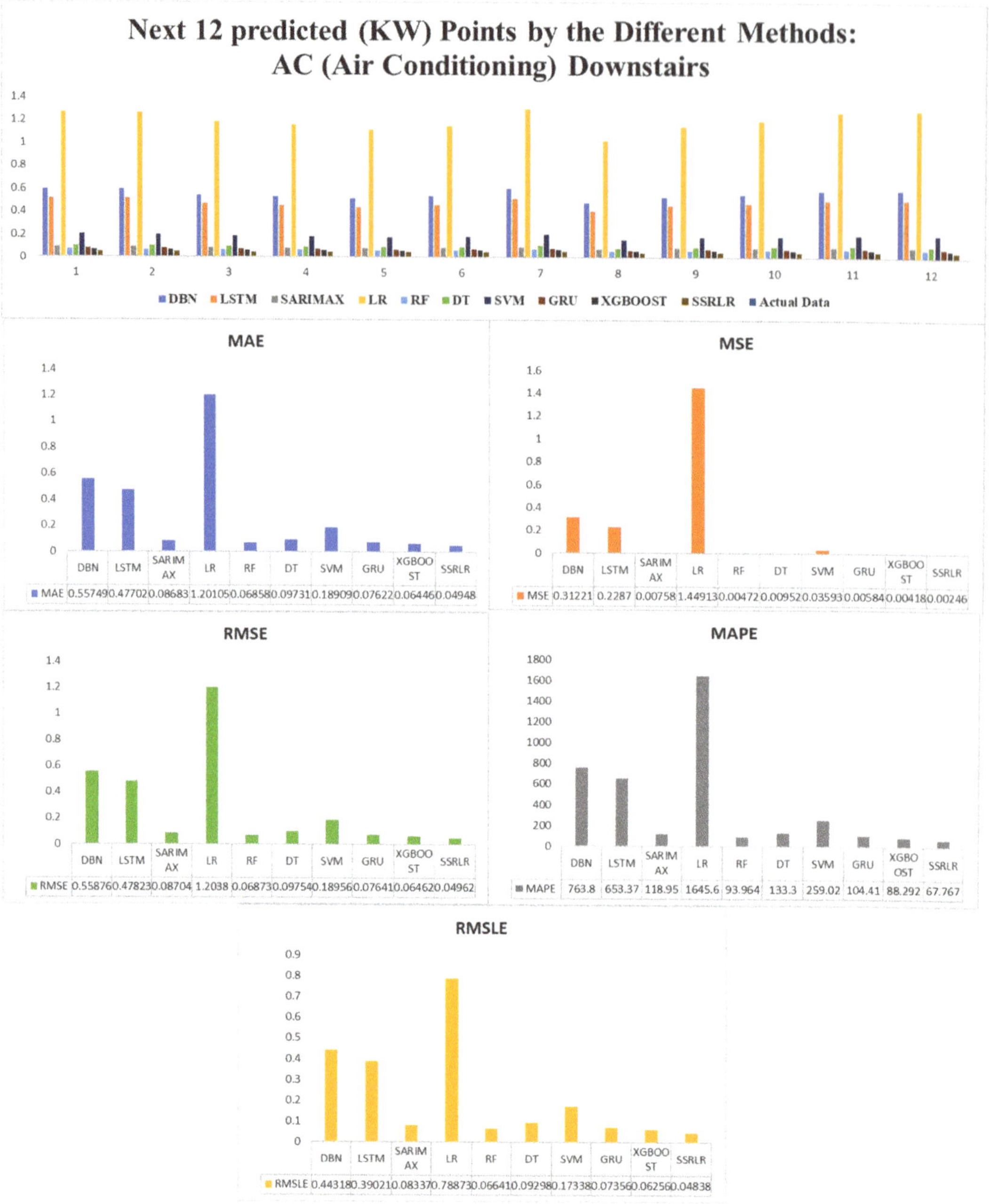

Figure 11. Comparison of models on AC downstairs consumption (12-step ahead).

Figure 12. Comparison of models on AC upstairs consumption (12-step ahead).

Figure 13. Comparison of models on refrigerator consumption (12-step ahead).

Figure 14. Comparison of models on laundry/dryer consumption (12-step ahead).

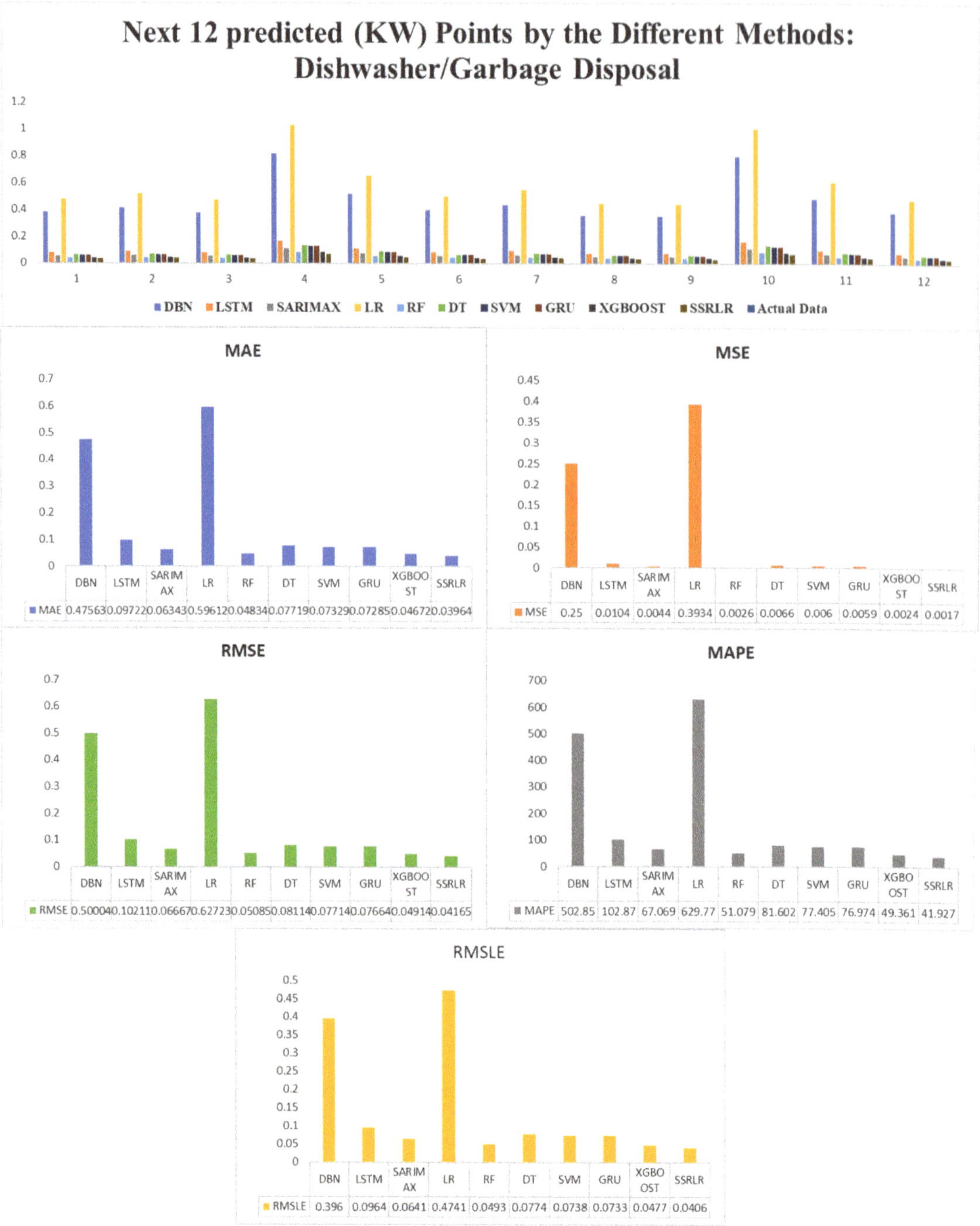

Figure 15. Comparison of models on dishwasher/garbage disposal consumption (12-step ahead).

Figure 16. Comparison of models on microwave consumption (12-step ahead).

Figure 17. Comparison of models on FAU upstairs consumption (12-step ahead).

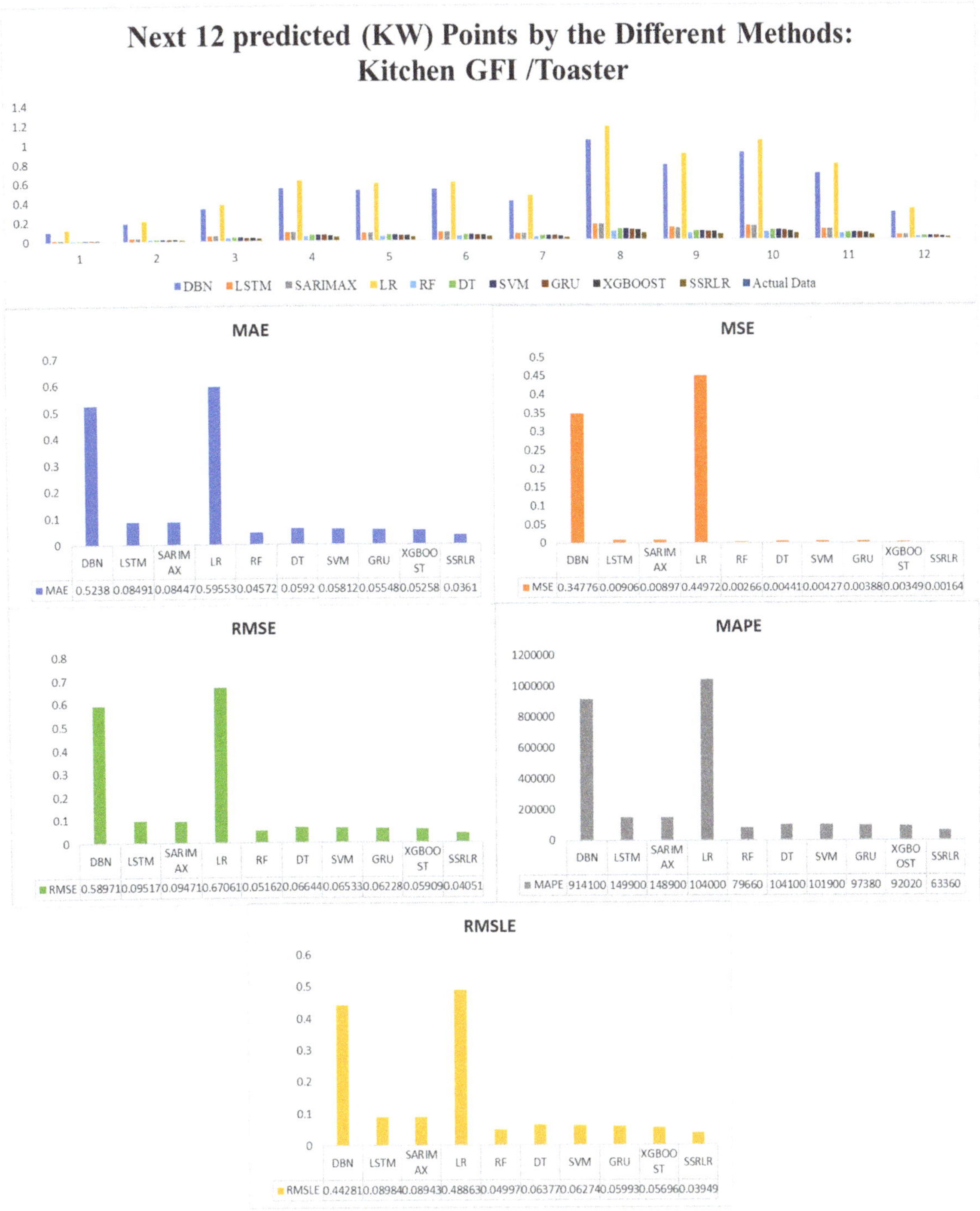

Figure 18. Comparison of models on kitchen GFI / toaster consumption (12-step ahead).

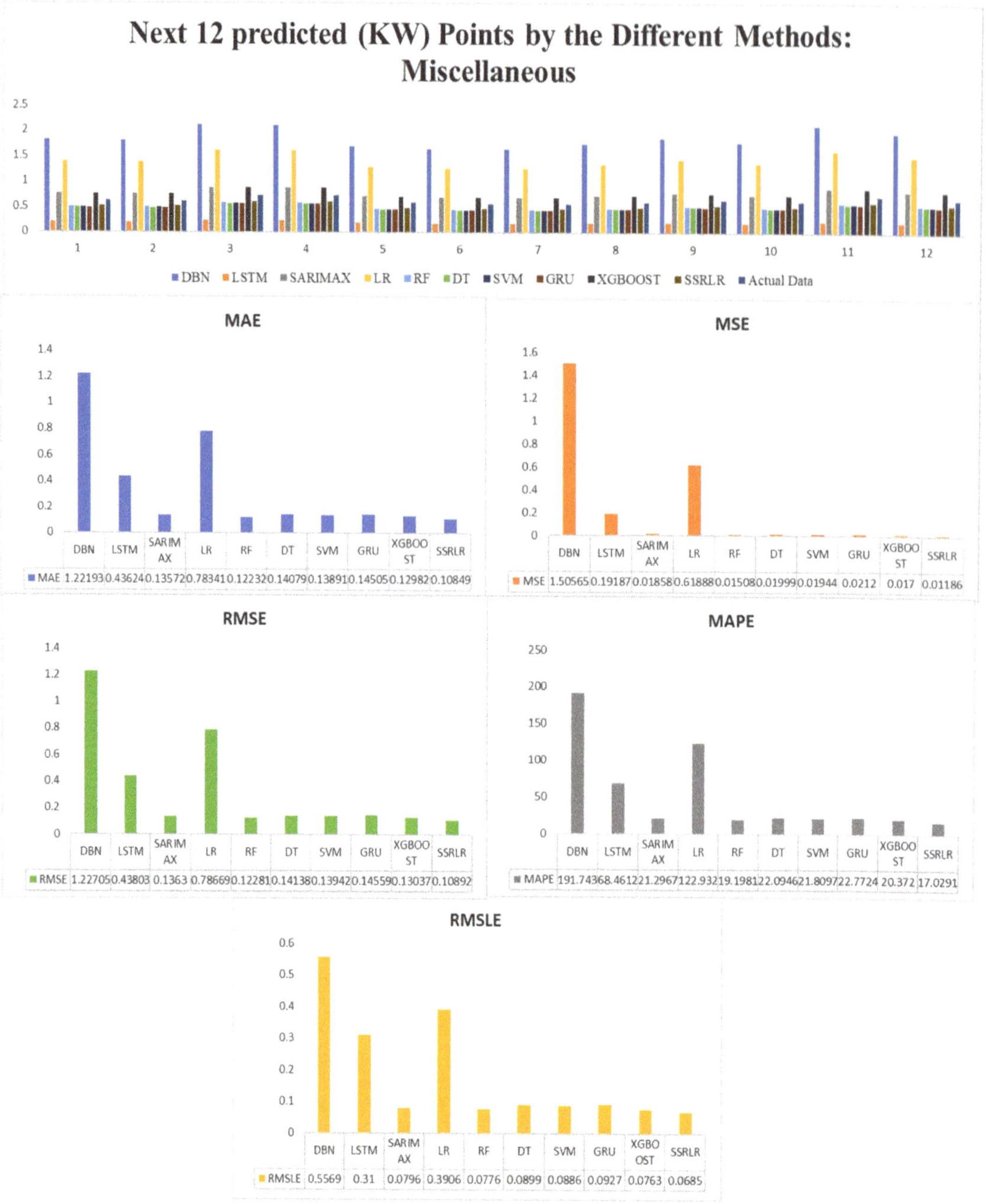

Figure 19. Comparison of models on miscellaneous consumption (12-step ahead).

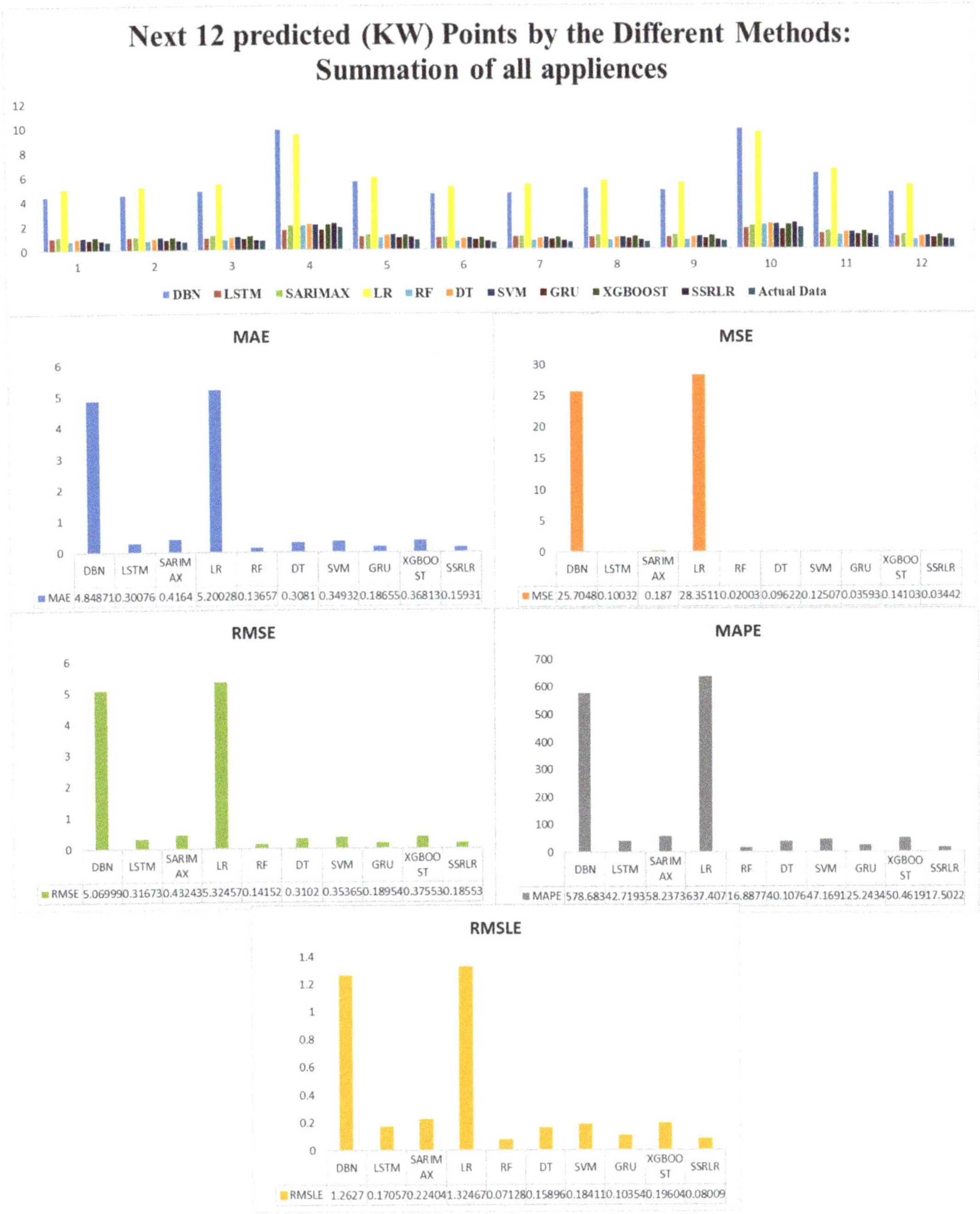

Figure 20. Comparison of models on summation of all appliance consumption (12-step ahead).

Table 5. Evaluation Metrics and Formulas.

Metric	Equation
MSE [41]	$MSE = \frac{1}{N}\sum_{i=1}^{N}(A_i - F_i)^2$
MAE [42]	$MAE = \frac{1}{N}\sum_{i=1}^{N}\lvert F_i - A_i \rvert$
MAPE [43]	$MAPE = \frac{1}{N}\sum_{i=1}^{N}\left\lvert \frac{A_i - F_i}{A_i} \right\rvert \times 100\%$
RMSE [44]	$RMSE = \sqrt{\frac{1}{N} \times \sum_{i=1}^{N}(F_i - A_i)^2}$
RMSLE [45]	$RMSLE = \sqrt{\frac{1}{N} \times \sum_{i=1}^{N}\left(\log(F_i + 1)^2 - log(A_i + 1)\right)^2}$

For the sum of appliance forecasts, the best model is RF with RMSLE = 0.07128 (MAE = 0.13657, RMSE = 0.14152) and second is the SSRLR with RMSLE = 0.08009. Comparing the two strategies best vs. best, AF: SSRLR (0.04311) beats sum of appliances RF (0.07128) by ~39.5% in RMSLE (the summer series is ~65% worse). This behavior is expected: when many appliances operate intermittently (e.g., microwave, dishwasher, small plugs), their individual errors can accumulate when the loads are summed, even if each model performs well.

In short, direct aggregate forecasting uses stable daily/weekly patterns in the total load, while the AL approach must first learn many on/off behaviors and then add them. Without hierarchical reconciliation (to force the sum to match the meter) or special handling of zeros/events, the AL sum can penalize small timing errors and stack relative errors.

Results of Per Appliance Forecasting:

Averaged across all appliances, SSRLR ranks first (avg RMSLE ≈ 0.042; RMSE/MAE show the same trend). RF and XGBoost are the strongest baselines. SSRLR is the best method for every single appliance; examples:

- FAU Downstairs: RMSLE 0.01783, RMSE 0.02536
- AC Downstairs: RMSLE 0.04838, RMSE 0.04962
- Refrigerator: RMSLE 0.04195, RMSE 0.04103
- AC Upstairs: RMSLE 0.04732, RMSE 0.04878
- Laundry/Dryer: RMSLE 0.04137, RMSE 0.04238
- Dishwasher/Garbage: RMSLE 0.04063, RMSE 0.04165
- Microwave: RMSLE 0.04116, RMSE 0.04207
- FAU Upstairs: RMSLE 0.04192, RMSE 0.07004
- Kitchen GFI/Toaster: RMSLE 0.03949, RMSE 0.04051
- General/Misc: SSRLR is best but errors are higher (RMSLE 0.06854, RMSE 0.10892) because this circuit mixes many small, irregular uses.

By the appliance type:

- Spiky/episodic loads (Microwave, Dishwasher/Garbage, Laundry) are handled well by SSRLR, which efficiently manages short bursts.
- HVAC loads (AC and FAU) also favor SSRLR, showing that it captures both seasonality and start/stop behavior.
- General/Misc is the hardest because it blends many small and random loads: SSRLR shows the best accuracy.

Model Ranking and Error Profiles:

A consistent ranking appears across aggregate, appliances, and the sum:

- SSRLR is best overall (aggregate and per appliance averages), strong on both relative (RMSLE) and absolute (RMSE/MAE) errors.

- RF/XGBoost are reliable runner ups; they capture nonlinear effects and are stable across devices.
- SARIMAX is a good seasonal baseline, but weaker on sharp spikes.
- LSTM and GRU do not dominate here, likely due to intermittency, short forecast horizons, and limited effective data per appliance.
- LR/DT/DBN/SVM perform poorly for short-term load forecasting in this setting.

This order holds for the aggregate and appliance averages. In the sum, RF wins, but it still trails the aggregate SSRLR by a wide margin.

Key Takeaways:

- Direct aggregate forecasting (SSRLR) is much more accurate than the sum of appliance forecasts on this dataset (~39.5% lower RMSLE, best vs. best).
- SSRLR is the most consistent model at both levels; RF/XGBoost are strong baselines.

It may look surprising that SSRLR wins for every appliance, yet RF wins after summing the results. The reason for that is the shift of the error pattern when summing consumption of many appliance forecasts: small positively correlated residuals from several SSRLR appliance models can add up for the same time points (for example, around evening peaks). RF tends to produce smoother appliance predictions with less shared timing error across devices, so when added some errors cancel each other. In addition, hierarchical reconciliation was not applied; therefore, perfect per-device calibration does not guarantee that the aggregated total matches the main meter. Because RMSLE on the summed series is sensitive to synchronized timing errors, RF ends up with a lower summed RMSLE (0.07128) than SSRLR (0.08009). Adding reconciliation and aggregate aware training would likely reduce and may remove this gap.

6. Conclusions

This paper introduced a cloud-enabled hybrid ML–statistical framework for 5 min residential load forecasting that blends Seasonal ARIMAX, Random Forest, and LSTM within a residual-correction architecture (SSRLR) and deploys redundant instances across multiple clouds with a two-out-of-three voting scheme. Using sub-metered data from a residential home in Summerlin, Las Vegas (2022), aggregate forecasting was evaluated against appliance-level forecasting. The evidence is clear: on the aggregate series, SSRLR is best across all metrics (RMSLE = 0.04311, RMSE = 0.07807, MAE = 0.05850), outperforming the next-best model (RF, RMSLE = 0.06610) by ~34.8%. For the appliance-level results, the RF model performs best (RMSLE = 0.07128), followed by the SSRLR model (RMSLE = 0.08009). However, the aggregate SSRLR model still outperforms the appliance-level results by up to approximately 39.5% in terms of RMSLE. This gap aligns with expectations for many intermittent end-uses: per appliance models can produce positively correlated residuals that accumulate when added, while the aggregate model exploits stable daily/weekly total load patterns. Importantly, RMSLE is the most informative headline metric in this setting because many appliances operate near zero and fairly scores small absolute errors; RMSE/MAE confirm the same ranking in absolute terms.

Practically, if the goal is maximum accuracy with minimal complexity, a single SSRLR model for the aggregate prediction is the preferred choice. The proposed approach supports real-time, fault-tolerant forecasting for smart-city energy management. Its cloud redundancy mechanism ensures operational continuity during single-cloud failures, making it suitable for deployment in distributed residential and micro-grid environments. The framework's modular design allows utilities to balance accuracy, latency, and cost by tuning cloud resources and model complexity.

Future work will extend the framework to multi-household, high-load buildings and cross-climate datasets, explore transfer learning for unseen dwellings, and integrate edge-cloud coordination for improved energy-efficiency and privacy. Further research will also focus on adaptive weighting and strengthen interpretability for decision-support systems.

Overall, this research demonstrates that hybrid multi-cloud forecasting can provide fine-grained, resilient, and scalable solutions for next-generation smart-residential infrastructures.

Author Contributions: Conceptualization, K.H.B., E.E.R. and Y.B.; methodology, K.H.B.; software, K.H.B.; validation, K.H.B., E.E.R. and Y.B.; formal analysis, K.H.B. and E.E.R.; investigation, K.H.B., E.E.R. and Y.B.; resources, K.H.B., E.E.R. and Y.B.; data curation, K.H.B. and Y.B.; writing original draft preparation, K.H.B., E.E.R. and Y.B.; writing review and editing, K.H.B., E.E.R. and Y.B.; visualization, K.H.B.; supervision, K.H.B., E.E.R. and Y.B.; project administration, K.H.B., E.E.R. and Y.B.; funding acquisition, K.H.B., E.E.R. and Y.B. All authors have read and agreed to the published version of the manuscript.

Funding: This research received no external funding.

Data Availability Statement: Data will be made available upon reasonable request.

Acknowledgments: The authors would like to express their sincere appreciation to Zeinab Farrokhi (Z.F) for her significant contributions during the revision process, including her substantial assistance in responding to the reviewers' comments and editing the manuscript.

Conflicts of Interest: The authors declare that they have no conflicts of interest to this work.

Abbreviations

The following abbreviations are used in this manuscript:

ML	Machine Learning
MSE	Mean Squared Error
LF	Load Forecasting
MAPE	Mean Absolute Percentage Error
MAE	Mean Absolute Error
STLF	Short-Term Load Forecasting
RMSLE	Root Mean Squared Logarithmic Error
MTLF	Mid-Term Load Forecasting
RMSE	Root Mean Squared Error
LTLF	Long-Term Load Forecasting
LR	Linear Regression
RF	Random Forest
DT	Decision Tree
GRU	Gated Recurrent Unit
SVM	Support Vector Machine
DBN	Deep Belief Network
ARIMA	Autoregressive Integrated Moving Average
SSRLR	Sparse SARIMAX, Random Forest, LSTM, and Remainder Prediction
SARIMAX	Seasonal Autoregressive Integrated Moving Average Exogenous
XGBoost	Extreme Gradient Boosting
LSTM	Long Short-Term Memory
CM	Correlation Matrix
MRMR	Minimum Redundancy Maximum Relevance
RP	Remainder Prediction
PP	Primary Prediction
AL	Appliance-level
AF	Aggregate forecasts

Appendix A

Table A1. Next 12 predicted (KW) points by different methods: overall consumption.

Method Point	DBN	LSTM	SARIMAX	LR	RF	DT	SVM	GRU	XGBOOST	SSRLR	Actual Data
1	1.569	0.911	0.592	1.634	0.853	0.951	0.621	0.938	0.871	0.782	0.650
2	2.104	0.887	0.545	1.610	0.703	0.279	0.602	0.923	0.699	0.775	0.646
3	2.104	1.040	0.516	1.576	0.643	0.513	0.569	0.901	0.637	0.768	0.752
4	1.569	0.941	1.294	1.941	1.963	3.180	0.665	1.112	1.968	1.730	1.840
5	2.104	0.884	0.812	1.658	0.892	0.946	0.637	0.953	0.89	0.883	0.846
6	1.569	0.888	0.580	1.573	0.643	0.513	0.565	0.899	0.643	0.612	0.590
7	1.569	0.920	0.561	1.577	0.631	0.513	0.570	0.902	0.632	0.612	0.590
8	1.569	0.927	0.535	1.624	0.854	0.788	0.613	0.932	0.833	0.708	0.625
9	2.639	1.050	0.535	1.634	0.853	0.951	0.621	0.938	0.865	0.795	0.664
10	2.104	0.958	1.247	2.004	1.708	1.031	0.642	1.143	1.71	1.714	1.732
11	2.639	0.902	0.789	1.689	0.988	1.019	0.655	0.972	0.988	0.989	0.989
12	1.569	0.911	0.575	1.596	0.738	0.794	0.589	0.914	0.737	0.701	0.699

Table A2. Next 12 predicted (KW) points by different methods: FAU (forced air unit) downstairs.

Method Point	DBN	LSTM	SARIMAX	LR	RF	DT	SVM	GRU	XGBOOST	SSRLR	Actual Data
1	0.017513	0.008902	0.007134	0.012425	0.006982	0.009112	0.007698	0.008235	0.007091	0.006785	0.00601
2	0.018425	0.009503	0.00721	0.012833	0.006991	0.009621	0.007895	0.008501	0.007254	0.006945	0.006043
3	0.017901	0.007625	0.007625	0.012332	0.006951	0.009432	0.00921	0.007193	0.008395	0.006851	0.005963
4	1.202331	0.731299	0.557128	0.983572	0.645221	0.71294	0.58932	0.638521	0.552118	0.501237	0.477047
5	0.391254	0.251332	0.173555	0.288413	0.210892	0.239144	0.199813	0.220335	0.176624	0.161225	0.145373
6	0.018311	0.009108	0.007324	0.012641	0.006914	0.009405	0.007809	0.00829	0.007114	0.006831	0.006037
7	0.018812	0.00961	0.007501	0.013001	0.007225	0.009832	0.008014	0.008502	0.007334	0.007009	0.006147
8	0.018501	0.009215	0.007254	0.012721	0.007001	0.009611	0.007921	0.008411	0.007209	0.006982	0.00611
9	0.0189	0.0097	0.007605	0.01322	0.007332	0.009955	0.008225	0.00872	0.007445	0.00712	0.00617
10	1.210422	0.740291	0.56321	0.990512	0.653222	0.721008	0.59512	0.642892	0.557231	0.503188	0.477707
11	0.39012	0.250992	0.172801	0.287134	0.209921	0.238551	0.198923	0.219832	0.175832	0.160534	0.145263
12	0.01825	0.00902	0.007111	0.01252	0.006952	0.009255	0.007715	0.008175	0.007003	0.006822	0.006033

Table A3. Next 12 predicted (KW) points by different methods: air conditioning downstairs.

Method Point	DBN	LSTM	SARIMAX	LR	RF	DT	SVM	GRU	XGBOOST	SSRLR	Actual Data
1	0.002301	0.001205	0.000924	0.00161	0.000851	0.001132	0.000967	0.001031	0.000912	0.000843	0.00078
2	0.002298	0.001211	0.000919	0.001599	0.000845	0.001127	0.000963	0.001028	0.000908	0.000839	0.00078
3	0.002115	0.001099	0.000842	0.001503	0.000799	0.001046	0.000892	0.00095	0.000838	0.000772	0.0007
4	0.002069	0.001067	0.000826	0.001469	0.00078	0.001021	0.00087	0.000928	0.00082	0.000755	0.00069
5	0.001986	0.001025	0.000793	0.001409	0.000748	0.000979	0.000834	0.00089	0.000787	0.000724	0.00066
6	0.002083	0.001074	0.00083	0.001453	0.000772	0.001009	0.00086	0.000919	0.000815	0.000747	0.000693
7	0.002352	0.001215	0.000938	0.001644	0.000869	0.001154	0.000985	0.001051	0.000933	0.000865	0.00081
8	0.001858	0.000961	0.000742	0.001298	0.000688	0.000906	0.000774	0.000827	0.000737	0.000676	0.000643
9	0.002067	0.001072	0.00083	0.00145	0.00077	0.001006	0.000858	0.000917	0.000813	0.000746	0.00071
10	0.002157	0.00112	0.000865	0.001515	0.000805	0.001051	0.000897	0.000958	0.000851	0.000781	0.00074
11	0.002276	0.001183	0.000914	0.001619	0.000861	0.001124	0.00096	0.001028	0.000913	0.000845	0.000773
12	0.002296	0.001196	0.000922	0.001635	0.000871	0.001137	0.00097	0.001039	0.000921	0.000852	0.000783

Table A4. Next 12 predicted (KW) points by different methods: AC upstairs.

Method Point	DBN	LSTM	SARIMAX	LR	RF	DT	SVM	GRU	XGBOOST	SSRLR	Actual Data
1	−0.00108	−0.00052	−0.00041	−0.0008	−0.00033	−0.00049	−0.00045	−0.00047	−0.00034	−0.0003	−0.00036
2	−0.00106	−0.00051	−0.0004	−0.00078	−0.00033	−0.00049	−0.00044	−0.00047	−0.00033	−0.00029	−0.00035
3	−0.00113	−0.00054	−0.00043	−0.00082	−0.00034	−0.00051	−0.00046	−0.00049	−0.00035	−0.00031	−0.00038
4	−0.00137	−0.00066	−0.00052	−0.00099	−0.00042	−0.00062	−0.00056	−0.00059	−0.00042	−0.00037	−0.00046
5	−0.00125	−0.0006	−0.00047	−0.0009	−0.00038	−0.00056	−0.00051	−0.00054	−0.00038	−0.00034	−0.00042
6	−0.00123	−0.00059	−0.00046	−0.00089	−0.00037	−0.00055	−0.0005	−0.00053	−0.00038	−0.00033	−0.00041
7	−0.00097	−0.00046	−0.00036	−0.0007	−0.00029	−0.00044	−0.00039	−0.00042	−0.0003	−0.00026	−0.00032
8	−0.00142	−0.00068	−0.00054	−0.00104	−0.00043	−0.00064	−0.00058	−0.00062	−0.00044	−0.00038	−0.00048
9	−0.00161	−0.00077	−0.00061	−0.00117	−0.00049	−0.00072	−0.00065	−0.00069	−0.00049	−0.00043	−0.00054
10	−0.00164	−0.00079	−0.00062	−0.00119	−0.0005	−0.00074	−0.00067	−0.00071	−0.0005	−0.00044	−0.00055
11	−0.00134	−0.00064	−0.00051	−0.00098	−0.00041	−0.00061	−0.00055	−0.00059	−0.00042	−0.00036	−0.00045
12	−0.00155	−0.00074	−0.00059	−0.00113	−0.00047	−0.0007	−0.00063	−0.00067	−0.00048	−0.00042	−0.00052

Table A5. Next 12 predicted (KW) points by different methods: refrigerator.

Method Point	DBN	LSTM	SARIMAX	LR	RF	DT	SVM	GRU	XGBOOST	SSXLR	Actual Data
1	0.018945	0.009118	0.007204	0.013375	0.007001	0.009402	0.008015	0.008465	0.007293	0.006981	0.006727
2	0.018832	0.009065	0.007166	0.013288	0.006942	0.009351	0.007984	0.008432	0.007259	0.006951	0.006703
3	0.01884	0.009071	0.00717	0.013296	0.006946	0.009356	0.007988	0.008437	0.007263	0.006955	0.006703
4	0.017858	0.008604	0.00681	0.01264	0.006601	0.008892	0.0076	0.008042	0.006897	0.006581	0.006343
5	0.018201	0.008771	0.00694	0.012894	0.00674	0.009086	0.007769	0.00823	0.00706	0.006732	0.00648
6	0.01826	0.0088	0.006962	0.012933	0.006761	0.009115	0.007794	0.008255	0.007081	0.006751	0.0065
7	0.018273	0.008807	0.006968	0.012944	0.006768	0.009122	0.0078	0.008261	0.007087	0.006756	0.0065
8	0.018284	0.008812	0.006971	0.01295	0.006771	0.009126	0.007804	0.008265	0.00709	0.006759	0.006503
9	0.018323	0.008832	0.006986	0.012978	0.006788	0.009145	0.00782	0.00828	0.007105	0.006772	0.006517
10	0.017671	0.008507	0.006718	0.012506	0.00655	0.008794	0.007527	0.007974	0.006833	0.006528	0.006193
11	0.018766	0.00902	0.007139	0.013233	0.006915	0.009312	0.007955	0.008407	0.007238	0.006931	0.006593
12	0.0189	0.009091	0.007201	0.01335	0.006986	0.009381	0.00801	0.00846	0.00729	0.006981	0.00664

Table A6. Next 12 predicted (KW) points by different methods: laundry/dryer.

Method Point	DBN	LSTM	SARIMAX	LR	RF	DT	SVM	GRU	XGBOOST	SSRLR	Actual Data
1	0.010825	0.005273	0.004242	0.007917	0.004008	0.005536	0.004821	0.005129	0.004181	0.003956	0.003603
2	0.01091	0.005316	0.004277	0.007988	0.004045	0.005588	0.004866	0.005175	0.004217	0.00399	0.003637
3	0.010789	0.005256	0.004227	0.007878	0.003993	0.005522	0.004804	0.005113	0.00416	0.003935	0.00358
4	0.010833	0.005278	0.004244	0.007913	0.004012	0.005539	0.004822	0.005133	0.004176	0.003951	0.003607
5	0.010537	0.005131	0.004127	0.00769	0.003898	0.005382	0.004686	0.004983	0.004058	0.003836	0.00351
6	0.010678	0.005202	0.004182	0.007824	0.003964	0.005464	0.004762	0.005063	0.004118	0.003893	0.003563
7	0.010579	0.005152	0.004142	0.007738	0.003921	0.005414	0.004718	0.005018	0.004079	0.003856	0.003527
8	0.010584	0.005154	0.004145	0.007744	0.003924	0.005418	0.004721	0.005021	0.004082	0.003858	0.003527
9	0.010606	0.005166	0.004154	0.00776	0.003934	0.00543	0.004731	0.005032	0.004091	0.003867	0.00352
10	0.010582	0.005154	0.004144	0.007742	0.003923	0.005417	0.00472	0.00502	0.004081	0.003857	0.003527
11	0.010919	0.00532	0.004281	0.008003	0.00405	0.005593	0.004872	0.005183	0.004224	0.003996	0.003637
12	0.010732	0.005227	0.004207	0.007853	0.003972	0.00549	0.004787	0.00509	0.004142	0.003919	0.003567

Table A7. Next 12 predicted (KW) points by different methods: dishwasher/garbage disposal.

Method Point	DBN	LSTM	SARIMAX	LR	RF	DT	SVM	GRU	XGBOOST	SSRLR	Actual Data
1	0.002256	0.001121	0.000874	0.001661	0.000752	0.001051	0.000942	0.001003	0.000789	0.000705	0.00075
2	0.002452	0.001222	0.000953	0.001799	0.000821	0.001152	0.00103	0.001098	0.000863	0.000773	0.000817
3	0.002224	0.001103	0.000859	0.001636	0.00074	0.001033	0.000926	0.000985	0.000775	0.000693	0.00074
4	0.004831	0.002374	0.001857	0.003555	0.001598	0.002243	0.002026	0.002155	0.001692	0.001515	0.001643
5	0.003074	0.001514	0.001186	0.002263	0.001022	0.001435	0.001296	0.001379	0.001082	0.000968	0.001047
6	0.002365	0.001174	0.000913	0.001719	0.000777	0.001093	0.00098	0.001049	0.000823	0.000739	0.000787
7	0.002599	0.001283	0.000997	0.001892	0.000857	0.0012	0.001077	0.001151	0.000902	0.000809	0.000863
8	0.002129	0.00105	0.000814	0.001552	0.000704	0.000985	0.000882	0.000944	0.00074	0.000664	0.000707
9	0.002097	0.001034	0.000802	0.001528	0.000693	0.00097	0.000869	0.00093	0.000729	0.000654	0.000697
10	0.004711	0.002321	0.001815	0.003466	0.001556	0.002184	0.001974	0.0021	0.00165	0.001477	0.00161
11	0.002857	0.001414	0.001106	0.002102	0.000945	0.001326	0.00119	0.001272	0.001	0.000897	0.000977
12	0.002264	0.00112	0.000874	0.001662	0.000752	0.00105	0.000942	0.001004	0.00079	0.000705	0.00075

Table A8. Next 12 predicted (KW) points by different methods: microwave.

Method Point	DBN	LSTM	SARIMAX	LR	RF	DT	SVM	GRU	XGBOOST	SSRLR	Actual Data
1	0.003218	0.001858	0.001458	0.002322	0.001188	0.001566	0.001501	0.001598	0.001382	0.001318	0.00108
2	0.003312	0.001927	0.00151	0.002387	0.001239	0.001645	0.001555	0.001678	0.001432	0.001362	0.0011266
3	0.003099	0.001774	0.001387	0.002224	0.001138	0.001512	0.001429	0.001535	0.001316	0.001252	0.0010433
4	0.003178	0.001847	0.001439	0.002299	0.001191	0.001568	0.001481	0.001589	0.001364	0.00131	0.0010733
5	0.003275	0.001898	0.001487	0.002364	0.001221	0.00161	0.001532	0.001654	0.00141	0.001343	0.00111
6	0.003214	0.001864	0.001459	0.002327	0.001206	0.00159	0.001503	0.001624	0.001395	0.001339	0.0010966
7	0.003329	0.00192	0.001507	0.002402	0.001228	0.001631	0.001552	0.001665	0.001428	0.001363	0.0011166
8	0.003112	0.001814	0.001417	0.002265	0.001169	0.001546	0.00146	0.001577	0.001342	0.001298	0.0010733
9	0.0032	0.001843	0.001432	0.002295	0.001184	0.001561	0.001474	0.001594	0.001369	0.001313	0.0010766
10	0.003252	0.001892	0.001481	0.002362	0.001225	0.001616	0.001536	0.001637	0.001403	0.001347	0.0011133
11	0.003284	0.00191	0.001497	0.002389	0.001228	0.001631	0.00154	0.001653	0.00142	0.001363	0.0011166
12	0.003286	0.001909	0.001499	0.002387	0.001232	0.001634	0.001543	0.001654	0.001421	0.001354	0.00111

Table A9. Next 12 predicted (KW) points by different methods: FAU upstairs.

Method Point	DBN	LSTM	SARIMAX	LR	RF	DT	SVM	GRU	XGBOOST	SSRLR	Actual Data
1	0.023895	0.01113	0.009172	0.017882	0.008612	0.012544	0.011276	0.011869	0.009442	0.008772	0.00794
2	0.02366	0.011016	0.00908	0.017737	0.008532	0.012452	0.011191	0.01178	0.009363	0.008698	0.007887

Table A9. *Cont.*

Method Point	DBN	LSTM	SARIMAX	LR	RF	DT	SVM	GRU	XGBOOST	SSRLR	Actual Data
3	0.02375	0.011062	0.009115	0.017801	0.008568	0.012493	0.011227	0.011821	0.009392	0.008728	0.007917
4	1.84365	0.869225	0.694118	1.385427	0.657311	0.951882	0.864113	0.910225	0.702447	0.635822	0.62757
5	0.320535	0.155773	0.124193	0.244153	0.115867	0.168152	0.152559	0.161037	0.12491	0.112967	0.109373
6	0.02404	0.011191	0.009218	0.017975	0.008665	0.012605	0.011332	0.011929	0.009491	0.008819	0.008013
7	0.02371	0.011043	0.009107	0.017768	0.008554	0.012476	0.011212	0.011802	0.009376	0.008713	0.007903
8	0.02399	0.011177	0.009207	0.01795	0.008655	0.012595	0.011323	0.01192	0.009481	0.00881	0.008007
9	0.02387	0.011118	0.00916	0.017868	0.008608	0.012544	0.011276	0.011871	0.009437	0.008771	0.007967
10	1.84782	0.871025	0.695452	1.388362	0.659014	0.954193	0.866174	0.912356	0.704123	0.63731	0.629273
11	0.317834	0.154462	0.123164	0.242273	0.114952	0.166883	0.151423	0.159822	0.123991	0.112142	0.10793
12	0.023565	0.010981	0.00905	0.017696	0.008513	0.012426	0.01117	0.011758	0.009338	0.008672	0.00785

Table A10. Next 12 predicted (KW) points by different methods: kitchen GFI (ground fault circuit interrupter)/toaster.

Method Point	DBN	LSTM	SARIMAX	LR	RF	DT	SVM	GRU	XGBOOST	SSRLR	Actual Data
1	0.000033	0.000018	0.000014	0.000027	0.000012	0.000019	0.000017	0.000018	0.000014	0.000013	0.0000133
2	0.00006	0.00003	0.000024	0.000048	0.000021	0.000033	0.000029	0.000031	0.000024	0.000022	0.00002
3	0.000108	0.000052	0.000041	0.000086	0.000038	0.000058	0.000051	0.000055	0.000043	0.000039	0.0000367
4	0.000176	0.000084	0.000067	0.00014	0.000061	0.000095	0.000084	0.00009	0.000071	0.000065	0.00006
5	0.000168	0.00008	0.000064	0.000134	0.000059	0.00009	0.000081	0.000087	0.000069	0.000063	0.0000567
6	0.00017	0.000081	0.000065	0.000136	0.000059	0.000091	0.000082	0.000088	0.00007	0.000064	0.0000567
7	0.00013	0.000062	0.00005	0.000104	0.000045	0.000069	0.000063	0.000067	0.000053	0.000048	0.0000433
8	0.00033	0.000155	0.000124	0.000264	0.000118	0.000176	0.000159	0.000168	0.000133	0.00012	0.00011
9	0.000249	0.00012	0.000095	0.0002	0.000089	0.000133	0.00012	0.000127	0.000101	0.000092	0.0000833
10	0.000289	0.000139	0.000111	0.000232	0.000103	0.000154	0.000139	0.000147	0.000116	0.000105	0.0000967
11	0.000219	0.000106	0.000084	0.000176	0.000079	0.000118	0.000106	0.000112	0.000089	0.000081	0.0000733
12	0.00009	0.000044	0.000035	0.000072	0.000032	0.00005	0.000045	0.000047	0.000037	0.000034	0.00003

Table A11. Next 12 predicted (KW) points by different methods: general (miscellaneous) household load.

Method Point	DBN	LSTM	SARIMAX	LR	RF	DT	SVM	GRU	XGBOOST	SSRLR	Actual Data
1	1.822104	0.91924	0.676118	1.392781	0.210125	0.534925	0.61302	0.57248	0.225194	0.196833	0.624233
2	1.807326	0.910481	0.669839	1.38035	0.208012	0.529689	0.607623	0.56747	0.223217	0.195084	0.619827
3	2.1154	1.06139	0.780626	1.61483	0.243584	0.620552	0.71228	0.665982	0.261883	0.229029	0.72587
4	2.10937	1.05824	0.778301	1.610066	0.242905	0.618736	0.710184	0.664031	0.260815	0.228099	0.723123
5	1.69062	0.848426	0.624025	1.29016	0.194717	0.495586	0.568477	0.530534	0.208309	0.182078	0.579567
6	1.64558	0.825861	0.60728	1.255844	0.189549	0.482486	0.553916	0.51662	0.203005	0.177666	0.56406
7	1.64375	0.82494	0.606606	1.254294	0.189313	0.481886	0.553226	0.515967	0.202685	0.177383	0.563447
8	1.74768	0.877021	0.644677	1.33329	0.201133	0.511807	0.587489	0.547783	0.21495	0.18815	0.59914
9	1.86149	0.933968	0.685921	1.42944	0.215455	0.547872	0.628909	0.586144	0.231409	0.202521	0.63872
10	1.78459	0.894606	0.657066	1.3691	0.206324	0.524664	0.602642	0.561501	0.220727	0.193142	0.612323
11	2.11404	1.06072	0.780099	1.61376	0.243448	0.620224	0.711905	0.665644	0.261753	0.228913	0.723473
12	1.96819	0.988808	0.726622	1.50408	0.226812	0.578106	0.663648	0.619833	0.243294	0.213253	0.673247

Table A12. Next 12 predicted (KW) points by different methods: overall consumption—summation of Tables A4–A12.

Method Point	DBN	LSTM	SARIMAX	LR	RF	DT	SVM	GRU	XGBOOST	SSRLR	Actual Data
1	4.402809	0.996231	1.087194	4.957531	0.753281	0.924844	1.001538	0.829909	1.033635	0.740985	0.650
2	4.527383	1.022295	1.101331	5.090235	0.760627	0.940943	1.015361	0.844171	1.040605	0.747044	0.646
3	4.838178	1.008571	1.225377	5.306084	0.844195	1.01274	1.083204	0.918764	1.162196	0.831497	0.752
4	9.84792	1.645598	1.992939	9.40809	2.022073	2.111469	2.054901	1.659253	2.045687	2.183267	1.840
5	5.580026	1.113722	1.2569	5.837724	0.982709	1.214348	1.236753	1.003334	1.209144	1.024185	0.846
6	4.546716	0.976813	1.05935	5.12311	0.727287	0.904705	0.968167	0.813176	0.986795	0.709672	0.590
7	4.611737	1.04493	1.069944	5.27021	0.740032	0.926468	1.005123	0.829526	0.997419	0.719465	0.590
8	4.975247	0.991344	1.151328	5.511958	0.778107	0.95712	1.009119	0.865246	1.056523	0.755695	0.625
9	4.854031	1.004369	1.159956	5.392624	0.789248	0.961816	1.023475	0.870084	1.078246	0.773757	0.664
10	9.811275	1.671429	1.906255	9.498042	1.959133	2.057394	1.998976	1.598688	1.939943	2.111271	1.732
11	6.211354	1.246643	1.465467	6.482264	1.119608	1.35283	1.388854	1.136581	1.403961	1.159455	0.989
12	4.606389	1.00431	1.149319	5.153972	0.791093	0.961008	1.034905	0.868796	1.091916	0.783885	0.699

References

1. Wang, S.; Deng, X.; Chen, H.; Shi, Q.; Xu, D. A bottom-up short-term residential load forecasting approach based on appliance characteristic analysis and multi-task learning. *Electr. Power Syst. Res.* **2021**, *196*, 107233. [CrossRef]

2. Shabbir, N.; Vassiljeva, K.; Nourollahi Hokmabad, H.; Husev, O.; Petlenkov, E.; Belikov, J. Comparative Analysis of Machine Learning Techniques for Non-Intrusive Load Monitoring. *Electronics* **2024**, *13*, 1420. [CrossRef]
3. Noroozi, Z.; Orooji, A.; Erfannia, L. Analyzing the impact of feature selection methods on machine learning algorithms for heart disease prediction. *Sci. Rep.* **2023**, *13*, 22588. [CrossRef]
4. Hassanpouri Baesmat, K.; Farrokhi, Z.; Chmaj, G.; Regentova, E.E. Parallel Multi-Model Energy Demand Forecasting with Cloud Redundancy: Leveraging Trend Correction, Feature Selection, and Machine Learning. *Forecasting* **2025**, *7*, 25. [CrossRef]
5. Farrokhi, Z. AI-Powered Grid Analytics for Sustainable Energy Management in ESG Frameworks. In *Data-Driven ESG Strategy Implementation Through Business Intelligence*; IGI Global Scientific Publishing: Hershey, PA, USA, 2026; pp. 279–318. [CrossRef]
6. Mohtasham, F.; Pourhoseingholi, M.; Hashemi Nazari, S.S.; Kavousi, K.; Zali, M.R. Comparative analysis of feature selection techniques for COVID-19 dataset. *Sci. Rep.* **2024**, *14*, 18627. [CrossRef] [PubMed]
7. Ali, M.; Islam, S.; Uddin, M.N.; Uddin, A. A conceptual IoT framework based on Anova-F feature selection for chronic kidney disease detection using deep learning approach. *Intell.-Based Med.* **2024**, *10*, 100170. [CrossRef]
8. Souza, F.; Premebida, C.; Araújo, R. High-order conditional mutual information maximization for dealing with high-order dependencies in feature selection. *Pattern Recognit.* **2022**, *131*, 108895. [CrossRef]
9. Qu, K.; Xu, J.; Hou, Q.; Qu, K.; Sun, Y. Feature selection using Information Gain and decision information in neighborhood decision system. *Appl. Soft Comput.* **2023**, *136*, 110100. [CrossRef]
10. Muthukrishnan, R.; Rohini, R. LASSO: A feature selection technique in predictive modeling for machine learning. In Proceedings of the 2016 IEEE International Conference on Advances in Computer Applications (ICACA), Coimbatore, India, 24 October 2016; pp. 18–20. [CrossRef]
11. Altarabichi, M.G.; Nowaczyk, S.; Pashami, S.; Mashhadi, P.S. Fast Genetic Algorithm for feature selection A qualitative approximation approach. *Expert Syst. Appl.* **2023**, *211*, 118528. [CrossRef]
12. Luong, H.H.; Tran, T.T.; Van Nguyen, N.; Le, A.D.; Nguyen, H.T.T.; Nguyen, K.D.; Tran, N.C.; Nguyen, H.T. Feature Selection Using Correlation Matrix on Metagenomic Data with Pearson Enhancing Inflammatory Bowel Disease Prediction. In *International Conference on Artificial Intelligence for Smart Community*; Ibrahim, R., Porkumaran, K., Kannan, R., Mohd Nor, N., Prabakar, S., Eds.; Lecture Notes in Electrical Engineering; Springer: Singapore, 2022; Volume 758. [CrossRef]
13. Rosidin, S.; Muljono; Fajar Shidik, G.; Zainul Fanani, A.; Al Zami, F.; Purwanto. Improvement with Chi Square Selection Feature using Supervised Machine Learning Approach on COVID-19 Data. In Proceedings of the 2021 International Seminar on Application for Technology of Information and Communication (iSemantic), Semarangin, Indonesia, 18–19 September 2021; pp. 32–36. [CrossRef]
14. Zhao, Z.; Anand, R.; Wang, M. Maximum Relevance and Minimum Redundancy Feature Selection Methods for a Marketing Machine Learning Platform. In Proceedings of the 2019 IEEE International Conference on Data Science and Advanced Analytics (DSAA), Washington, DC, USA, 5–8 October 2019; pp. 442–452. [CrossRef]
15. Jamali, H.; Dascalu, S.M.; Harris, F.C. AI-Driven Analysis and Prediction of Energy Consumption in NYC's Municipal Buildings. In Proceedings of the 2024 IEEE/ACIS 22nd International Conference on Software Engineering Research, Management and Applications (SERA), Honolulu, HI, USA, 30 May–1 June 2024; pp. 277–283. [CrossRef]
16. Baesmat, K.H.; Shokoohi, F.; Farrokhi, Z. SP-RF-ARIMA: A sparse random forest and ARIMA hybrid model for electric load forecasting. *Glob. Energy Interconnect.* **2025**, *8*, 486–496. [CrossRef]
17. Baesmat, K.H.; Regentova, E.E.; Baghzouz, Y. A Hybrid Machine Learning–Statistical Based Method for Short-Term Energy Consumption Prediction in Residential Buildings. *Energy AI* **2025**, *21*, 100552. [CrossRef]
18. Alam, M.M.; Latifi, S. A Systematic Review of Techniques for Early-Stage Alzheimer's Disease Diagnosis Using Machine Learning and Deep Learning. *J. Data Sci. Intell. Syst.* **2025**. Vol. 00(00) 1–13. Available online: https://ojs.bonviewpress.com/index.php/jdsis/article/view/5037/1573 (accessed on 24 November 2025). [CrossRef]
19. Shaon, M.A.R.; Baghzouz, Y. Day-Ahead Residential Customer Load Forecasting Using Prophet. In Proceedings of the 2023 IEEE International Conference on Environment and Electrical Engineering and 2023 IEEE Industrial and Commercial Power Systems Europe (EEEIC/I&CPS Europe), Madrid, Spain, 6–9 June 2023; pp. 1–4. [CrossRef]
20. Sina, L.B.; Secco, C.A.; Blazevic, M.; Nazemi, K. Hybrid Forecasting Methods A Systematic Review. *Electronics* **2023**, *12*, 2019. [CrossRef]
21. Fattah, J.; Ezzine, L.; Aman, Z.; El Moussami, H.; Lachhab, A. Forecasting of demand using ARIMA model. *Int. J. Eng. Bus. Manag.* **2018**, *10*, 1847979018808673. [CrossRef]
22. Kumari, S.; Muthulakshmi, P. SARIMA Model: An Efficient Machine Learning Technique for Weather Forecasting. *Procedia Comput. Sci.* **2024**, *235*, 656–670. [CrossRef]
23. Ram, C.S.; Raj, M.; Chaturvedi, R.K. Boosting Time-Series Forecasting Accuracy with SARIMAX Seasonal Interval Automation. *Procedia Comput. Sci.* **2025**, *260*, 814–821. [CrossRef]
24. Li, J.; Lei, Y.; Yang, S. Mid-long term load forecasting model based on support vector machine optimized by improved sparrow search algorithm. *Energy Rep.* **2022**, *8* (Suppl. S5), 491–497. [CrossRef]

25. Nooruldeen, O.; Baker, M.R.; Aleesa, A.M.; Ghareeb, A.; Shaker, E.H. Strategies for predictive power: Machine learning models in city-scale load forecasting, e-Prime—Advances in Electrical Engineering. *Electron. Energy* **2023**, *6*, 100392. [CrossRef]
26. Wang, Y.-F.; Wang, M.Y.-F.; Tu, L.-Y. An Evaluation of Machine Learning Models for Forecasting Short-Term U.S. Treasury Yields. *Appl. Sci.* **2025**, *15*, 6903. [CrossRef]
27. Kanagaraj, R.; Revanth, N.; Vishnu, K.; Saxena, R. Demand Forecasting with XGBoost and LSTM: An Exploratory Analysis. In Proceedings of the 2025 International Conference on Inventive Computation Technologies (ICICT), Kirtipur, Nepal, 23–25 April 2025; pp. 1216–1220. [CrossRef]
28. Abbasimehr, H.; Shabani, M.; Yousefi, M. An optimized model using LSTM network for demand forecasting. *Comput. Ind. Eng.* **2020**, *143*, 106435. [CrossRef]
29. Gomez-Rosero, S.; Capretz, M.A.M.; Mir, S. Transfer Learning by Similarity Centred Architecture Evolution for Multiple Residential Load Forecasting. *Smart Cities* **2021**, *4*, 217–240. [CrossRef]
30. Li, Y.; Gao, Z.; Zhou, Z.; Zhang, Y.; Guo, Z.; Yan, Z. Abnormal Load Variation Forecasting in Urban Cities Based on Sample Augmentation and TimesNet. *Smart Cities* **2025**, *8*, 43. [CrossRef]
31. Aldossary, M. Enhancing Urban Electric Vehicle (EV) Fleet Management Efficiency in Smart Cities: A Predictive Hybrid Deep Learning Framework. *Smart Cities* **2024**, *7*, 3678–3704. [CrossRef]
32. Bin Kamilin, M.H.; Yamaguchi, S.; Bin Ahmadon, M.A. Radian Scaling and Its Application to Enhance Electricity Load Forecasting in Smart Cities Against Concept Drift. *Smart Cities* **2024**, *7*, 3412–3436. [CrossRef]
33. Richter, L.; Lenk, S.; Bretschneider, P. Advancing Electric Load Forecasting: Leveraging Federated Learning for Distributed, Non-Stationary, and Discontinuous Time Series. *Smart Cities* **2024**, *7*, 2065–2093. [CrossRef]
34. Malashin, I.; Tynchenko, V.; Gantimurov, A.; Nelyub, V.; Borodulin, A. Applications of Long Short-Term Memory (LSTM) Networks in Polymeric Sciences: A Review. *Polymers* **2024**, *16*, 2607. [CrossRef] [PubMed]
35. Islam, M.S.; Alam, M.M.; Ahamed, A.; Meerza, S.I.A. Prediction of Diabetes at Early Stage using Interpretable Machine Learning. In Proceedings of the SoutheastCon 2023, Orlando, FL, USA, 13–16 April 2023; pp. 261–265. [CrossRef]
36. Alharbi, F.R.; Csala, D. A Seasonal Autoregressive Integrated Moving Average with Exogenous Factors (SARIMAX) Forecasting Model-Based Time Series Approach. *Inventions* **2022**, *7*, 94. [CrossRef]
37. Jamali, H.; Karimi, A.; Haghighizadeh, M. A new method of cloud-based computation model for mobile devices: Energy consumption optimization in mobile-to-mobile computation offloading. In Proceedings of the 6th International Conference on Communications and Broadband Networking, Singapore, 24–26 February 2018. [CrossRef]
38. Chetry, B.P.; Kar, B. Online Signature Verification Using Chi-Square ($\chi 2$) Feature Selection and SVM Classifier. In Proceedings of the 2024 6th International Conference on Energy, Power and Environment (ICEPE), Shillong, India, 20–22 June 2024; pp. 1–6. [CrossRef]
39. Hermo, J.; Bolón-Canedo, V.; Ladra, S. Fed-mRMR: A lossless federated feature selection method. *Inf. Sci.* **2024**, *669*, 120609. [CrossRef]
40. Forghani, R.; Savadjiev, P.; Chatterjee, A.; Muthukrishnan, N.; Reinhold, C.; Forghani, B. Radiomics and Artificial Intelligence for Biomarker and Prediction Model Development in Oncology. *Comput. Struct. Biotechnol. J.* **2019**, *17*, 995–1008. [CrossRef]
41. Farrokhi, Z.; Baesmat, K.; Regentova, E. Enhancing Urban Intelligence Energy Management: Innovative Load Forecasting Techniques for Electrical Networks. *J. Power Energy Eng.* **2024**, *12*, 72–88. [CrossRef]
42. Zhao, Q.; Xiang, W.; Huang, B.; Wang, J.; Fang, J. Optimised extreme gradient boosting model for short term electric load demand forecasting of regional grid system. *Sci. Rep.* **2022**, *12*, 19282. [CrossRef] [PubMed]
43. Zhao, M.; Guo, G.; Fan, L.; Han, L.; Yu, Q.; Wang, Z. Short-term natural gas load forecasting based on EL-VMD-Transformer-ResLSTM. *Sci. Rep.* **2024**, *14*, 20343. [CrossRef]
44. Joshi, B.; Singh, V.K.; Vishwakarma, D.K.; Ghorbani, M.A.; Kim, S.; Gupta, S.; Chandola, V.K.; Rajput, J.; Chung, I.-M.; Yadav, K.K.; et al. A comparative survey between cascade correlation neural network (CCNN) and feedforward neural network (FFNN) machine learning models for forecasting suspended sediment concentration. *Sci. Rep.* **2024**, *14*, 10638. [CrossRef] [PubMed]
45. Sarker, B.; Chakraborty, S.; Čep, R.; Kalita, K. Development of optimized ensemble machine learning-based prediction models for wire electrical discharge machining processes. *Sci. Rep.* **2024**, *14*, 23299. [CrossRef] [PubMed]

energies

MDPI

Review

Power Consumption Analysis of the Power System in a Gigafactory: A Review

Manzar Ilyas *, Paolo Guglielmi and Andrea Mazza

Department of Energy "Galileo Ferraris", Politecnico di Torino, 10129 Torino, Italy;
paolo.guglielmi@polito.it (P.G.); andrea.mazza@polito.it (A.M.)
* Correspondence: manzar.ilyas@polito.it

Abstract

Recent decades have seen substantial growth in the demand for lithium-ion batteries (LIBs); as a result, the number of gigafactories is increasing. The power systems of these gigafactories are the most important parts of these installations, as they are directly responsible for energy efficiency, operational cost, and sustainable growth of the LIB industry. This necessitates the need for comprehensive studies of power consumption in a gigafactory during the LIB manufacturing process. This paper presents a detailed review of the state-of-the-art of different parts and components of power systems in gigafactories, and power consumption estimation during cell production. This research analyzes the existing components of a power system, including power sources, different power distribution mechanisms, various power equipment, thermal management strategies, failure analysis methods, and several technologies for regenerative functions. The analysis of the above-mentioned components, systems, and technologies will enable us to understand the cumulative power consumption profile of the gigafactory, including the power consumption at each step of the production process, for normal non-production operations, like powering and lighting the facility, and for the complex and highly sophisticated power distribution system. The outcomes of this research paper highlight the importance of an optimized power system for the gigafactory, with maximum possible efficiency and minimal power losses during transmission, distribution, operational stages, and the cell formation process. This paper also helps to understand the shortcomings in existing systems and technologies, suggests improvements, and provides targets for future research directions.

Keywords: lithium-ion batteries; gigafactory; cell formation process; power systems; power distribution; power equipment; failure analysis; regenerative function

1. Introduction

Large-scale manufacturing facilities for lithium-ion batteries (LIBs) with a total manufacturing capability of more than 1 GWh are generally referred to as gigafactories. These gigafactories are fundamental to meeting the ever-increasing demand for LIBs in a rapidly digitalizing world. The production of LIBs has been projected to increase about sevenfold to reach 2200 GWh by 2030 [1]. These gigafactories consume a massive amount of electrical energy for LIB production, particularly during the cell formation process. For instance, the manufacturing of a 24 kWh battery pack requires 16.31 MWh [2,3]. This makes an energy-efficient power system a critical necessity for a modern, reliable, and sustainable gigafactory [4]. This research analyzes power consumption during cell manufacturing. The main goal is to identify different power consumption patterns and trends, establish key areas of power losses, and propose improvements to enhance the energy efficiency of the power system for future research directions.

Figure 1 provides brief information about the key elements of the power system analysis in a gigafactory. Historically, power consumption analysis was overlooked. But in recent times, with a growing focus on sustainable and energy-efficient power systems for industrial plants, the field's importance has increased, and the available literature has expanded significantly. This review paper explains the critical components and methodologies for power consumption analysis in gigafactories. These include power sources, distribution systems, power equipment, thermal management systems, failure analysis, regenerative functions, and other critical energy-consumption metrics of cell formation systems.

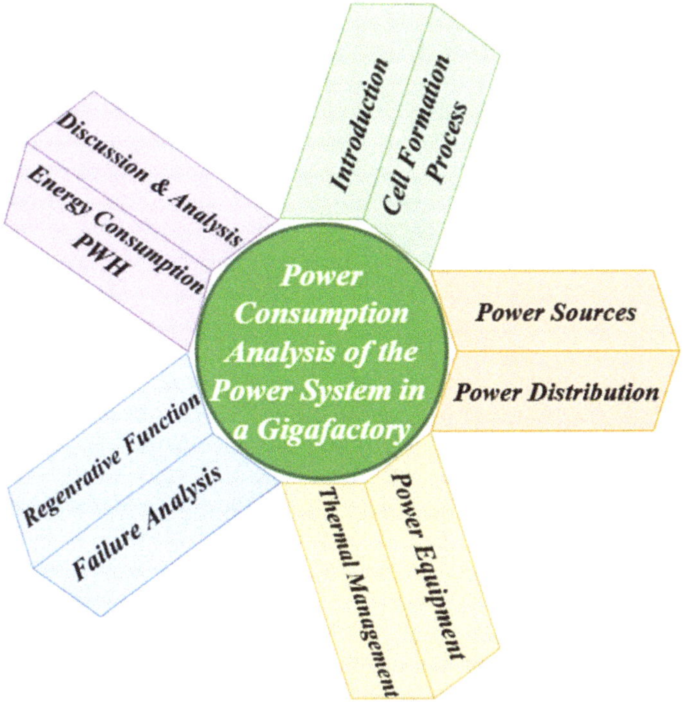

Figure 1. Key elements of the power system analysis.

Exploring the diversification of the power sources and their integration into the gigafactory's existing power systems is fundamental. An increase in the production of renewable energy sources, such as solar and wind, may support the gigafactories' reliance on the traditional grid power. The integration of renewable power into the existing power systems of industrial plants results in considerable reductions in operational costs and carbon emissions, as highlighted in [5]. For example, Tesla's gigafactory in Nevada has installed solar panels on site and uses this renewable power in combination with grid power, which has reduced its dependency on fossil fuels and other non-renewable power sources and increased its sustainability [6].

The thermal management of a power system is crucial for preventing equipment from overheating and maintaining optimal operating conditions [7]. A large amount of heat is generated by the electrical equipment in a gigafactory power system and released into the surrounding environment. These thermal loads have been effectively reduced by using a liquid cooling system in combination with heat exchangers in industrial plants.

https://doi.org/10.3390/en19051345

The advantages of this cooling method have been explained in [8]. In addition to cooling liquids and heat exchangers, improvements in the design and use of advanced materials for cooling systems and heat sinks can effectively enhance cooling system efficiency by managing the generated heat [9,10].

Optimization and reduction in operational costs of the industrial plant can be achieved with regenerative functions used as a power-saving measure. Such regenerative systems are particularly advantageous in high-power applications, where frequent braking and acceleration cycles occur. Regenerative systems can effectively be used to recover and reuse heat energy that would otherwise be lost as waste heat [11]. Integration of regenerative systems into existing industrial processes can result in significant energy savings and a reduction in operational costs [12]. Another important factor for power system analysis is understanding and mitigating power system failures. Mitigation of these failures is critical for maintaining an uninterruptible power supply (UPS) and the smooth functioning of gigafactories (or any other industrial plant). Commonly identified failure points in an industrial plant's power system are switches, transformers, and power transmission lines, as explained in [13]. The use of predictive maintenance and real-time monitoring is necessary to prevent power failures [14]. It is vital to avoid or decrease power failures in the power system through the implementation of robust maintenance schedules and high-quality switching and control components [15].

For an effective and efficient analysis of the power system in the gigafactory, per-watt-hour power consumption is considered a critical parameter. Comparison of different power consumption metrics from several gigafactories can help us identify best practices for power saving and also indicate the critical areas for improvement [16]. Power management systems, specific operational practices, and advanced technologies are important factors for enhancing power efficiency. As an instance, the overall power consumption can be considerably reduced by implementing the strategy of optimized production schedules and energy-efficient lighting systems in industrial plants [17].

The reminder of the paper is as follows: Section 2 presents the cell formation process; Section 3 focuses on power sources; Section 4 analyses the power distribution system, including earthing systems; Section 5 focuses on power equipment for the cell formation process; Section 6 describes all the thermal management approaches; Section 7 covers the generative function; Section 8 covers the failure analysis in gigafactory. All of the above aspects are discussed in Section 10, whereas Section 11 presents the concluding remarks.

2. Cell Formation Process

The most important stage in the gigafactory of LIBs is the cell formation process. This process electrochemically activates newly manufactured LIB cells. The LIB cells are electrochemically activated through rigorous and continuous charging and discharging cycles. Different charging and discharging topologies and C-rates are used to achieve desired characteristics. This makes the cell formation process one of the most energy-intensive stages in the gigafactory, consuming more power compared to any other stage in the LIB manufacturing process.

The formation area of the cell manufacturing facility consists of formation towers made up of chambers or individual racks with an independent control channel for each cell in both topologies [18].

The power system required for the cell formation process comprises numerous power equipment, including battery testers, AC/DC and DC/DC converters, cell chargers, and monitoring and protection systems. The performance of the cell formation process can be improved by using higher efficiency and innovative power systems.

https://doi.org/10.3390/en19051345

Cell formation includes important processes, critical for normal cell operation, including Solid Electrolyte Interphase (SEI) formation, Cathode Electrolyte Interphase (CEI) formation, and structural changes in active materials [19,20]:

- The cell formation process is responsible for the formation of the SEI layer on the surface of the anode, so that direct contact between the anode and electrolyte can be avoided during normal cell operation [19].
- Just as SEI is formed at the anode layer, CEI is formed at the cathode surface, and generally the thickness of the CEI layer is less than the SEI layer [21].
- Cell formation process is also responsible for establishing stable electrical contact between the electrolyte and the active materials [22].

The target end of the cell formation process is according to the parameter called Columbic efficiency (CE): the cell formation process is thought to be completed when the value of CE reaches 100% [19].

Figure 2 shows the flowchart of the critical stages in the cell formation process of LIBs. These stages are explained in the following paragraphs. During the first stage (precharging), the newly manufactured LIBs are prepared for formation cycling, leaving an open hole at the top of the cell. The cells are charged to a very low level with very low charging current rates, i.e., (1/20)C [23].

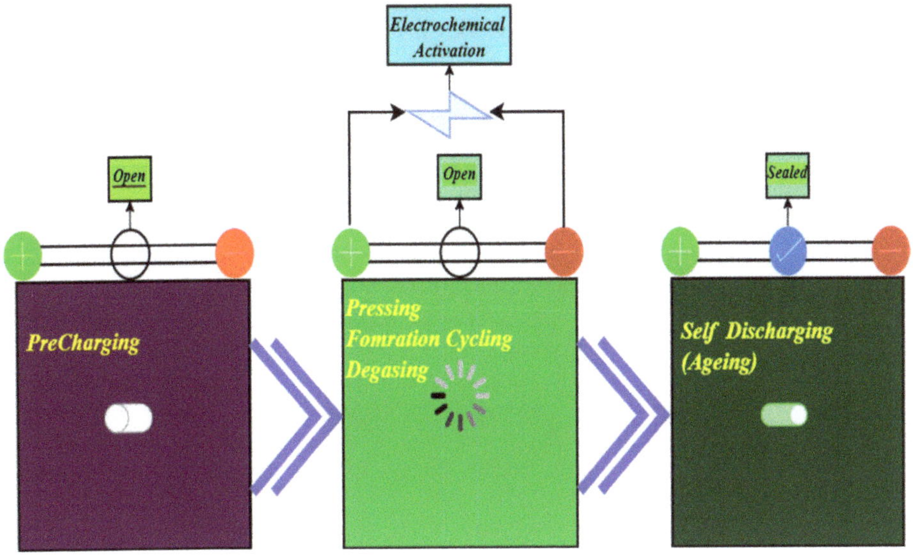

Figure 2. Basic flowchart of the cell formation process.

In the second stage (cell assembly), which consists of pressing, formation cycling, and de-gassing, rigorous formation cycles are carried out on these cells under considerable pressure to avoid swelling during the initial charging period. Cells are charged and discharged according to different protocols depending on the charging rates and current and voltage levels. Different chemical materials in the cells produce gases during this charging process, which are sucked out of the cells through the open hole at the top of the cases.

When the formation cycling reaches the desired levels, the third stage (Self Discharging or Aging) starts: after degassing any remaining gases from the cell, the hole is sealed. Then these cells are moved to the shelves for the aging test (self-discharging) to analyze the results of the formation process and the quality of the produced cells.

https://doi.org/10.3390/en19051345

The cell formation process impacts cell performance and determines the characteristics and performance of parameters such as aging behavior, power capability, energy consumption, cost, and safety.

Figure 3 shows the different stages of the cell formation process, consisting of pre-charging, de-gassing, formation cycling, and aging (self-discharging) [24]. The important steps of the cell formation process are briefly explained below:

- **Pre-charging**: In the pre-charging phase, the LIB cells are charged for the first time to form the SEI layer on the anode and create a good electrical connection among active materials and electrolyte.
- **De-gassing**: In the de-gassing phase, the LIB cells are compressed under suitable pressure to avoid swelling and to remove any gases generated during the pre-charging stage.
- **Formation Cycling**: In the formation cycling phase, the LIB cells are charged according to the manufacturer's guidelines. Usually, charging starts with a lower current and is increased to a certain level set by the cell manufacturer. At this stage, the cells are charged up to 90–95% using constant-current (CC), constant-voltage (CV), or constant-current–constant-voltage (CC-CV) charging techniques. Some novel and innovative protocols are also available as alternatives, such as MS-CC (multi-stage constant-current), PC (pulse charging), CP-CV (constant-power–constant-voltage), and BC (boost charging). However, they are not as widely used as CC-CV. Most gigafactories still prefer the CC-CV protocol due to its obvious advantages. CC-CV offers steady current rates and a simpler power electronics setup. It generates less heat and reduces the power consumption to maintain the cooling of the formation chambers and towers. It also avoids noise and spikes, ensuring a smooth ramp curve during formation cycles. CC-CV provides a higher CE as compared with other protocols and allows virgin cells to be fully charged without overstressing the structure (cell chemistry). Although, when considered in isolation, these innovative formation protocols might show comparatively less power consumption during the formation process, the lifecycle assessment shows that CC-CV still provides the best solution as a cell formation protocol [25]. At the end of the formation cycling stage, the cell case is properly sealed to avoid any leakage.
- **Aging (Self-discharging)**: In the aging (self-discharging) phase, 90–95 percent charged cells are stored in specially designed cell storage chambers, where these cells undergo the self-discharging process. This is important to understand the quality of newly produced cells.

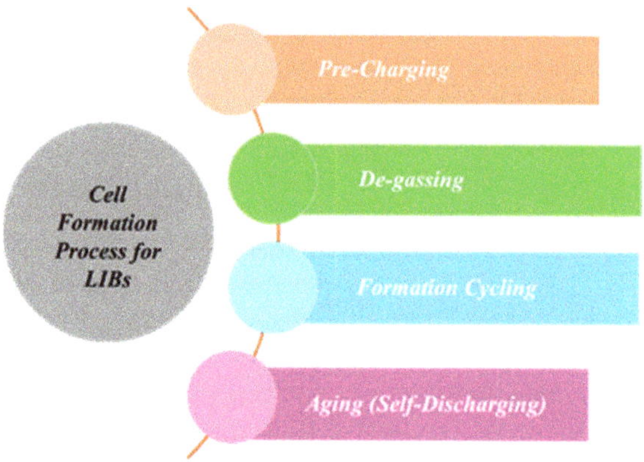

Figure 3. Important stages of the cell formation process [23].

https://doi.org/10.3390/en19051345

3. Power Sources

Power sources are critical for a gigafactory producing LIBs, as these power sources ensure the efficient, sustainable, and economical operation of the gigafactory [26]. Usually, a combination of different power sources is used in a gigafactory to manage the huge energy requirements for battery production [27]. This section analyzes common power sources used in a gigafactory (including grid power, renewable energy systems, and backup power supply systems), as well as the characteristics of each power source. In addition to this, it also analyzes the importance of energy storage systems in order to ensure a robust and reliable power supply for the gigafactory [28].

Figure 4 shows the approximate estimation of the percentages of different power sources, combined to meet the power requirement of Tesla's gigafactory located in Nevada, based on the data published in the 2023 annual impact report [29].

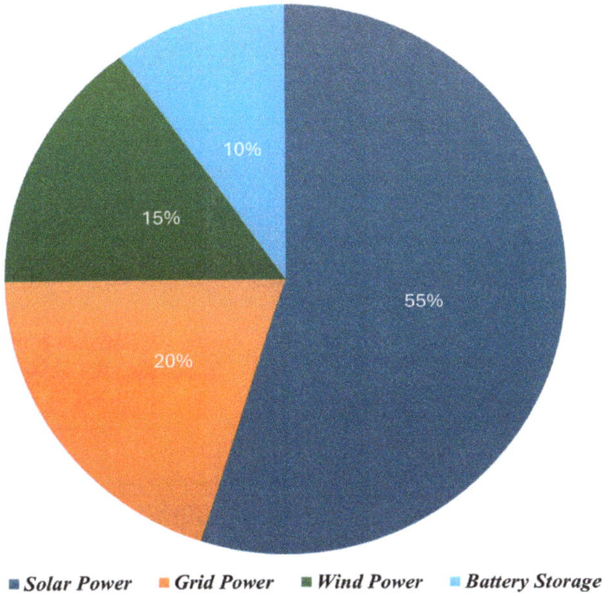

Figure 4. Power sources used in Tesla's factory in Nevada.

Among these power sources, grid power remains the largest for most gigafactories. To maintain an optimal production process, the grid sources supply continued power to the gigafactory. However, dependence on grid power only has its own drawbacks: it can be expensive due to the change in electricity prices, and it can also face outages due to grid instability [30].

Gigafactories are signing green and stable energy contracts with private electricity providers to manage these risks, including controlling electricity prices and minimizing power outages. These contracts may usually include clauses about demand response programs: in that case, gigafactories agree to minimize their power usage during peak demand hours, and, in return, these electricity suppliers give them reduced prices and other benefits [28].

Gigafactories manufacturing LIB cells require a large amount of stable and uninterrupted electricity. Renewable energy sources play an important role in meeting overall energy requirements. During grid-sourced power outages, these systems can also serve as backup power sources for critical sections of the gigafactory [31]. Apart from supporting power supply, renewable energy sources also help the LIB industry to optimize cost and

https://doi.org/10.3390/en19051345

reduce CO_2 emissions during the production process, as reported in the annual reports of famous LIB manufacturers such as CATL and LGES [32,33]. Usual renewable sources adopted in a gigafactory include the following:

- **Solar power**: In recent times, gigafactories have begun to install a large number of solar panels on rooftops and empty lands in the surroundings of the main structure, to support grid power, reduce operational costs, and minimize the environmental impact of the gigafactory during manufacturing processes [27].
- **Hydropower**: If a gigafactory is located in a suitable location with hydroelectric potential due to the presence of a water reservoir, it can be used to generate hydropower from this water resource. This hydroelectric power can serve as a backup power source to the traditional grid sources. This is quite rare and location-dependent, as gigafactories usually prefer to use on-site power generation to reduce external dependence [34].
- **Cogeneration systems**: They are often called combined heat and power (CHP) units as backup power sources. These CHP units generally consist of an electrical generator equipped with the capability to recover and use the generated heat for power generation [35].

  Such CHP systems are quite efficient and useful due to their dual-generation capability, producing both thermal energy and electricity. The electricity is produced by using the extra heat generated during the process of electricity generation [36,37].

  During grid outages, CHPs are usually more efficient and quicker than renewable energy sources as backup power sources. CHPs use natural gas as a generation source. They can also efficiently support primary sources during peak demand hours in case of deficiency in terms of power. However, the main drawback of these CHPs is that they use natural gas as fuel, resulting in increased carbon emissions from facilities [38].
- **Electrical storage systems**: Gigafactories need an uninterrupted, balanced, and reliable power supply throughout the manufacturing process. This gives rise to the use of energy storage systems in order to make intermittent energy from renewable energy sources, such as solar and wind [39], consistently available. Gigafactories are established to manufacture LIBs. An efficient and reliable energy storage system can be built by using similar batteries produced in the same gigafactory. These energy storage systems can be used to store additional renewable energy produced during low-demand hours. Later, this energy can be released during peak-demand hours, or at a time when the potential for renewable power generation is low due to atmospheric conditions [27]. The prominent examples of such energy storage systems are Tesla's Powerpack and Megapack, which are currently being used in Tesla's facilities and also by other manufacturers around the globe [40].
- **Flywheels**: Flywheel technology can be used to store power in the form of kinetic energy. In case of a change in power demand inside the gigafactory, it can provide a rapid response [41,42].
- **Thermal storage systems**: To balance the heating and cooling requirements in gigafactories, thermal energy storage systems are used to store extra heat and thermal energy. This technology can be used effectively and reliably in CHP units and conjunction systems (a combination of storage and backup).

All gigafactories must have a sophisticated and robust Energy Management System (EMS) to control the power generated from these diverse power sources. An EMS is used to optimize energy consumption, maximize efficiency, and minimize cost by monitoring real-time energy production, consumption, and storage levels. This system ensures the most suitable available power source is used based on the prediction of energy production and

https://doi.org/10.3390/en19051345

demand. An EMS makes this prediction based on production schedules and environmental constraints [27,43].

Furthermore, a gigafactory requires the implementation of smart grid technologies. They are used for the integration of power from these diverse renewable power sources and storage systems with existing traditional power sources and the grid [44]. This integration of smart grid technology makes the power system more efficient and reliable through the implementation of real-time monitoring, supply and demand balance, and automated control [45].

4. Power Distribution

Gigafactories have quite a complicated power distribution network. The productivity and efficiency of a gigafactory can be considerably enhanced by using an efficient power distribution system. This system ensures that the power is optimally supplied to all the equipment in the factory. It also helps to reduce operational costs by minimizing disruptions and losses during power supply and distribution [46]. This section reports the details of the different components of power distribution systems, including transformers, distribution networks, smart grids, and automation technologies [47].

A. *Power distribution system overview*

The power distribution system of the gigafactory can be divided into three sections, namely, high-voltage (HV), medium-voltage (MV), and low-voltage (LV) sections.

The HV section represents the power supply from sources (the utility grid and the backup power sources) to the gigafactory. The MV section represents the internal power distribution system, whereas the LV section represents the process unit of the gigafactory where the LIBs are produced. This division of the power distribution system enables us to understand and analyze each section and piece of power equipment. Appropriate safety and protection devices and equipment are designed and proposed according to the voltage level of the particular section. These safety measures and devices protect power equipment and personnel working in proximity to high-voltage equipment.

An overview of the power distribution system in a gigafactory is shown in Figure 5: it can be divided into three sections, namely, the high-voltage (HV) section, medium-voltage (MV) section, and low-voltage (LV) section, based on the operating voltage range of each section.

The HV section of the power distribution system is shown in Figure 6; this section generally refers to the power supply section of the gigafactory, consisting of the utility grid supply, backup power systems such as generators, high-voltage switchgears, and the main transformer of the manufacturing facility. This section is generally operated at 220 kV and 100 MVA. Due to these high operational voltage ranges, this section is classified as a high-voltage section. Then, a step-down power transformer is used to step down this power supply to the MV section at 20 kV.

Figure 7 shows the MV section of the power distribution system. This section generally refers to the internal power distribution. It consists of the MV switchgear, a step-down transformer to step down voltage from MV to LV, capacitor banks, renewables, and energy storage systems for additional backup power for the gigafactory.

Figure 8 shows the LV section of the power distribution system. This section represents the process section of the gigafactory, where the cell production line is located. It is the most complex section and consists of production line equipment, power conversion systems, motor control systems, uninterruptible power supply systems for the control and monitoring of manufacturing equipment, and formation chambers. The operational voltage range of the LV section extends up to 480 V and 2.5 MVA.

https://doi.org/10.3390/en19051345

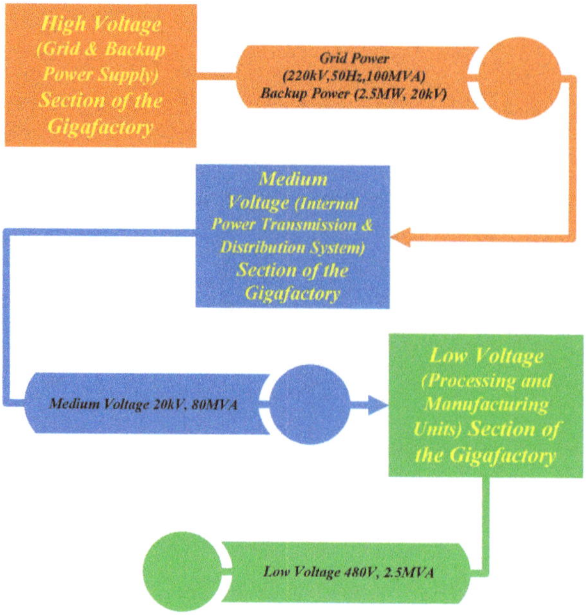

Figure 5. Overview of the power distribution network of the gigafactory.

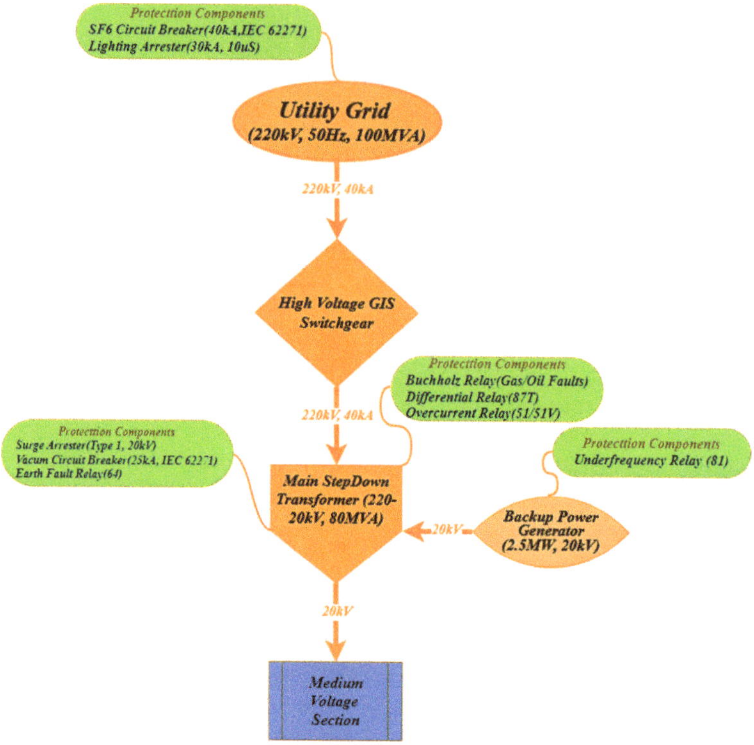

Figure 6. HV section of the power distribution network.

https://doi.org/10.3390/en19051345

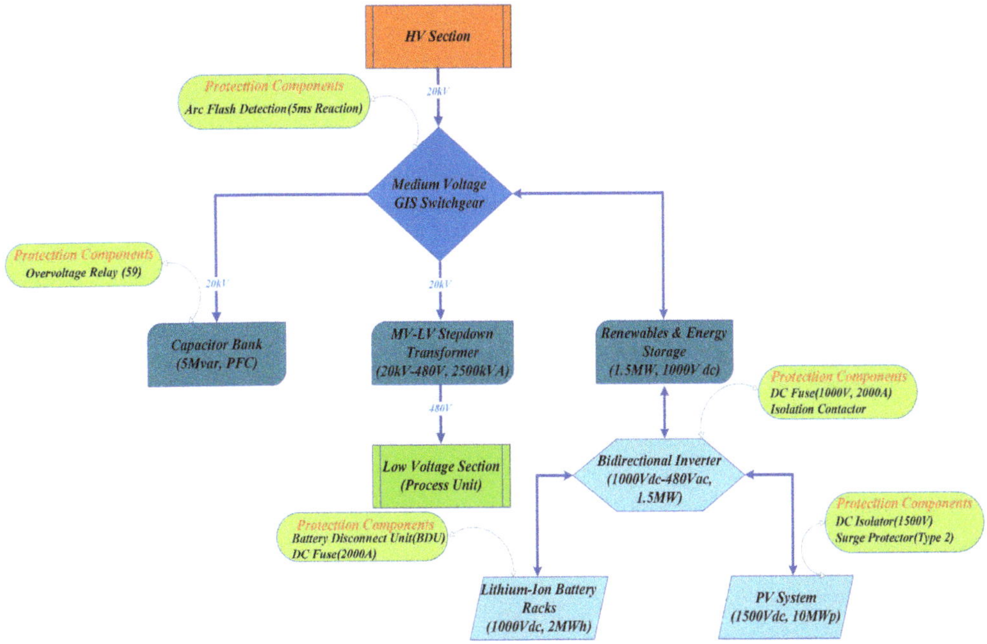

Figure 7. MV section of the power distribution network.

The three sections described above are combined to form a very complex power distribution network (PDN) of a gigafactory. This power distribution network is explained in detail in the following sections of this paper.

B. *Electricity interconnections and protections*

Generally, electricity is supplied to the gigafactory at an HV level from the electricity grid. So, this voltage is stepped down using transformers, making it suitable for factory usage. This makes transformers one of the basic components of the distribution network: it is crucial to select transformers with optimal characteristics and install them at suitable locations to maintain a stable voltage for different voltage sections, ensuring that power losses are minimized [48].

The electrical power distribution is optimized by the strategic and careful placement of transformers at optimized locations in the manufacturing facility [13,49]. These strategic locations include the main intake station, where HV (220 kV) power is received from the electric grid and stepped down to suitable MV (20 kV) and low-voltage (400/480 V) levels. The locations of these transformers must not be too far from the production lines, heating, ventilation, and air conditioning (HVAC) systems, and battery storage systems, which are major power-consuming parts of the factory [50].

The basic function of the switchgears is to control, isolate, and protect electrical equipment. They consist of circuit breakers, switches, and fuses. These switchgears are carefully designed to manage electrical flow and protect the power system. They minimize faults to ensure the safe and reliable operation of the entire power distribution network. These switchgears are placed at both HV and MV levels [51].

Busbars enable efficient power distribution among different electrical paths for different components. These are metallic strips or bars used to conduct electricity within a switchboard, distribution board, substation, or other electrical apparatus. Usually, bus-bars are used to conduct electricity over short distances.

https://doi.org/10.3390/en19051345

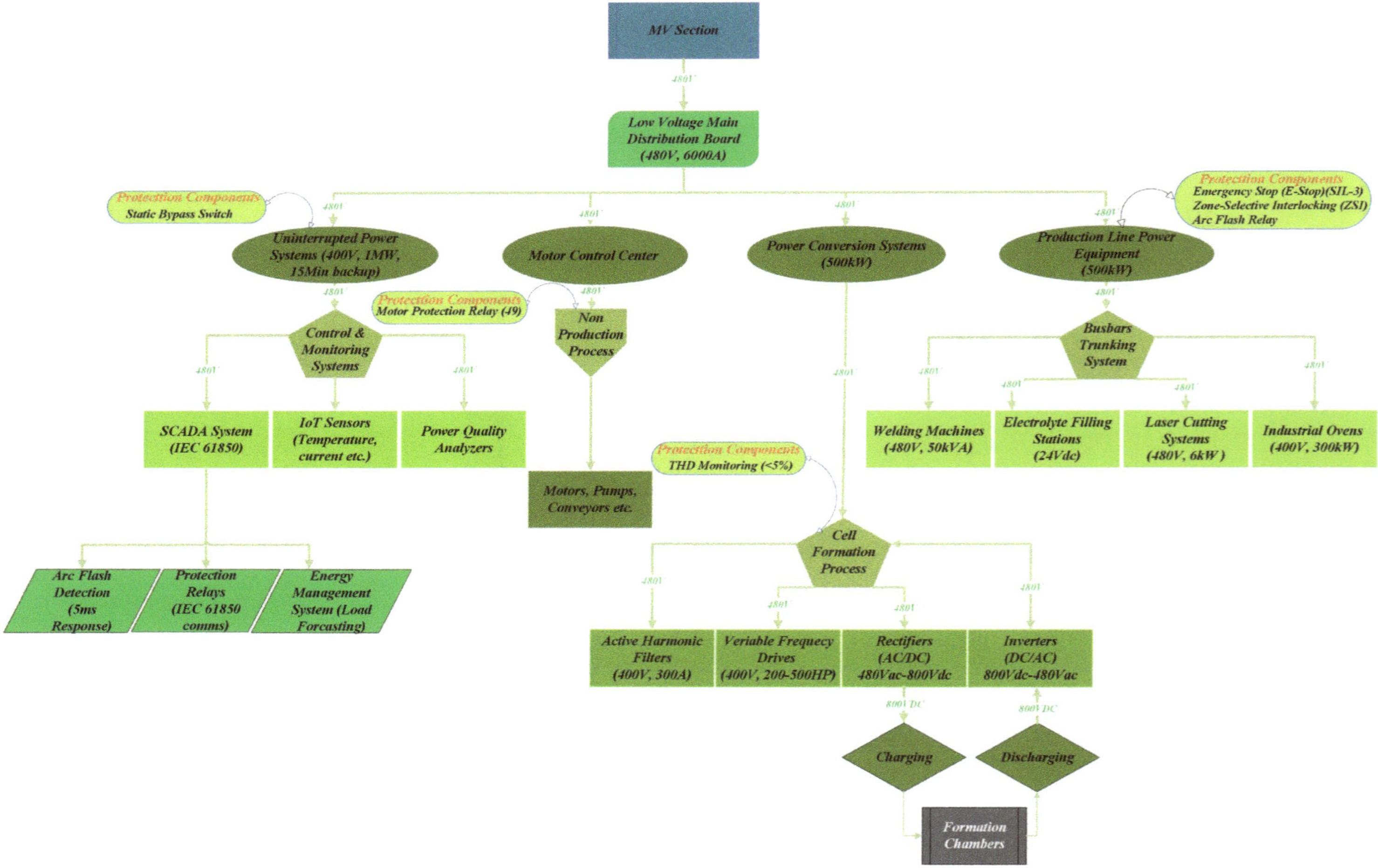

Figure 8. LV section of the power distribution network.

https://doi.org/10.3390/en19051345

Cabling is a critical part of an efficient power distribution network. The cabling should be high-quality and well-insulated to ensure safety and minimize transmission losses. The cabling system layout is carefully designed to minimize the distance between the power distribution boards and the consumption points, enhance transmission efficiency, and reduce energy losses. The efficiency and reliability of the power distribution network are enhanced by making improvements to the design layout of the cabling network and using cables made from advanced and high-efficiency materials [52].

C. *Power management system*

The development of smart grid technology has transformed the approach for managing the PDN of a gigafactory. Smart grids optimize the flow of electricity in the gigafactory by incorporating advanced sensors, communication networks, and automated control systems [53]. The key features of smart grid technology are as follows:

- The real-time monitoring of electricity consumption, generation, and storage is done by smart grid technology. This real-time monitoring enables the identification of deficiencies, the detection of faults, and the optimization of power flow across the distribution network.
- The automated control system is used to adjust the power distribution based on real-time demand of the facility, reduce losses, and enhance efficiency. For example, power conservation is achieved by temporarily shutting down non-critical processes during peak demand hours [54,55].
- EMS is another important part of smart grid technology. It provides a comprehensive analysis of energy consumption by integrating data from various sections. EMS ensures optimal distribution and energy saving by automating power distribution based on predictive analytics [56].
- It is critical to maintain an uninterrupted and reliable power supply throughout the manufacturing procedure. Various redundancy measures are usually incorporated into the power distribution systems of the gigafactory [57].
- Multiple meshed distribution pathways are laid for electricity distribution to ensure an uninterrupted supply of power. In case of the failure of one path, the other can take over. These can also be used for load sharing to avoid the overloading of one path.
- Advanced control systems are required to manage the variability and intermittency of the power if generated by renewable energy sources like solar and wind. Energy storage systems, such as large-scale batteries, are used to store excess energy generated during peak production hours of renewable sources. Then, it can be released during periods of low renewable generation or high-demand hours [58].

These integrated smart grid systems are used for dynamic load balancing to ensure a stable and continuous power supply. This is achieved by buffering the variability of renewable energy sources through sophisticated grid management and energy storage solutions [59].

Backup power generators and battery storage systems are installed on the site in addition to the primary power sources to provide emergency power during grid outages or system failures [60].

In its gigafactories, Tesla enhances the efficiency and reliability of the power distribution systems of the factories by utilizing the advanced EMS, and by integrating real-time monitoring and smart grid technologies into the PDN, making these factories a prime example of manufacturing units with modern and advanced power systems [40].

D. *Earthing and Grounding*

The grounding and earthing of electrical connections and equipment are critical for optimal and reliable operation in large industrial plants such as gigafactories. These

https://doi.org/10.3390/en19051345

grounding measures are necessary for the safety and protection of the human resources and power equipment inside the gigafactory.

Generally, there are two types of systems used in large-scale industrial plants, such as gigafactories. These are known as the TN system and the IT system. The choice of the system is made based on the scale of the industrial plant, the nature of the operation, and industrial standards for such operations [61,62].

- **TN System**: In a TN system, a separate earth conductor PE (Protective Earth) runs through from the transformer to the main distribution board. The transformer's neutral terminal is connected to the exposed chassis of power equipment by a PE conductor. Then, the whole system is grounded using this connection point of the main transformer [63].
- **IT System**: An IT system is generally used at the secondary-level distribution network for the critical equipment of a gigafactory. In this system, the exposed conductors of the power equipment are separately earthed. While the secondary transformer is either not grounded at all or is earthed with a fault limiter, having a high impedance value [63].

In the gigafactory, it is critical to choose a suitable earthing and grounding system in accordance with the power equipment, operational voltage ranges, and the critical nature of the process. In the important stages of battery manufacturing, including the cell formation process, a combination of TN and IT systems is used to achieve optimal results.

The PDN of the gigafactory can be divided into two stages, i.e., the transmission and distribution stage and the manufacturing and processing stage. In the first stage, which includes the HV and MV sections of the PDN, the TN system is used to achieve optimal results. In the second stage (LV section of the PDN), power supply interruptions are minimized by implementing an IT system for the earthing and grounding of power equipment. This stage consists of the manufacturing and distribution section of the gigafactory.

Figure 9 shows the types and implementation of earthing and grounding systems. It shows the sections of the gigafactory where each earthing and grounding system is implemented. It also indicates two categories of power components according to the earthing and grounding system implementation.

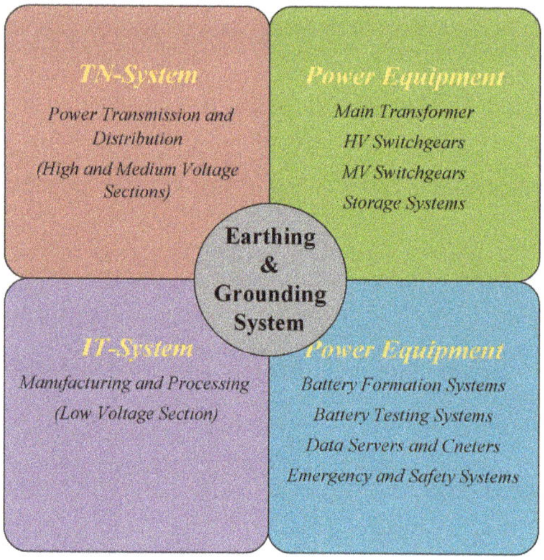

Figure 9. Earthing and grounding system.

https://doi.org/10.3390/en19051345

5. Power Equipment

Different types of power equipment are used in a gigafactory, which are critical for the efficient and reliable working of the factory [64]. For the power system concerned, the cell formation process is the most critical part of a gigafactory. This section discusses the types of power equipment used in the gigafactory for the cell formation process [65,66]. The key power equipment in the gigafactory is explained in the following sections of this article.

A. *Process-related power equipment*

The power equipment specifically used in the cell formation process includes the following:

- Cell formation chargers are the most critical devices in the formation process. These chargers are used to charge the battery cells for the very first time at the beginning of the formation process. This process usually takes several hours [67]. Battery performance and longevity are the main purposes of the cell formation process. These were achieved by using high-precision chargers, which are mandatory to ensure a uniform charging profile across all cells inside a formation chamber. Voltage and current profiles are adjusted according to the real-time monitoring of cells through advanced formation chargers with unique charging and discharging circuitries [68].
- The devices used to perform continuous charging–discharging cycles for the cell formation process are called cyclers [69].
  These devices track cell performance and capacity during charging and discharging cycles. These are equipped with precise control and real-time monitoring. The cells are passed through rigorous cycles with high levels of power, voltage, and current, and must prove their quality by showing consistent performance throughout the formation cycles [23].
- The battery management system (BMS) is installed with formation equipment to acquire real-time data accurately during the cell formation process. Real-time voltage, current, temperature, and other parameters for each cell are monitored by this BMS [70]. BMS is used to optimize the formation process to enhance performance and efficiency by using the data collected to identify any cells that do not meet quality standards [71]. Robust safety systems are mandatory in the cell formation process because of the potential risks of thermal runaway and cell failure. The instances of thermal runaway over the years are listed in [72]. These safety systems are activated during anomalies and consist of fire suppression systems, gas detection sensors, and automated shutdown mechanisms [73]. To achieve greater efficiency and reliability, it is vital to integrate these power equipment systems into a cohesive and automated setup.
- Advanced EMSs, used in modern gigafactories to monitor and control power equipment, are used to ensure optimal performance and energy consumption [74].
- Automation technologies, including robotics and machine learning algorithms, are used to enhance the quality control and productivity of the manufacturing process [18].
- This automation process is also used for predictive maintenance to predict and prevent failures even before their occurrence, and the data retrieved from equipment is continuously analyzed, enhancing the component lifespan and reducing the maintenance costs [75].

B. *Auxiliary power equipment*

A considerable amount of total energy in a gigafactory is consumed by auxiliary power equipment. This power consumption may range from anywhere between 6.33% and 8.89% of the total power consumption of the plant [76]. These components are crucial for controlling the speed and torque of machines used in various manufacturing processes, including electric motors and variable-frequency drives (VFDs). A considerable amount of

https://doi.org/10.3390/en19051345

energy is saved, and higher production efficiency is achieved in the manufacturing process by using these high-efficiency motors and VFDs [65].

The Uninterruptible Power Supply (UPS) is also a crucial piece of power equipment in a gigafactory. UPS ensures uninterrupted operation during power failures and protects crucial devices and data by providing backup power.

HVAC systems are also important power equipment inside the gigafactory. HVAC systems are used to create optimal environmental conditions by controlling temperature, air, and humidity within the manufacturing facility. Technologically advanced HVAC systems can improve power efficiency and decrease operational costs by using smart controls and heat recovery units [9].

Energy-efficient lighting systems are another necessary type of power-consuming equipment in a gigafactory. Electricity consumption can be significantly reduced by using LED lighting systems with smart controls.

Power consumption efficiency can be further improved by using automated lighting systems that can adjust the intensity of lighting based on occupancy and ambient light conditions [77].

6. Thermal Management

The thermal management of the heat generated during the manufacturing of LIBs in a gigafactory is a very critical aspect of its optimal operation. An effective thermal management system is mandatory to have an efficient, reliable, and safe manufacturing process [78]. This section explores the various techniques and technologies used in a gigafactory for thermal management. The thermal management system of a gigafactory covers several aspects, including waste heat recovery, the cooling of the manufacturing equipment and devices, and HVAC systems for the facility [79]. Key factors of the thermal management system are briefed in the following paragraphs.

A thermal management system is mandatory for the safety of the cells during the cell formation process. LIBs are highly sensitive to temperature fluctuations and can be severely affected if the temperature inside the formation chambers is not controlled and managed properly [80]. If cells get overheated during the formation process, it can result in thermal runaway, fires, or even explosions. So, thermal management and control are of the utmost importance, keeping the temperature within a suitable range to enable cells to prevent such incidents [81].

Apart from the safety of the cells, the performance and lifecycle of the LIBs are also influenced by the temperature of the manufacturing environment, specifically the formation chambers. It is mandatory to have consistent and suitable thermal conditions to achieve the optimal quality and efficiency of the batteries manufactured in the gigafactory [82].

In addition, a proper thermal management system also helps to enhance the overall energy efficiency of the factory as it optimizes the heating and cooling process of the facility, which results in a reduction in energy consumption [83,84].

Air handling units (AHUs) are another type of component used for thermal management in the gigafactory. The basic purpose of the AHUs is to circulate air throughout the factory. AHUs consist of heating and cooling coils, filters, and humidifiers to achieve the desirable air quality and temperature in the factory premises. They maintain a suitable working environment inside the gigafactory [85].

These AHUs are also used to cool down manufacturing equipment and processes. Chillers in the AHUs use coolant to remove heat. Then, the heat is dissipated by the cooling towers. AHUs are fundamental for maintaining low temperatures in critical areas of the factory, such as formation towers.

https://doi.org/10.3390/en19051345

Ventilation is also an important component of a thermal management system. An efficient ventilation system is critical to maintain the desired air quality and remove excess heat from the factory. The air inside the gigafactory is polluted by several contaminants and chemicals released during the manufacturing process, and the ventilation system ensures the removal of these contaminants from various zones of the factory [86].

Temperature optimization is critical during production, specifically in the cell formation process. The presence of effective cooling methods is crucial for the cell formation process. These methods are summarized in Figure 10.

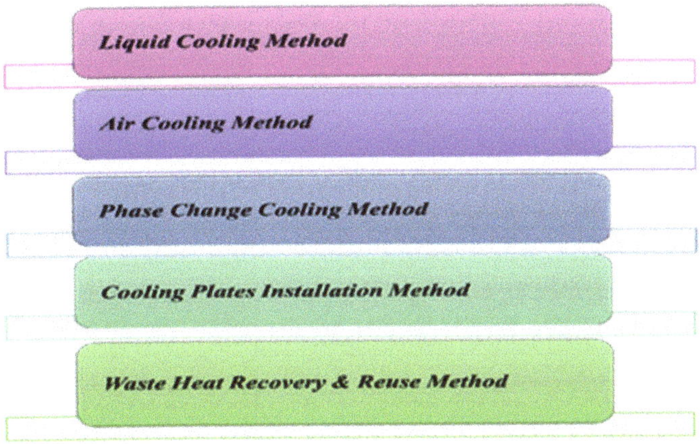

Figure 10. Common cooling techniques used in gigafactories.

Among the most advanced cooling techniques used in existing gigafactories are liquid cooling systems. The liquid cooling system is highly efficient and effective in managing the large thermal loads generated by high-power equipment and manufacturing processes [87]. Coolant is continuously circulated through the cooling system's heat exchangers around heat-intensive power equipment. Heat exchangers absorb and dissipate generated heat. The liquid cooling systems are considerably more efficient than the other cooling systems used in gigafactories [88].

The air-cooling system is another type of cooling technique commonly used in gigafactories. This system circulates air in the place of coolants to dissipate heat from heat-generating equipment and systems by using fans and blowers. Although this cooling method is less efficient than a liquid cooling system, it has a simpler working principle, and it is quite suitable for systems that generate relatively low thermal loads [89].

Phase Change Materials (PCMs) are also used in some places of the gigafactory for thermal management. PCMs act as effective thermal buffers. They absorb and release thermal energy while they transition from one phase to another. These materials are specifically used at places where it is crucial to have very precise temperature control, like formation chambers and dry rooms [90].

The cell formation process is a highly thermally intensive stage during the manufacturing of lithium-ion batteries. Considerable heat is generated due to the continuous charging and discharging of the lithium-ion cells [91].

It is fundamental to ensure that the temperature of the cell formation chambers is uniformly maintained to achieve optimal safety and performance. Advanced control and monitoring systems are installed along with these chambers to monitor and adjust temperature in real time during the formation process [92].

One of the most commonly used thermal management techniques during the formation process is the implementation of cooling plates. These plates directly dissipate heat

https://doi.org/10.3390/en19051345

generated from cells. Plates are integrated and installed in combination with the liquid cooling systems to achieve the desired thermal conditions in the formation chambers [93]. A combination of thermal imaging cameras and sensors is used to monitor the temperature distribution across the cells in the formation chambers. The presence of the hotspots is identified in the formation chambers that need cooling [94].

Another important technique for thermal management in a gigafactory is to use waste heat recovery systems. These systems can enhance the overall energy efficiency of the factory by capturing generated heat and reusing it in various industrial processes [95]. There are several ways to use waste heat recovery in a gigafactory, e.g., this recovered heat can be used to preheat materials or to heat some specific space by transferring it from exchangers, exhaust gases, or coolant.

The waste recovery heat can also be used as an additional source of electrical power for the factory by using Organic Rankine Cycle (ORC) systems to convert this low-grade waste heat into electrical power [96]. The power generated from these ORC systems can be used to run cooling processes and reduce the requirement for electrical energy in HVAC systems [97].

7. Regenerative Function

This section explores the details of the methodologies and techniques used for the implementation of the regenerative function in a gigafactory [98].

The LIB cell manufacturing process incurs significant energy losses. Capturing and reusing this wasted energy improves the overall energy efficiency of the manufacturing process [99]. In addition to reducing the price, the application of the regenerative function also minimizes the environmental impact of the gigafactory and makes it more sustainable and environmentally friendly [100].

One of the most common applications of regenerative function is the implementation of automated guided vehicle (AGV) conveyor systems, used for transportation and material handling in the gigafactory. Regenerative braking systems installed in such transportation systems help to generate electrical power by capturing the kinetic energy produced by the braking of AGVs and conveyor belts.

The regenerative braking system accumulates energy produced by the slowing down of AGVs or conveyors. Batteries and capacitors are used to store the generated electrical power so it can be used to run other components of the power system. The power requirements from external power sources are reduced as a result [101].

Another system that can be incorporated with regenerative functioning is the HVAC system inside the gigafactory. It is implemented through heat recovery ventilators (HRVs) installed in the HVAC system. This system reduces the energy requirement for heating or cooling of the facility, as these HRVs capture the heat from the outgoing exhaust air and transfer it to the incoming fresh air [102].

Energy Recovery Ventilators (ERVs) are another example of systems that can use regenerative functions to manage humidity levels inside the gigafactory, working on similar principles as the HRVs. These systems are installed at places where it is crucial to control humidity levels. Power consumption requirements are reduced by implementing this system [103].

The cell formation process for LIBs has considerable potential for energy regeneration, as it is the most energy-intensive process inside the gigafactory, requiring about 33% of the total power consumed during the manufacturing of lithium-ion cells [16]. Once the cells are fully charged and passed through the discharging cycles, there is a huge potential to recapture the energy during these discharging cycles, which can be used to charge other cells or even run other operations in the facility [104]. In modern gigafactories, advanced

https://doi.org/10.3390/en19051345

bidirectional power converters are used with both formation chambers and formation towers. These converters can operate in both directions and enable these devices to feed the captured discharging energy back into the grid or store it in the central energy storage system [105].

Furthermore, a large number of production machines in a gigafactory are involved in cell production, such as presses, motors, and robotics. There can be a substantial reduction in energy consumption if regenerative functions are incorporated during the design of these high-power electrical machines. Such electrical machines are equipped with special drives that convert kinetic energy into electrical power while decelerating. This power can be used to run other power equipment in the gigafactory, or it can be supplied back to the grid [106].

It is compulsory to install a centralized energy storage system to make the application of regenerative functions more effective and efficient. This central energy storage system consists of high-capacity batteries or supercapacitors to store the recovered energy coming from different processes and equipment [107]. These energy storage systems can act as an additional power source during peak demand hours or as an alternative power source during outages [108].

8. Failure Analysis

A. *Failure analysis procedure*

Effective failure analysis is imperative for enhanced production quality and reduction in shutdown time, improving the reliability, safety, and efficiency of the overall manufacturing process in the gigafactory [109].

Failures can occur during several processes or at different stages. As explained in Figure 11, failures can occur at two levels: either at the plant and power system level or at the process and product level. Plant and system failures can be further divided into mechanical and electrical faults, which usually occur in the power and auxiliary equipment. Likewise, product and process faults can be further categorized into thermal and chemical issues. Product and process faults generally occur during the manufacturing and formation of LIBs.

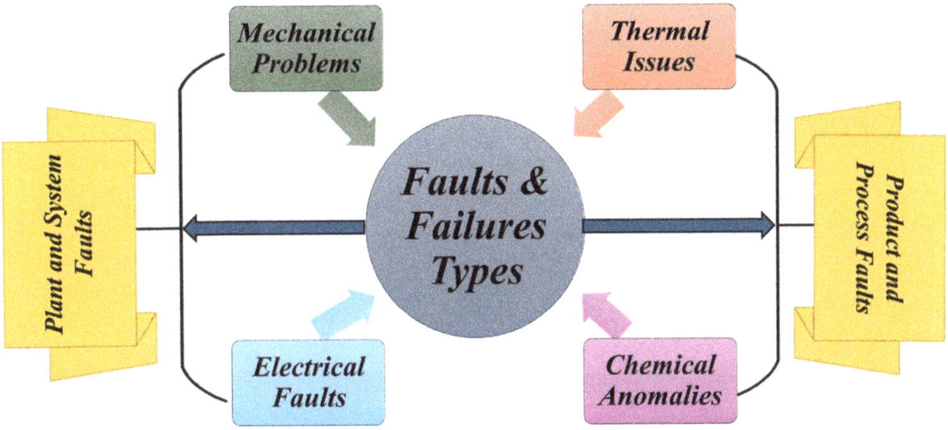

Figure 11. Failures and fault types in the gigafactory power system.

Two of the most common types of plant and system failures are mechanical and electrical failures in the equipment used to operate the power system smoothly. These failures can occur in motors, drives, transformers, or components of the HVAC systems.

https://doi.org/10.3390/en19051345

These failures can prove fatal and can result in the disruption or complete shutdown of operations [110].

Another type of failure is a fault, which occurs during the manufacturing processes of LIB cells. This failure occurs during the formation process due to improper formation protocols, defective thermal management, or wrong material handling. Generally, these failures can be more costly because they go unnoticed and may result in defective production. These are usually called individual defects and do not usually result in a complete shutdown of the plant, but can be fixed by tweaking parameters or through the replacement of small electronic components [111].

Historical failure data is gathered from the gigafactory over a certain time period. For example, this data can be collected from sensors installed at critical locations, operational logs, maintenance records, and quality control reports. In modern gigafactories, real-time data collection can also be facilitated by advanced monitoring systems and Internet of Things (IoT) devices [112].

Failures can be identified and categorized into different failure modes, such as mechanical breakdowns, electrical faults, thermal issues, or chemical reactions [113].

Some of the prominent techniques used in failure analysis to inspect components without causing any harm to these parts are X-ray imaging, SCADA, ultrasonic testing, and infrared thermography [114].

In addition to the above-mentioned techniques, several additional methods can be adopted to analyze microstructural defects in materials. Such techniques involve Scanning Electron Microscopy (SEM) and Energy-Dispersive X-ray Spectroscopy (EDS) [115].

Various analysis models and techniques are used to identify the root causes of failures in the gigafactory: Failure Mode and Effects Analysis (FMEA), Fault Tree Analysis (FTA), and root cause analysis (RCA) are prominent examples [116,117].

B. *Root cause identification and Preventive measures*

Identifying root causes is essential for implementing preventive measures that minimize failure recurrence and ensure uninterrupted gigafactory operations.

(1) Root Cause Identification

The most important part of failure analysis is the root cause identification of failures to facilitate failure elimination. It can also prevent accidents and improve the safety of personnel and equipment inside the gigafactory [118]. Understanding these failure mechanisms also enhances the quality and quantity of production. Reduction in maintenance costs and the eradication of recurring problems can also be achieved by the effective management of these mechanisms. It can also improve production quality standards and lead to the complete elimination of failures during the manufacturing process [119].

The root causes of failures are eliminated by using root failure analysis: in the case of electrode degradation or separator failure, different techniques, such as electrochemical impedance spectroscopy (EIS) and differential scanning calorimetry (DSC), can be used. The problems identified using these techniques are addressed by implementing changes in the electrode manufacturing process, material sourcing, and cell assembly techniques [120].

The effectiveness of these suggested corrective actions can be validated by continuous monitoring and thorough testing to validate the analysis and prevent recurring failures [121].

(2) Preventive Measures

Predictive analysis or protection can be assessed by the effects of these failures on the system's reliability. This is achieved using different computational and simulation models. Weibull Analysis and Reliability Block Diagrams (RBDs) are examples of statistical methods that are used to analyze the failure data and reliability of the models [122].

https://doi.org/10.3390/en19051345

After the suggested corrective measures have been implemented, extensive testing of the modified cells is conducted. The effectiveness and reliability of these measures in overall cell performance are checked under various test conditions.

Moreover, preventive measures are also implemented to avoid future failures. This preventive analysis is carried out using real-time monitoring and machine learning methods to collect data from critical equipment and prevent any failure from occurring during the manufacturing process [121,123].

9. Energy Consumption per Watt-Hour

The energy consumption per watt-hour is a critical parameter, and aims to estimate the exact amount of energy required to produce LIBs with a capacity of approximately one watt-hour. The required energy is estimated from the complete battery manufacturing process, starting from the processing of raw materials to the cell formation process of the cell assembly [4]. This section elaborates on the key factors that influence energy usage and the methods developed for measurements.

Energy consumption is important for a gigafactory, as it highlights the strategies for energy efficiency in the manufacturing process and key areas where energy can be saved. The energy used in the battery manufacturing process adds to production cost, and understanding energy consumption helps to reduce the operational costs of LIB cells. In addition to reducing the operational cost of the manufacturing process, this also decreases energy consumption, improves environmental sustainability, and minimizes the carbon footprint of the gigafactory [16].

Figure 12 shows the percentages of power consumption during different stages of the cell manufacturing process. The most power is consumed during the cell finishing stage, which includes cell formation, aging, and the handling of the newly manufactured cells. It accounts for 33% of total power consumed, closely followed by the electrode production stage at 32%, the dry room at 26%, and cell assembly at 9%.

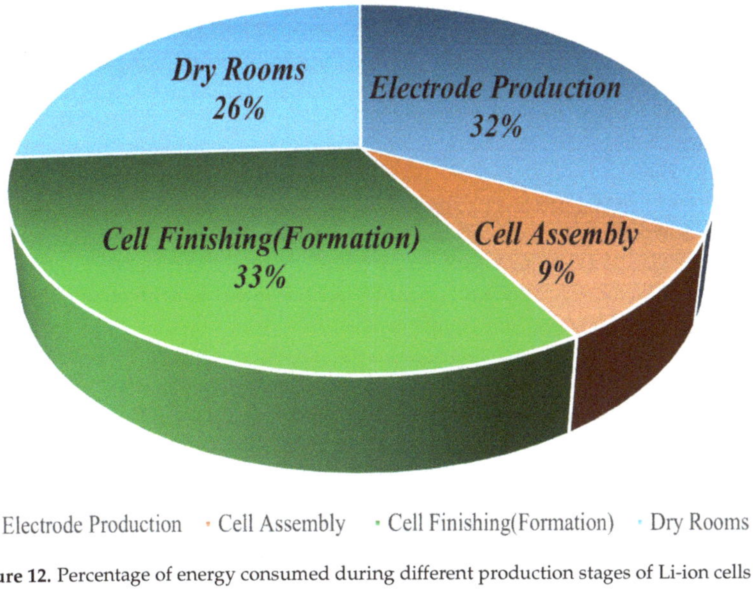

Figure 12. Percentage of energy consumed during different production stages of Li-ion cells [16].

The total power consumed during the LIB cell manufacturing process for each kWh of capacity is 41.43 kWh. Table 1 shows the breakdown of power consumption at each

https://doi.org/10.3390/en19051345

important stage of the LIB manufacturing process, further divided into substages based on the activities performed at that particular stage of cell manufacturing.

Table 1. Power consumption at different process stages of the cell production of LIBs [16].

Process	Stage	Power Consumed per kWh
Electrode Production	Mixing	0.13
	Coating and drying	11.02
	Calendaring	0.53
	Slitting	0.16
	Vacuum drying	1.61
Subtotal		**13.45**
Cell Assembly	Winding	0.25
	Assembly	1.59
	Washing	1.98
Subtotal		**3.82**
Cell Finishing	Formation	10.17
	Aging and Testing	1.39
	Handling	1.98
Subtotal		**13.54**
Dry Rooms	Drying	10.62
Subtotal		**10.62**
Grand Total		**41.43**

In addition to the process, the energy consumption of different equipment, including mixers, coaters, and formation chambers, is also crucial for assessing energy consumption per watt-hour in the gigafactory [124].

Energy consumption can be further reduced by implementing advanced control and automation techniques to achieve an effective thermal management system for heating and cooling. Another way to improve efficiency in the LIB manufacturing process is to use high-quality raw materials [125].

Energy consumption per watt-hour is measured throughout the manufacturing process by monitoring and analyzing energy consumption. Energy consumption patterns and areas for improvement are analyzed by conducting comprehensive energy audits. These surveys are conducted by measuring the energy consumed during specific stages of the manufacturing process [4]. In advanced gigafactories, IoT sensors and smart meters are installed for the real-time monitoring of energy consumption. The data collected from the above-mentioned techniques provides a granular view of energy consumption. It helps in making well-informed decisions for energy consumption optimization.

The total energy consumption is assessed throughout battery life by performing a lifecycle analysis from raw material extraction to battery disposal. Figure 13 shows some crucial stages at which this holistic approach is adopted to measure and optimize energy consumption [16]:

- **Electrode Production**: The process of electrode production is quite energy-intensive and involves the mixing of active materials with binders. The energy used during this process can be reduced by optimizing mixer efficiency and decreasing batch sizes. Energy consumption can be further reduced by using advanced drying technologies. This process is quite energy-intensive because it involves coating electrodes onto the substrates.

https://doi.org/10.3390/en19051345

- **Cell Assembly**: Cell assembly requires significant amounts of energy at different stages, such as hermetic cell sealing, laser welding, and other joining processes, which require precise energy control to ensure high-quality products. Energy efficiency can be improved at the cell assembly stage by using advanced laser-cutting techniques, sealing technologies, and materials.
- **Cell Formation Process**: The cell formation process is the most energy-intensive stage during the manufacturing process, as cells go through multiple and continuous charging and discharging cycles. The energy consumption during the cell formation process can be significantly reduced by using energy-efficient power systems and optimizing the formation protocols involving cycle duration, power supplied, chamber configuration, etc.
- **Thermal Management**: The thermal management system of the gigafactory requires a considerable amount of energy as it is responsible for maintaining optimal temperatures in key production areas of the facility. Energy consumption during this process can be reduced by implementing ERVs and high-efficiency HVAC systems. Overall energy savings can be enhanced by installing efficient cooling systems for machinery and other process equipment. This reduces the requirement for external energy input to maintain optimal thermal conditions inside the factory.

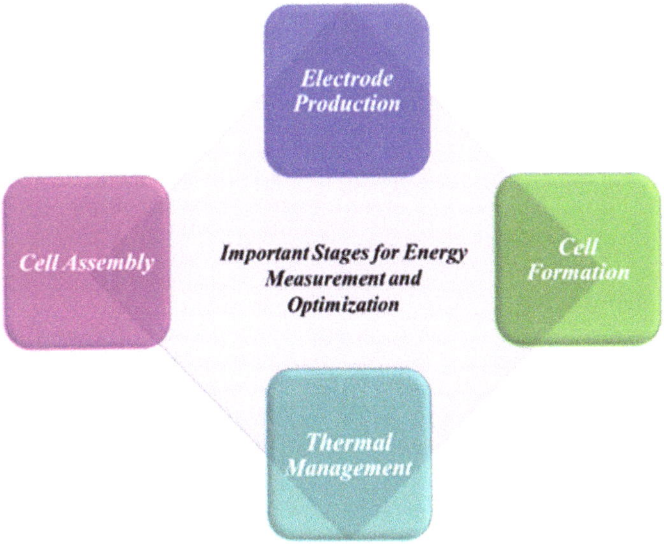

Figure 13. Key stages for energy measurement and optimization.

Energy consumption can be further decreased by adopting various strategies.

- **Process Optimization**: This can be achieved by implementing lean manufacturing principles to eliminate energy waste and improve process efficiency. Continuous improvements like Six Sigma [126] are implemented to identify and eliminate inefficiencies in the manufacturing process.
- **Advanced Technologies**: Advanced automation systems are installed to optimize process control and decrease energy usage. The integration of smart manufacturing technologies, such as IoT and AI, helps in real-time monitoring and optimization of energy consumption.
- **Employee Training and Engagement**: Energy efficiency awareness training programs must be conducted among employees to educate them about the importance of energy

https://doi.org/10.3390/en19051345

savings and to encourage them to contribute to the implementation of incentive training programs.

10. Discussions and Analysis

This section comprehensively discusses and analyzes the critical elements of power systems, based on the collective energy profile shape of a gigafactory.

Gigafactories generally require a significant amount of power for their operation and require a diverse mix of power sources. It is recommended to integrate renewable energy (such as solar and wind) with the grid power. This integration can help reduce load on the electricity grid and also make the factory sustainable. If the renewable resources produce excess power, it is suggested to use large-scale battery banks to create centralized energy storage systems. These centralized energy storage systems can be used as a supplementary or backup power source during power outages.

The paramount goal is a significant reduction in power losses and a steady power supply to all the equipment and processes in the facility. It is suggested that modern gigafactories should implement setups. These setups enable them to maintain a dynamic power flow and real-time monitoring. That helps in energy distribution optimization based on real-time demand and supply situations and reduces power losses. Overall energy efficiency can be enhanced by using data-driven insights, using information such as production schedules, equipment failures, or fluctuations in renewable energy generation, as well as by using low-loss transformers.

The energy consumption profile of the gigafactory can also be greatly affected by the selection and optimization of the power equipment. A large amount of specialized power equipment is extensively used in the cell formation process. All devices used in the cell formation process must be carefully designed and controlled to optimize the charging and discharging cycles for high-quality production. Moreover, predictive maintenance strategies are implemented to monitor equipment health and performance using sensors and data analytics so that optimal power flow and efficient operations are maintained, sudden unexpected failures can be prevented, and spikes are avoided.

It is recommended to use modern thermal management solutions in gigafactories, applying a combination of heat recovery, advanced cooling techniques, and efficient HVAC systems. Heat recovery systems capture heat and reuse it for various processes. The heat generated during the cell formation stage can be used to preheat the raw material or to maintain the temperature. This technique can result in energy savings and reduce operational costs.

Similar to the heat recovery system, it is recommended to use regenerative functions to maximize energy efficiency in the gigafactory, as the energy that would otherwise be wasted can be captured and reused. Regenerative systems should be further complemented by the installation of central energy storage systems to store recovered energy, to reduce energy costs, and to reduce dependence on external power sources.

11. Concluding Remarks

The power consumption analysis described in this paper provides a comprehensive analysis of the various factors that can affect the overall energy efficiency of a gigafactory.

The strategies suggested in this paper help to reduce operational costs and enable environmentally friendly production.

The largest amount of energy is consumed during the cell formation process of the LIB cell production; hence, the authors are developing innovative protocols for formation chambers to minimize energy losses at this stage of production by adopting innovative and energy-efficient techniques. Further research directions will focus on the develop-

https://doi.org/10.3390/en19051345

ment of more integrated and holistic approaches towards power consumption analysis in gigafactories to address critical issues like optimization in power consumption, reduction in operational costs, and the minimization of environmental effects by reducing CO_2 emissions during production processes [18].

The optimal placement of AC/DC and DC/DC converters and the consequent choice of which segment of the internal distribution system could operate in either AC or DC will be an interesting research topic, even though it can be difficult to obtain detailed public information on the AC/DC and DC/DC energy consumption trade-off in the gigafactories. However, authors are currently working on an innovative formation system in which one AC/DC converter is used for the formation tower, and a minimal number of DC/DC converters are utilized for the formation chambers by implementing novel and innovative chamber configurations.

Funding: This research is co-funded by the European Union—Next Generation EU, Mission 4 Component 2-ECS00000036-CUP E14D23002000004 and MUR DM 117/Comau—Innovative formation system for lithium-ion cells project being carried out by a research collaboration between Politecnico di Torino and COMAU S.P.A. ITALY.

Data Availability Statement: No new data were created or analyzed in this study. Data sharing is not applicable to this article.

Conflicts of Interest: The authors declare no conflicts of interest. The funders had no role in the design of the study; in the collection, analyses, or interpretation of data; in the writing of the manuscript; or in the decision to publish the results.

References

1. Bechberger, M.; Vorholt, F.; Bünting, A.; Oehl-Schalla, N.; Arnold-Triangeli, L.; Beermann, V.; Bilge, P.; Giacomini, M.; Steven, J.M.; Neupert, P.; et al. *Sustainability of Battery Cell Production in Europe*; VDI/VDE Innovation + Technik GmbH: Berlin, Germany, 2021.
2. Yuan, C.; Deng, Y.; Li, T.; Yang, F. Manufacturing Energy Analysis of Lithium Ion Battery Pack for Electric Vehicles. *CIRP Ann. Manuf. Technol.* **2017**, *66*, 53–56. [CrossRef]
3. Kallitsis, E. On the Energy Use of Battery Gigafactories. *J. Clean. Prod.* **2022**, *364*, 132573. [CrossRef]
4. Davidsson Kurland, S. Energy Use for GWh-Scale Lithium-Ion Battery Production. *Environ. Res. Commun.* **2020**, *2*, 012001. [CrossRef]
5. Zemite, L.; Kozadajevs, J.; Jansons, L.; Bode, I.; Dzelzitis, E.; Palkova, K. Integrating Renewable Energy Solutions in Small-Scale Industrial Facilities. *Energies* **2024**, *17*, 2792. [CrossRef]
6. Tesla Impact Report; Austin, 2024. Available online: https://www.tesla.com/ns_videos/2023-tesla-impact-report-highlights.pdf (accessed on 25 January 2026).
7. Choe, H.; Choi, Y.; Kim, J.H. Threshold Cell Diameter for High Thermal Insulation of Water-Blown Rigid Polyurethane Foams. *J. Ind. Eng. Chem.* **2019**, *73*, 344–350. [CrossRef]
8. Liu, J.; Chen, H.; Huang, S.; Jiao, Y.; Chen, M. Recent Progress and Prospects in Liquid Cooling Thermal Management System for Lithium-Ion Batteries. *Batteries* **2023**, *9*, 400. [CrossRef]
9. Olatunde, T.M.; Okwandu, A.C.; Akande, D.O.; Sikhakhane, Z.Q. Review of Energy-Efficient HVAC Technologies for Sustainable Buildings. *Int. J. Sci. Technol. Res. Arch.* **2024**, *6*, 12–20. [CrossRef]
10. Jouhara, H.; Khordehgah, N.; Almahmoud, S.; Delpech, B.; Chauhan, A.; Tassou, S.A. Waste Heat Recovery Technologies and Applications. *Therm. Sci. Eng. Prog.* **2018**, *6*, 268–289. [CrossRef]
11. Jouhara, H.; Olabi, A.G. Editorial: Industrial Waste Heat Recovery. *Energy* **2018**, *160*, 1–2. [CrossRef]
12. Diez, L.; Marangé, P.; Levrat, É. Regeneration Management Tool for Industrial Ecosystem. *IFAC-Pap.* **2017**, *50*, 12950–12955. [CrossRef]
13. Toren, M. Optimization of Transformer Parameters at Distribution and Power Levels with Hybrid Grey Wolf-Whale Optimization Algorithm. *Eng. Sci. Technol. Int. J.* **2023**, *43*, 101439. [CrossRef]
14. Molęda, M.; Małysiak-Mrozek, B.; Ding, W.; Sunderam, V.; Mrozek, D. From Corrective to Predictive Maintenance—A Review of Maintenance Approaches for the Power Industry. *Sensors* **2023**, *23*, 5970. [CrossRef] [PubMed]
15. Meriem, H.; Nora, H.; Samir, O. Predictive Maintenance for Smart Industrial Systems: A Roadmap. *Procedia Comput. Sci.* **2023**, *220*, 645–650. [CrossRef]

https://doi.org/10.3390/en19051345

16. Degen, F.; Schütte, M. Life Cycle Assessment of the Energy Consumption and GHG Emissions of State-of-the-Art Automotive Battery Cell Production. *J. Clean. Prod.* **2022**, *330*, 129798. [CrossRef]
17. Gao, K.; Huang, Y.; Sadollah, A.; Wang, L. A Review of Energy-Efficient Scheduling in Intelligent Production Systems. *Complex Intell. Syst.* **2020**, *6*, 237–249. [CrossRef]
18. Potter, K.; Blessing, E.; Oluwaseyi, J. Enhancing Industrial Automation Through the Integration of Deep Learning and Robotics. 2024. Available online: https://www.researchgate.net/publication/378522207_Enhancing_Industrial_Automation_through_the_Integration_of_Deep_Learning_and_Robotics_Authors (accessed on 25 January 2026).
19. Xu, J. CEI and SEI Formation in Li-Ion Batteries. In *Corrosion and Degradation in Fuel Cells, Supercapacitors and Batteries*; Saji, V.S., Ed.; Springer Nature: Cham, Switzerland, 2024; pp. 307–324. ISBN 978-3-031-57012-4.
20. Jinasena, A.; Burheim, O.S.; Strømman, A.H. A Flexible Model for Benchmarking the Energy Usage of Automotive Lithium-Ion Battery Cell Manufacturing. *Batteries* **2021**, *7*, 14. [CrossRef]
21. Wood, D.L.; Li, J.; An, S.J. Formation Challenges of Lithium-Ion Battery Manufacturing. *Joule* **2019**, *3*, 2884–2888. [CrossRef]
22. Tornheim, A.; Trask, S.E.; Zhang, Z. Evaluation of Electrolyte Oxidation Stability on Charged $LiNi_{0.5}Co_{0.2}Mn_{0.3}O_2$ athode Surface through Potentiostatic Holds. *J. Electrochem. Soc.* **2016**, *163*, A1717–A1722. [CrossRef]
23. Schomburg, F.; Heidrich, B.; Wennemar, S.; Drees, R.; Roth, T.; Kurrat, M.; Heimes, H.; Jossen, A.; Winter, M.; Cheong, J.Y.; et al. Lithium-Ion Battery Cell Formation: Status and Future Directions towards a Knowledge-Based Process Design. *Energy Environ. Sci.* **2024**, *17*, 2686–2733. [CrossRef]
24. Robben, T.; Offermanns, C.; Heimes, H.; Kampker, A. Advancements in Battery Cell Finalization: Insights from an Expert Survey and Prospects for Process Optimization. *World Electr. Veh. J.* **2024**, *15*, 219. [CrossRef]
25. Weng, A.; Mohtat, P.; Attia, P.M.; Sulzer, V.; Lee, S.; Less, G.; Stefanopoulou, A. Predicting the Impact of Formation Protocols on Battery Lifetime Immediately after Manufacturing. *Joule* **2021**, *5*, 2971–2992. [CrossRef]
26. Knehr, K.W.; Kubal, J.J.; Yoon, S.; Jeon, H.; Roh, W.J.; Ahmed, S. Energy Consumption of Lithium-Ion Pouch Cell Manufacturing Plants. *J. Clean. Prod.* **2024**, *468*, 143050. [CrossRef]
27. Prakash, R.B.R.; Srinivasa Varma, P.; Ravikumar, C.; Vijay Muni, T.; Srinivasulu, A.; Bagadi, K.; Rajesh, A.; Sathish, K. Intelligent Energy Management for Distributed Power Plants and Battery Storage. *Int. Trans. Electr. Energy Syst.* **2023**, *2023*, 6490026. [CrossRef]
28. Yasmin, R.; Amin, B.M.R.; Shah, R.; Barton, A. A Survey of Commercial and Industrial Demand Response Flexibility with Energy Storage Systems and Renewable Energy. *Sustainability* **2024**, *16*, 731. [CrossRef]
29. Tesla Impact Report 2018; Nevada, 2019. Available online: https://www.tesla.com/ns_videos/tesla-impact-report-2019.pdf (accessed on 25 January 2026).
30. Caetano, N.S.; Martins, F.F.; Oliveira, G.M. Life Cycle Assessment of Renewable Energy Technologies. In *The Renewable Energy-Water-Environment Nexus*; Elsevier: Amsterdam, The Netherlands, 2024; pp. 37–79.
31. Department for Business, Energy & Industrial Strategy. *Combined Heat and Power-Technologies*; The National Archives: London, UK, 2021.
32. Fraunhofer Research Institution for Battery Cell Production FFB. Time Production Output Mastering Ramp-up of Battery Production 2. Available online: https://www.ffb.fraunhofer.de/content/dam/ipt/forschungsfertigung-batteriezelle/Bilder/presse-medien/whitepaper-ramp-up/Whitepaper_Mastering%20Ramp-up%20of%20Battery%20Production.pdf (accessed on 25 January 2026).
33. ESG IT. LG Energy Solution ESG REPORT; Seoul, 2024. Available online: https://www.lgensol.com/upload/file/sustainability/LG_Energy_Solution_2024_ESG_Report_EN.pdf (accessed on 25 January 2026).
34. Kumar, A.; Schei, T.; Ahenkorah, A.; Caceres Rodriguez, R.; Devernay, J.-M.; Freitas, M.; Hall, D.; Liu, Z.; Killingtveit, Å. Hydropower. In *IPCC Special Report on Renewable Energy Sources and Climate Change*; Cristobal Diaz, M., Krug, T., Eds.; Cambridge University Press: Cambridge, UK, 2011; pp. 1–60.
35. Rodríguez Martínez, J.H.; Alcaraz Calderón, A.; Ruiz Flores, L.I.; Valdez Vargas, R. Technical and Economic Analysis of Cogeneration Systems for Refinery Power Plant Applications. In *Proceedings of the ASME 2010 Power Conference, Chicago, IL, USA, 13–15 July 2010*; ASMEDC: Chicago, IL, USA, 2010; pp. 499–507.
36. De Souza, R.; Casisi, M.; Micheli, D.; Reini, M. A Review of Small–Medium Combined Heat and Power (CHP) Technologies and Their Role within the 100% Renewable Energy Systems Scenario. *Energies* **2021**, *14*, 5338. [CrossRef]
37. Giaccone, L.; Canova, A. Economical Comparison of CHP Systems for Industrial User with Large Steam Demand. *Appl. Energy* **2009**, *86*, 904–914. [CrossRef]
38. Bagherian, M.A.; Mehranzamir, K.; Pour, A.B.; Rezania, S.; Taghavi, E.; Nabipour-Afrouzi, H.; Dalvi-Esfahani, M.; Alizadeh, S.M. Classification and Analysis of Optimization Techniques for Integrated Energy Systems Utilizing Renewable Energy Sources: A Review for Chp and Cchp Systems. *Processes* **2021**, *9*, 339. [CrossRef]

https://doi.org/10.3390/en19051345

39. Wu, M.; Su Lee Ming, E.; Zheng, Y.; Dong, B.; Fai Yeong, C.; Holderbaum, W. Green Automation: Empowering a Sustainable Future with Energy-Efficient Autonomous Mobile Robots in Manufacturing. 2024. Available online: https://neric.salford.ac.uk/wp-content/uploads/sites/216/2024/09/Green-Automation-Empowering-a-Sustainable-Future-with-Energy-Efficient-Autonomous-Mobile-Robots-in-Manufacturing.pdf (accessed on 25 January 2026).
40. Tesla. Power Pack System. 2022. Available online: https://impulsoragdl.com/wp-content/uploads/2020/09/Ficha-Tecnica-Power-Pack.pdf (accessed on 25 January 2026).
41. Yulong, P.; Cavagnino, A.; Vaschetto, S.; Feng, C.; Tenconi, A. Flywheel Energy Storage Systems for Power Systems Application. In *Proceedings of the 2017 6th International Conference on Clean Electrical Power (ICCEP), Santa Margherita Ligure, Italy, 27–29 June 2017*; IEEE: New York, NY, USA, 2017; pp. 492–501.
42. Bamisile, O.; Zheng, Z.; Adun, H.; Cai, D.; Ting, N.; Huang, Q. Development and Prospect of Flywheel Energy Storage Technology: A Citespace-Based Visual Analysis. *Energy Rep.* **2023**, *9*, 494–505. [CrossRef]
43. Liang, N.; Zhang, J.; Zheng, F.; Pan, Z. A Multi-Time Scale Source–Load Scheduling Method for Power Systems with Wind Power Considering Typical Industrial Production Processes. *J. Electr. Eng. Technol.* **2024**, *19*, 1033–1049. [CrossRef]
44. Bouchard, P.; Voß, S.; Heilig, L.; Shi, X. A Case Study on Smart Grid Technologies with Renewable Energy for Central Parts of Hamburg. *Sustainability* **2023**, *15*, 15834. [CrossRef]
45. Ali Syed, A.; Kumar Pundhir, S. Study of Different Distribution Models for Industrial Consumables and Applicability in India Limitations. *EPRA Int. J. Multidiscip. Res.* **2020**, *6*, 145–153. [CrossRef]
46. Büschel, S. *Industrial Plants Applications for Electric Power Distribution*; Siegbert, H., Ed.; Siemens AG, Smart Infrastructure and Distribution Systems: Berlin, Germany, 2021; Volume 1.
47. Amoiralis, E.I.; Tsili, M.A.; Kladas, A.G. Transformer Design and Optimization: A Literature Survey. *IEEE Trans. Power Deliv.* **2009**, *24*, 1999–2024. [CrossRef]
48. Pinzón, S.; Yánez, S.; Ruiz, M. Optimal Location of Transformers in Electrical Distribution Networks Using Geographic Information Systems. *Enfoque UTE* **2020**, *11*, 84–95. [CrossRef]
49. Amoiralis, E.I.; Tsili, M.A.; Kladas, A.G. Global Transformer Design Optimization Using Deterministic and Non-Deterministic Algorithms. In *Proceedings of the 2012 XXth International Conference on Electrical Machines*; IEEE: New York, NY, USA, 2012; pp. 2323–2331.
50. Chakraborty, S. Components of Distribution System Distribution System. In *Transmission & Distribution of Power*; EE/JCGP; 2020; pp. 1–7. Available online: https://www.jcgpolytechnic.in/wp-content/uploads/2020/04/DEE-S4-Transmission-distribution-Components-of-Distribution-system.pdf (accessed on 25 January 2026).
51. Fan, L.; Si, W.; Xuan, Y.; Sun, Z.; Zhao, J.; Xu, B.; Gu, Q. Multi-Objective Optimal Configuration of Multiple Switchgear Considering Distribution Network Fault Reconfiguration. *IEEE Access* **2021**, *9*, 69905–69912. [CrossRef]
52. GCTT Industrial and Construction Cables Reference Guide; Highland Heights, 2016. Available online: https://na.prysmian.com/sites/na.prysmian.com/files/media/documents/INS-0177-0916%20Industrial%20Cable%20Product%20Reference%20Guide.pdf (accessed on 25 January 2026).
53. Souran, D.M.; Safa, H.H.; Moghadam, B.G.; Ghasempour, M.; Heravi, P.T. Smart Grid Technology in Power Systems. In *Advances in Intelligent Systems and Computing*; Springer: Berlin/Heidelberg, Germany, 2016; Volume 357, pp. 1367–1381.
54. Favuzza, S.; Ippolito, M.; Massaro, F.; Musca, R.; Riva Sanseverino, E.; Schillaci, G.; Zizzo, G. Building Automation and Control Systems and Electrical Distribution Grids: A Study on the Effects of Loads Control Logics on Power Losses and Peaks. *Energies* **2018**, *11*, 667. [CrossRef]
55. Liu, Y. Research on Electrical Automation Devices under Embedded Real-Time Operating System Design. *Appl. Math. Nonlinear Sci.* **2024**, *9*, 1–19. [CrossRef]
56. Alam, M.S.; Arefifar, S.A. Energy Management in Power Distribution Systems: Review, Classification, Limitations and Challenges. *IEEE Access* **2019**, *7*, 92979–93001. [CrossRef]
57. Lakota, B.; Exnar, Z.; Matejcek, M.; Sostronek, M. Backup Power Supply System Analysis. *Adv. Electr. Electron. Eng.* **2015**, *13*, 115–119. [CrossRef]
58. Faraji, J.; Babaei, M.; Bayati, N.; Ahejazi, M. A Comparative Study between Traditional Backup Generator Systems and Renewable Energy Based Microgrids for Power Resilience Enhancement of a Local Clinic. *Electronics* **2019**, *8*, 1485. [CrossRef]
59. Das, C.K.; Bass, O.; Kothapalli, G.; Mahmoud, T.S.; Habibi, D. Overview of Energy Storage Systems in Distribution Networks: Placement, Sizing, Operation, and Power Quality. *Renew. Sustain. Energy Rev.* **2018**, *91*, 1205–1230. [CrossRef]
60. Kumar, Y.V.P.; Bhimasingu, R. Key Aspects of Smart Grid Design for Distribution System Automation: Architecture and Responsibilities. *Procedia Technol.* **2015**, *21*, 352–359. [CrossRef]
61. Gardner, J.; Alanis, J. A Comparison of International Wiring and Grounding Systems with North American Methods. In *Proceedings of the 2008 55th IEEE Petroleum and Chemical Industry Technical Conference*; IEEE: New York, NY, USA, 2008; pp. 1–7.
62. Panethiere, M.A. *Analysis of Grounding and Bonding for Power Distribution Systems in Commercial, Industrial, and Institutional Buildings*; Capitol Technology University: Laurel, MD, USA, 2022.

https://doi.org/10.3390/en19051345

63. Lu, W.-H.; Yao, W.; Luo, J.; Cheng, Z.; Cheng, J.-Z.; Zhang, P.-H.; Li, S.-S.; Wang, S.-R. Comparative Study of the Earthing Systems for Micro-Grid. *Dianli Xitong Baohu Yu Kongzhi/Power Syst. Prot. Control* **2012**, *40*, 103–108.
64. Sajjadi, M.; Seifi, H. A New Approach for Parameter Estimation of Power System Equipment Models. In *Proceedings of the 2019 International Power System Conference (PSC)*; IEEE: Tehran, Iran, 2019; pp. 530–535.
65. Ibekwe, K.I.; Fabuyide, A.; Hamdan, A.; Ilojianya, V.I.; Etukudoh, E.A. Energy Efficiency through Variable Frequency Drives: Industrial Applications in Canada, USA, and Africa. *Int. J. Sci. Res. Arch.* **2024**, *11*, 730–736. [CrossRef]
66. Zhao, H.; Zhang, W.; Zheng, K.; Xu, W. A Novel High-Efficient Lithium-Ion Battery Serial Formation System Scheme Based on Partial Power Conversion. *J. Energy Storage* **2024**, *97*, 112350. [CrossRef]
67. Patnaik, L.; Praneeth, A.V.J.S.; Williamson, S.S. A Closed-Loop Constant-Temperature Constant-Voltage Charging Technique to Reduce Charge Time of Lithium-Ion Batteries. *IEEE Trans. Ind. Electron.* **2019**, *66*, 1059–1067. [CrossRef]
68. Matsuda, M.; Matsumoto, S. Modeling of Manufacturing Equipment for Conguring a Virtual Production Line to Simulate Electric Power-Consumption. *Trans. Inst. Syst. Control Inf. Eng.* **2017**, *30*, 97–103. [CrossRef]
69. Anseán, D.; García Fernández, V.; Gonzalez, M.; Carlos Viera, J.; Anseán, D.; García, V.; González, M.; Viera, J.; Antón, J.; Blanco, C. Efficient Fast-Charging Strategies for Li-Ion Batteries. In Proceedings of the EVS28 International Electric Vehicle Symposium and Exhibition, Goyang, Republic of Korea, 3–6 May 2015; pp. 1–11.
70. Challoob, A.F.; Bin Rahmat, N.A.; Ramachandaramurthy, V.K.A.; Humaidi, A.J. Energy and Battery Management Systems for Electrical Vehicles: A Comprehensive Review & Recommendations. *Energy Explor. Exploit.* **2024**, *42*, 341–372. [CrossRef]
71. Schnell, J.; Reinhart, G. Quality Management for Battery Production: A Quality Gate Concept. *Procedia CIRP* **2016**, *57*, 568–573. [CrossRef]
72. Huhn, E.; Braxtan, N.; Chen, S.E.; Bombik, A.; Zhao, T.; Ma, L.; Sherman, J.; Roghani, S. Lithium-Ion Battery Thermal Runaway Suppression Using Water Spray Cooling. *Energies* **2025**, *18*, 2709. [CrossRef]
73. Kompass, K.; Königs, S.; Florian, E.; Euhus, F.; Fuchs, F.; Zimmermann, S.; Schöneburg, R.; Rainer, J.; Saúl, M.; Genís, M. Safety of Electro Mobility-White Paper of the Fisita Intelligent Safety Working Group. 2023. Available online: https://www.researchgate.net/publication/371991285_Safety_of_Electro_Mobility_-_White_Paper_of_the_FISITA_Intelligent_Safety_Working_Group (accessed on 25 January 2026).
74. Bruneo, D.; Cucinotta, A.; Minnolo, A.L.; Puliafito, A.; Scarpa, M. Energy Management in Industrial Plants. *Computers* **2012**, *1*, 24–40. [CrossRef]
75. Musthopa; Harsanto, B.; Yunani, A. Electric Power Distribution Maintenance Model for Industrial Customers: Total Productive Maintenance (TPM), Reliability-Centered Maintenance (RCM), and Four-Discipline Execution (4DX) Approach. *Energy Rep.* **2023**, *10*, 3186–3196. [CrossRef]
76. Gellings, C.W. Back to Basics: Enhancing Efficiency in the Generation and Delivery of Electricity. In *Energy Efficiency: Towards the End of Demand Growth*; Elsevier: Amsterdam, The Netherlands, 2013; pp. 375–417. ISBN 978-0-12397-887-5.
77. Rajesh, T.; Aravindhan, S.; Sowmiya, M.; Thenmozhi, S. Energy Efficient Lighting System in Industries. *Int. J. Appl. Sci. Eng. Manag.* **2021**, *2*, 214–226.
78. Ortiz, Y.; Arévalo, P.; Peña, D.; Jurado, F. Recent Advances in Thermal Management Strategies for Lithium-Ion Batteries: A Comprehensive Review. *Batteries* **2024**, *10*, 83. [CrossRef]
79. Ghalkhani, M.; Habibi, S. Review of the Li-Ion Battery, Thermal Management, and AI-Based Battery Management System for EV Application. *Energies* **2022**, *16*, 185. [CrossRef]
80. Chang, C.-H.; Gorin, C.; Zhu, B.; Beaucarne, G.; Ji, G.; Yoshida, S. Lithium-Ion Battery Thermal Event and Protection: A Review. *SAE Int. J. Electrified Veh.* **2023**, *13*, 357–398. [CrossRef]
81. Madani, S.; Hajihosseini, M.; Ziebert, C.; Nandhakumar, I. Battery Thermal Management System for Greater Lifetime and Safety: Mini-Review. In *Proceedings of the Engineering for Rural Development*; Latvia University of Life Sciences and Technologies: Jelgava, Latvia, 2024; Volume 23, pp. 1073–1089.
82. Ma, S.; Jiang, M.; Tao, P.; Song, C.; Wu, J.; Wang, J.; Deng, T.; Shang, W. Temperature Effect and Thermal Impact in Lithium-Ion Batteries: A Review. *Prog. Nat. Sci. Mater. Int.* **2018**, *28*, 653–666. [CrossRef]
83. Kapp, S.; Choi, J.-K.; Kissock, K. Toward Energy-Efficient Industrial Thermal Systems for Regional Manufacturing Facilities. *Energy Rep.* **2022**, *8*, 1377–1387. [CrossRef]
84. Mawson, V.J.; Hughes, B.R. Thermal Modelling of Manufacturing Processes and HVAC Systems. *Energy* **2020**, *204*, 117984. [CrossRef]
85. Lun, Y.H.V.; Tung, S.L.D. Air Handling Unit. In *Green Energy and Technology*; Springer Science and Business Media Deutschland GmbH: Berlin/Heidelberg, Germany, 2020; pp. 51–64.
86. Chen, C.; Lai, D.; Chen, Q. Energy Analysis of Three Ventilation Systems for a Large Machining Plant. *Energy Build.* **2020**, *224*, 110272. [CrossRef]
87. Boshkova, I.; Titlov, A.; Volgusheva, N.; Kolesnychenko, N.; Sagala, T. Designing a System of Liquid Cooling for Industrial Microwave Installations. *East.-Eur. J. Enterp. Technol.* **2019**, *4*, 65–70. [CrossRef]

https://doi.org/10.3390/en19051345

88. Cheli, L.; Carcasci, C. Modelling and Analysis of a Liquid-Cooled System for Thermal Management Application of an Electronic Equipment. *E3S Web Conf.* **2020**, *197*, 10008. [CrossRef]
89. Memar, S.; Napoles, O.M.; Hofland, B.; Melling, G. Simulation of Parallel Layered Air Cooling Thermal Management System for Li-Ion Batteries. In *Proceedings of the 31st European Safety and Reliability Conference, ESREL 2021*; Research Publishing: Singapore, 2021; Volume 1, pp. 2525–2532.
90. Bianco, V.; De Rosa, M.; Vafai, K. Phase-Change Materials for Thermal Management of Electronic Devices. *Appl. Therm. Eng.* **2022**, *214*, 118839. [CrossRef]
91. Bhattacharjee, A.; Mohanty, R.K.; Ghosh, A. Design of an Optimized Thermal Management System for Li-Ion Batteries under Different Discharging Conditions. *Energies* **2020**, *13*, 5695. [CrossRef]
92. Kumar, K.; Sarkar, J.; Mondal, S.S. Assessment of Newly-Designed Hybrid Nanofluid-Cooled Micro-Channeled Thermal Management System for Li-Ion Battery. *J. Electrochem. Energy Convers. Storage* **2024**, *21*, 011011. [CrossRef]
93. Napa, N.; Agrawal, M.K.; Tamma, B. Design of Novel Thermal Management System for Li-Ion Battery Module Using Metal Matrix Based Passive Cooling Method. *J. Energy Storage* **2023**, *73*, 109119. [CrossRef]
94. Sequino, L.; Capasso, C.; Costanza, G.; Tata, M.E. Experimental Investigation on Thermal Effects of a Metal Foam-Based Frame Application for Lithium-Ion Cells. In *Proceedings of the SAE International Journal of Advances and Current Practices in Mobility*; SAE International: Warrendale, PA, USA, 2023; Volume 6, pp. 2141–2150.
95. Mohammadaliha, N.; Amani, M.; Bahrami, M. Thermal Performance of Heat and Water Recovery Systems: Role of Condensing Heat Exchanger Material. *Clean. Eng. Technol.* **2020**, *1*, 100024. [CrossRef]
96. Slauch, I.M.; Deceglie, M.G.; Silverman, T.J.; Ferry, V.E. Optical Approaches for Passive Thermal Management in C-Si Photovoltaic Modules. *Cell Rep. Phys. Sci.* **2021**, *2*, 100430. [CrossRef]
97. Kim, S.H.; Doh, Y.; Heo, T.-W.; Lee, I.W. Economic Analysis of a Redox Flow Batteries-Based Energy Storage System for Energy Savings in Factory Energy Management System. *Batteries* **2023**, *9*, 418. [CrossRef]
98. Khartchenko, N.V.; Kharchenko, V.M. *Advanced Energy Systems*, 2nd ed.; CRC Press: Boca Raton, FL, USA, 2010.
99. Sarode, R.P.; Vinchurkar, S.M. An Approach to Recovering Heat from the Compressed Air System Based on Waste Heat Recovery: A Review. *Energy Sources Part A Recovery Util. Environ. Eff.* **2023**, *45*, 9465–9484.
100. Ferreira, F.J.T.E.; de Almeida, A.T. Overview on Energy Saving Opportunities in Electric Motor Driven Systems—Part 1: System Efficiency Improvement. In *Proceedings of the 2016 IEEE/IAS 52nd Industrial and Commercial Power Systems Technical Conference (I&CPS)*; IEEE: New York, NY, USA, 2016; pp. 1–8.
101. Ferreira, F.J.T.E.; de Almeida, A.T. Overview on Energy Saving Opportunities in Electric Motor Driven Systems—Part 2: Regeneration and Output Power Reduction. In *Proceedings of the 2016 IEEE/IAS 52nd Industrial and Commercial Power Systems Technical Conference (I&CPS)*; IEEE: New York, NY, USA, 2016; pp. 1–8.
102. Šebo, J.; Prester, J.; Šebová, M. The Role of Environmental Management Systems and Energy Management Systems in the Adoption of Energy Recuperation Technologies in Central European Manufacturing Companies. *Sustainability* **2023**, *15*, 16913. [CrossRef]
103. Moro, A.; Kwakye-Boateng, P. Heat Recovery Analysis of a Fixed Plate Energy Recovery Ventilator. *Sustain. Energy Res.* **2024**, *11*, 29. [CrossRef]
104. Orozco, L.; Liao, W. Power Efficient Battery Formation/Testing System with Energy Recycling. *System* **2015**, *512*, 768.
105. Chen, Y.-C.; Chang, Y.-C.; Cheng, J.-F.; Yu, W.-C.; Lin, C.-L. Regenerative Braking-Driving Control System. In *Proceedings of the 2018 13th IEEE Conference on Industrial Electronics and Applications (ICIEA)*; IEEE: New York, NY, USA, 2018; pp. 887–892.
106. Badhoutiya, A.; Darokar, H.; Verma, R.P.; Saraswat, M.; Devaraj, S.; Raj, V.H.; Abdulhussain, Z.N. Regenerative Manufacturing: Crafting a Sustainable Future through Design and Production. *E3S Web Conf.* **2023**, *453*, 01038. [CrossRef]
107. Potrč, S.; Nemet, A.; Čuček, L.; Varbanov, P.S.; Kravanja, Z. Synthesis of a Regenerative Energy System—Beyond Carbon Emissions Neutrality. *Renew. Sustain. Energy Rev.* **2022**, *169*, 112924. [CrossRef]
108. Ferreira, F.J.T.E.; de Almeida, A.T. Reducing Energy Costs in Electric-Motor-Driven Systems: Savings Through Output Power Reduction and Energy Regeneration. *IEEE Ind. Appl. Mag.* **2018**, *24*, 84–97. [CrossRef]
109. Durham, M.O.; Robert Durham, C.A.; Rosemary Durham, C.; Jason Coffin, C. Electrical Failure Analysis for Fire & Incident Investigations Lightning Excerpts; Durham, 2011. Available online: https://www.scribd.com/document/127712615/Electrical-Failure-Analysis-for-Fire-and-Incident-Investigation (accessed on 25 January 2026).
110. Zitouni, A. Power Transformer Differential Relay Reliability Assessment Using False Trip Root Cause Analysis. In *Proceedings of the 2020 International Conference on Electrical Engineering (ICEE), Istanbul, Turkey, 25-27 September 2020*; IEEE: New York, NY, USA, 2020; pp. 1–5.
111. Nikolaev, A.; Maklakov, A.; Bulanov, M.; Gilemov, I.; Denisevich, A.; Afanasev, M. Current Electromagnetic Compatibility Problems of High-Power Industrial Electric Drives with Active Front-End Rectifiers Connected to a 6–35 KV Power Grid: A Comprehensive Overview. *Energies* **2023**, *16*, 293. [CrossRef]

https://doi.org/10.3390/en19051345

112. Bauermann, L.P.; Mesquita, L.V.; Bischoff, C.; Drews, M.; Fitz, O.; Heuer, A.; Biro, D. Scanning Acoustic Microscopy as a Non-Destructive Imaging Tool to Localize Defects inside Battery Cells. *J. Power Sources Adv.* **2020**, *6*, 100035. [CrossRef]
113. Kılıç, L. Assessing and Improving Recommendations for Local Power Quality Efficiency for Industrial Plants with the Help of Real Data. *Int. J. Energy Smart Grid* **2019**, *4*, 12–20. [CrossRef]
114. Përvetica, R.; Përvetica, N.; Rexhepi, V.; Purellku, M. An Analysis of the Failures in the Power System Using the SCADA/EMS Database. In Proceedings of the 1st International Conference, CIGRE, Pristina, Kosovo, 6–11 January 2020; pp. 1–5.
115. Girão, A.V.; Caputo, G.; Ferro, M.C. Application of Scanning Electron Microscopy–Energy Dispersive X-Ray Spectroscopy (SEM-EDS). *Compr. Anal. Chem.* **2017**, *75*, 153–168. [CrossRef]
116. Cristea, G.; Constantinescu, D.M. A Comparative Critical Study between FMEA and FTA Risk Analysis Methods. *IOP Conf. Ser. Mater. Sci. Eng.* **2017**, *252*, 012046. [CrossRef]
117. Pazhayattil, A.B.; Sharma, S. Root Cause Analysis (RCA) Methods. In *Pharmaceutical Manufacturing Deviation and Failure Investigations: Principles, Practices, and Case Studies*; Pazhayattil, A.B., Sharma, S., Eds.; Springer Nature: Cham, Switzerland, 2025; pp. 11–31. ISBN 978-3-031-86504-6.
118. Rau, C.A.; Becker, D.G.; Besuner, P.M.; Cipolla, R.C.; Egan, G.R.; Gupta, P.; Johnson, D.P.; Tetelman, A.S.; Rettig, T.W.; Peters, D.C. Failure Analysis and Failure Prevention in Electric Power Systems. *Nucl. Eng. Des.* **1977**, *43*, 1–97. [CrossRef]
119. Sun, K.; Hou, Y.; Sun, W.; Qi, J. *Power System Control Under Cascading Failures Understanding, Mitigation, and System Restoration*, 1st ed.; John Wiley & Sons: Hoboken, NJ, USA; IEEE: New York, NY, USA, 2019; ISBN 201-8-04229-7.
120. Ariyoshi, K.; Siroma, Z.; Mineshige, A.; Takeno, M.; Fukutsuka, T.; Abe, T.; Uchida, S. Electrochemical Impedance Spectroscopy Part 1: Fundamentals. *Electrochemistry* **2022**, *90*, 102007. [CrossRef]
121. Evans, D.; Luc, P.-M.; Tebruegge, C.; Kowal, J. Detection of Manufacturing Defects in Lithium-Ion Batteries-Analysis of the Potential of Computed Tomography Imaging. *Energies* **2023**, *16*, 6958. [CrossRef]
122. Balasivanandha Prabu, S.; Karunamoorthy, L. Microstructure-Based Finite Element Analysis of Failure Prediction in Particle-Reinforced Metal–Matrix Composite. *J. Mater. Process. Technol.* **2008**, *207*, 53–62. [CrossRef]
123. Hamed, M.; Soliman, A. Analyzing Failure to Prevent Problems. *Ind. Manag.* **2014**, *5*, 10–15. [CrossRef]
124. Yang, Z.G. The Vanadium Redox-Flow Battery Is Big and Long-Lived, and It Won't Catch Fire Is this the Ultimate Grid Battery. *IEEE Spectr.* **2017**, *54*, 36–41. [CrossRef]
125. Hintsala, M. *Industrial Management Production Requirements for 35 Gwh Lithium-Ion Battery Factory*; University of VAASA, Faculty of Technology: Vaasa, Finland, 2018.
126. Lee, J.; Yuvamitra, K.; Guiberteau, K.; Kozman, T.A. Six-Sigma Approach to Energy Management Planning. *Strateg. Plan. Energy Environ.* **2014**, *33*, 23–40. [CrossRef]

https://doi.org/10.3390/en19051345

processes

MDPI

Article

Application of a Multi-Objective Optimisation (MOO) via Pareto Front to the Energy Performance of a Domestic Oven

Simona Rustico [1,*], Beatrice Bonfanti Pulvirenti [1] and Marco Reguzzoni [2]

1 Department of Industrial Engineering (DIN), Alma Mater Studiorum Università di Bologna, Viale Risorgimento 2, 40136 Bologna, Italy; beatrice.pulvirenti@unibo.it
2 Research and Development Department, SMEG S.p.A., Via Leonardo da Vinci 4, 42016 Guastalla, Italy; marco.reguzzoni@smeg.it
* Correspondence: simona.rustico2@unibo.it

Abstract

The growing demand for environmentally sustainable technologies is driving the adoption of increasingly stringent energy regulations across Europe. The residential sector is particularly impacted, not only through requirements for highly insulated buildings but also through stricter standards for household appliances. Among these, domestic ovens represent a critical target, requiring manufacturers to develop technologies that support laboratory testing while reducing energy consumption. This work proposes a tool to support manufacturers during laboratory testing by applying a multi-objective optimisation approach using the Pareto front method. The code was developed in MATLAB® and aims to minimise final consumption by acting exclusively on the management of the heating element. The results obtained from the code are first tested in the Simulink® digital model of the oven and then through experimental testing. The results demonstrate that the proposed tool, specifically tailored for these systems, provides outcomes consistent with real operating conditions, while enabling a substantial reduction in laboratory testing time.

Keywords: multi-objective optimisation; Pareto front; energy consumption

1. Introduction

According to Eurostat, most of the EU's final energy consumption in the residential sector is covered by electricity (24.6%). Household appliances represent the most significant source of consumption. This awareness emphasises the development of more effective technologies to minimise energy consumption. European standards on the energy consumption of ovens standardise the techniques for classifying the energy efficiency of these appliances. Due to regulatory requirements, companies have invested resources in research to develop new technologies aimed at reducing consumption. These new technologies, such as pulsed electric fields (PEF) [1], ohmic heating [2], are very promising techniques. However, they almost never reached the market due to relative technological issues (e.g., production cost). For this reason, while many researchers are focusing on new technologies, others are more interested in optimising processing using consolidated technologies [3]. For example, this includes quantifying the energy demand for food and the process-related energy consumption, allowing optimisation of the total energy absorbed by the product [4]. The numerical approach is certainly widespread among the various methods, and dynamic models are essential tools for design and control. Usually, a lumped-parameter-based methodology is considered to develop a low-order dynamic model, reducing computational cost and run-time [5–7]. In these works, an analysis of energy consumption was carried out in accordance with EN 60350 [8]. By implementing a brick model and adopting alternative formulations of the evaporation term, we lay the foundation for the development of an accurate numerical model. In one approach, evaporation was modelled through a linear relationship between oven temperature and load temperature [6], while in another, it was expressed as a function of vapour pressure, which depends on the temperature measured at the brick surface [7]. In this type of system, energy consumption is typically associated with the oven's heating elements. Further studies focused on the product's energy demand, accounting for sensible and latent heat contributions [9]. A more detailed analysis was carried out by examining the heater usage algorithm governing the cooking process in a domestic oven. Numerical simulations and experimental tests were performed to examine heat distribution and the primary mechanisms of heat loss. The fidelity of the baking-process model was further improved by modelling evaporation and condensation phenomena within the thermal properties of a cupcake [10]. Among the various optimisation techniques, the multi-objective technique is also included as an algorithm aimed at reducing energy consumption. These studies represent the largest category (47%) and are most frequently implemented using MATLAB (32%) [11]. Multi-objective optimisation turns out to be a very useful tool for critical challenges in advanced energy systems, such as thermal energy storage devices based on phase change materials and heat exchangers, where optimisation strategies are used to simultaneously improve heat transfer and system efficiency [12]. Similarly, hybrid methodologies combining Computational Fluid Dynamics and Design of Experiments have been employed to identify optimal operating conditions for industrial equipment, such as centrifugal fans used in industrial ovens [13]. Furthermore, it is used to achieve cleaner, more sustainable systems, thereby reducing CO_2 emissions [14]. Moreover, this method can be used to develop a model based on active power losses and voltage deviation. Within the food and cooking sector, optimisation frameworks have been adopted to improve thermal processes and product quality. For instance, model-based multi-objective optimisation has been applied to beef roasting to balance energy consumption, cooking time, and final product characteristics [15]. In addition, multi-objective optimisation frameworks have been proposed for energy management in commercial baking processes to simultaneously optimise energy use and thermal performance during bread baking [16]. Within this framework, the Pareto front serves as the solution tool for the multi-objective optimisation problem [17]. However, they are also used to identify the optimal locations for installing power-quality monitors, aiming to minimise investment costs while ensuring adequate observability of power-quality disturbances, particularly voltage sags [18]. The variety of available multi-objective optimisation algorithms continues to expand. Recent studies have extended these methods to complex mechanical systems, such as six-axis manipulators, to simultaneously optimise motion paths and control energy consumption [19]. The use of the Pareto front and its variants remains a widely explored topic in the literature, with studies focusing on adapting these algorithms to a broad range of engineering applications [13,20–23]. Multi-objective optimisation methods are frequently employed in thermal and energy systems to negotiate conflicting goals, including efficiency, cost, and performance. Most research focuses on temperature uniformity or component design rather than on reducing electrical energy use; therefore, their application to domestic cooking appliances is limited. Using the Pareto front approach, this study develops a multi-objective optimisation framework and applies it to the thermal control of a heating cavity system typical of a domestic oven. Using a multi-objective approach with a physics-based digital model, the method reduces energy and time demands, thereby enabling the discovery of optimal operating settings. First examined with the dynamic model created in MATLAB/Simulink (version R2024a, MathWorks), results are then verified via lab experiments.

https://doi.org/10.3390/pr14060979

2. Digital Model

A digital model of the system, namely a physical model that reproduces the dynamics of the real system and its interactions with the environment, is needed to test the control strategy and compare it with the real system. In this section, the oven considered in this analysis is presented, and a brief description is given of the approach used for modelling the different components, with a special emphasis on the control system.

2.1. System Description

The system considered is an electromechanical domestic oven with a volume of 70 L. The oven has an enamelled steel outer shell designed to withstand high temperatures. The front section includes a door fitted with low-emissivity, double-layer tempered glass to ensure both thermal insulation and user safety. The cooking chamber is coated with nonstick materials and equipped with support guides for racks and drip pans. Heating is provided by electric elements positioned at the top and bottom of the cavity. Figure 1 shows the dimensions and the main components of the oven. Two heating elements, called Grill and Cielo, are mounted on the upper wall of the cavity. The bottom element, called Suola, is located in a drawer below the cooking chamber. A fan, covered by a baffle, is mounted on the rear wall, allowing forced-convection operation during ventilated cooking modes. The baffle has vents used to recirculate the air inside the cavity. The lateral walls (left and right), the back wall, the bottom surface, and the door define the thermal boundaries of the cavity. This configuration supports natural convection (static mode) and forced convection (ventilated mode), depending on the selected operating condition.

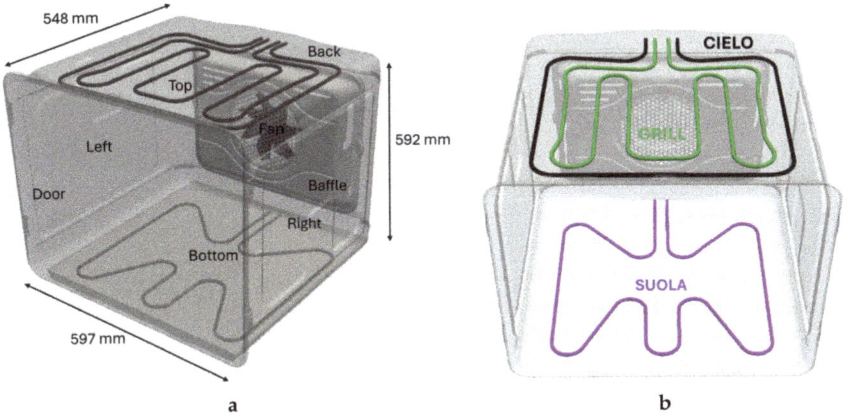

Figure 1. Overview of the oven considered in this analysis: (**a**) Names of the boundaries and the dimensions of the cavity walls; (**b**) Names of the three heating elements present in this model and their placement.

The control panel is equipped with three knobs:

- Temperature selector: regulates the oven temperature from 50 °C to 250 °C.
- Cooking function selector: allows the user to choose among several operating modes, including Static, Ventilated, Grill, and Eco.
- Timer: sets the desired cooking duration.

The digital model was developed in the MATLAB/Simulink environment and is based on a set of non-linear equations that describe the dominant heat-transfer mechanisms in the oven. It adopts a lumped-parameter formulation, detailed in a previous work [24].

https://doi.org/10.3390/pr14060979

In Table 1, all the nodes are listed and described. The energy balance for each node is expressed by:

$$C_i \frac{dT_i}{d\tau} = \sum_j Q_{conv\, i,j} + \sum_j Q_{rad\, i,j} + P_{el,i} + Q_{loss,i}, \tag{1}$$

where $P_{el\, i}$ represents the heat source term, which is non-zero only for the nodes corresponding to the heating elements. The oven walls are represented using thermal-block components, allowing the thermal properties of each layer to be specified. Convective, conductive, and radiative heat exchanges are implemented through dedicated physical blocks, each based on its corresponding physical laws and connections. As mentioned in [24], the heat transfer coefficients obtained from CFD analysis under steady state conditions are used as input in the dynamic model. The heating elements are described as a volume energy source, with an emissivity of 0.85 [25,26]. The interior elements, oven cavity, and baffle are prescribed with an emissivity of 0.9 [25–27]. The parameters do not depend on the temperature.

Table 1. Nodes considered for the lumped parameter model.

Nodes	$Q_{conv\, i,j}$	$Q_{rad\, i,j}$	$P_{el\, i}$	$Q_{loss\, i}$
Oven Cavity air (OC)	CW, GH and Brick	-	-	-
Grill heater (GH)	OC	CW and Brick	$P_{el,GH}$	-
Cavity walls (CW)	OC	CW and GH	-	$Q_{loss,CW}$
Bottom wall (BW)	OC and AIR BH	CW, BH, GH, BOW, and Brick	-	$Q_{loss,BW}$
Bottom heater (BH)	AIR BH	BW and BOW	$P_{el,BH}$	-
Bottom case wall (BOW)	AIR BH	BW and AIR BH	-	-

The dynamic system also incorporates a brick model to evaluate energy consumption in accordance with the standard [8]. The standard explains how to carry out the test in laboratories. The brick undergoes a preparation phase: it is soaked in cold water with two thermocouples inserted into the brick's core. Once it reaches approximately 5 °C, it is positioned on a rack in the centre of the oven. During the test, the sensor's temperature, the Brick's core temperature, and consumption are monitored. The test ends when the brick reaches +55 K. In the dynamic model, the brick is defined by two nodes, one associated with the external part BrS Equation (2) and one associated with the internal part BrC Equation (3) of the brick. According to [7] the relationship is defined as:

$$\begin{aligned} r(m_{dry}c_{dry} + m_w c_w)\frac{dT_{BrS}}{d\tau} &= Q_{OCBrS} + Q_{BrCBrS} + Q^{rad}_{THBrS} + Q^{rad}_{TWBrS} + \\ & Q^{rad}_{BWBrS} + Q^{rad}_{RWBrS} + Q^{rad}_{LWBrS} + Q^{rad}_{BaBrS} + Q^{rad}_{DBrS} - \dot{m}_w \Delta h_{lv}, \end{aligned} \tag{2}$$

$$\left[(1-r)(m_{\mathrm{dry}}c_{\mathrm{dry}} + m_w c_w) - m_{w\,ev}\, c_w\right]\frac{dT_{\mathrm{BrC}}}{dt} = Q_{BrC\, BrS}. \tag{3}$$

In this configuration, radiative heat transfer constitutes the dominant mechanism between the oven cavity and the brick surface. Additionally, evaporation plays a significant role during the heating phase and must be accounted for in the thermal modelling.

2.2. Electromechanical Control

Temperature control in an electromechanical oven is via a mechanical thermostat that regulates the heating elements. This liquid-expanding thermostat senses the temperature and turns the heating on and off. A thermometer bulb filled with heat-expanding fluid is placed in the cooking chamber. The bulb is connected to a capillary tube filled with the same fluid, which reaches the thermostat. As the temperature rises, the fluid in the

https://doi.org/10.3390/pr14060979

bulb expands, exerting pressure on a diaphragm in the thermostat. This pressure actuates an electrical contact, interrupting or reactivating the resistance circuit. The bulb in the dynamic model is represented as follows:

$$C_{bu}\frac{dT_{bu}}{d\tau} = \dot{Q}_{ocbu} + \dot{Q}_{TWbu} + \dot{Q}^{rad}_{TWbu} + \dot{Q}^{rad}_{THbu}. \quad (4)$$

The thermostat regulates the switching on and off of the heating elements. When the temperature drops below the set value, the thermostat closes the circuit and energises the heating elements. When the temperature is reached, the thermostat opens the circuit, cutting off the power. Temperature control is carried out in a cyclic on–off mode: the oven alternates between active heating phases and turning off the heating elements to maintain the desired temperature. The cycle does not maintain a perfectly constant temperature but fluctuates around the set value by about ±5 °C. A logic block represented the on/off behaviour of the resistors, which was called the switch. This has three inputs and one output:

- Threshold: a scalar signal or constant that determines the selection criterion.
- First option: the signal that will be routed if the condition is true.
- Second option: the signal that will be routed if the condition is false.
- Output: the signal selected according to the condition set.

The threshold input is calculated as the difference between the set-point temperature and the probe's temperature. If the measured temperature is below the set threshold, the thermostat closes the circuit and the heating elements start heating. When the temperature reaches the set value, the thermostat opens the circuit and turns off the heating elements. A gain block is used to express the power of each heater. The output is then used as input to a controlled heat flow that feeds the heating elements, modelled using a thermal block. Given that the goal is to determine the configuration that achieves minimal energy consumption, it is necessary to evaluate all possible combinations of the heating elements. In the experimental tests, we manually evaluated the operation of the heating elements in series and parallel configurations to assess the response of the electromechanical control system. Each setup was created by connecting the resistances based on the chosen circuit design and then operating the thermostat as described. In the series setup, the heating elements were powered one after another, with the thermostat managing the total circuit. This allowed the observation of the combined effect on the chamber temperature. In the parallel configuration, the elements were energised simultaneously, and the thermostat regulated their combined output, enabling the comparison of the temperature rise and the stability between the two setups. Figure 2 shows all the possible combinations implemented in the oven: in red, the parallel function, and in blue, the series function. During these manual tests, the oven's temperature cycles were monitored with the sensor bulb, confirming the expected on/off behaviour of the heating elements and the ±5 °C fluctuations around the set point. These tests validated the model assumptions and the logic implemented in the switch and gain blocks, ensuring that the digital representation of the heating control accurately reproduces the actual electromechanical behaviour for both series and parallel configurations.

The configurations considered in this study are representative of standard household electric oven architectures. Commercial appliances of this type are designed and certified according to internationally harmonised safety standards (IEC60335-2-6 [28] for stationary electric cooking ranges and ovens), which address electrical insulation, protection against electric shock, surface temperature limits, and mechanical safety. While the present work focuses on the modelling and optimisation of thermal performance, any practical imple-

https://doi.org/10.3390/pr14060979

mentation or deployment in a commercial product would require compliance with these safety standards by components and configurations.

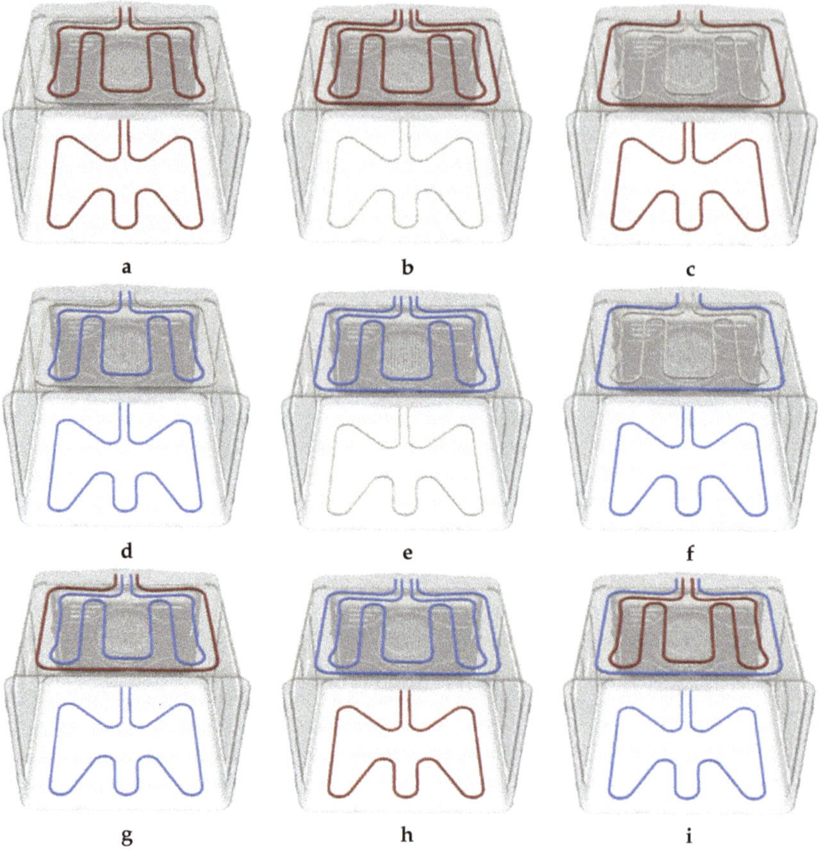

Figure 2. All the possible combinations implemented in the oven. In red: the parallel connection; in blue: the series connection. (**a**) Grill and Suola in parallel; (**b**) Grill and Cielo in parallel; (**c**) Suola and Cielo in parallel, (**d**) Grill and Suola in series; (**e**) Grill and Cielo in series; (**f**) Suola and Cielo in series; (**g**) Grill and Suola in series, with Cielo in parallel; (**h**) Grill and Cielo in series, with Suola in parallel; (**i**) Suola and Cielo in series, with Grill in parallel.

3. Optimisation Strategy

The approach used to achieve a lower energy consumption than the baseline relies on managing the power supplied to the heating elements. The standard operating mode plays the most significant role in determining the oven's energy class. According to the production standard, the grill element and the bottom heating element operate in parallel. The study began by identifying all possible heater configurations, in series and in parallel, available within the oven's electrical layout. As mentioned in Section 2.2, testing each configuration requires technicians to manually modify the wiring connections between heating elements. This procedure is not only labour-intensive but also time-consuming, as every wiring adjustment must be followed by a complete energy consumption measurement.

3.1. Generation and Power Constrained Selection of Heating Element Configurations

The first step in the optimisation process was the development of an initial script to compute all possible combinations of heating element configurations. Each electrical

https://doi.org/10.3390/pr14060979

resistance is defined and associated with its nominal power rating (in watts). The code then generates all feasible combinations of heaters operating simultaneously. A for loop is used in conjunction with the nchoosek function to produce every combination involving two or more resistors. For each resulting configuration, the corresponding power output is calculated. In parallel configurations, the total power is simply the sum of the individual resistor powers. In series configurations—applicable only to pairs of heaters—the total power is computed using the appropriate electrical relationship for series circuits:

$$\frac{(x \cdot y)}{(x + y)}, \tag{5}$$

where x and y are two generic power heaters. With an if loop, the constraint on the maximum acceptable power is applied. Only configurations with power <3000 W are accepted. The same process is carried out for the calculation of "mixed" combinations, i.e., those involving the resistors in series and one in parallel. Combinations of three active resistors are iterated to calculate all possible mixed connections. Finally, with an fprint, the valid combinations are printed, specifying the type: Parallel, Series, or Series + Parallel.

3.2. Development of the Thermal Model for Oven Performance Evaluation

The second part of the code set up a model for the oven's thermal behaviour. Table 2 shows all the parameters defined in the code.

Table 2. Thermal properties and operating parameters set for modelling the system's thermal behaviour.

Parameters	Value	Unit
Oven volume	70	L
Brick mass	2	kg
Brick heat capacity	2263.66	J/kg °C
Brick initial temperature	5	°C
Oven initial temperature	23	°C
Brick target temperature	+55 °C	°C
Efficiency heaters	0.7	-

Most model parameters were derived from the manufacturer, reference standards, and physical property correlations. Experimental tests were therefore used only for model validation, ensuring that the assessment of the model performance is independent from the parameter identification process and avoiding circular validation. The volume of the oven and the brick mass obtained from the datasheet. The brick heat capacity was calculated using the standard volume-based mixing rule:

$$c_s = \varphi_w c_w + (1 - \varphi_w) c_{b,dry}. \tag{6}$$

The initial and target temperatures for the brick and the oven were defined in accordance with the relevant reference standards. The efficiency of the heating elements was calculated as the ratio between the energy used for water evaporation and the total electrical energy supplied during the simulation. The heater combinations generated by the initial script are imported into this second code for further optimisation. A for loop is then executed, and for each valid configuration, it computes the energy and time required to heat the oven, transfer thermal energy to the brick, and produce the resulting temperature increase within the brick.

https://doi.org/10.3390/pr14060979

Polynomial Fitting of the Oven–Brick Thermal Transfer Function

To obtain the most realistic possible model of brick heating, the code incorporates a function that defines the relationship between the brick and oven temperatures. This relationship is nonlinear and was derived experimentally through laboratory testing. The experiment was conducted under the production-standard operating mode, meaning the Grill and Suola in parallel. The brick was prepared according to the reference procedure explained in Section 2.1. During the test, the brick's core temperature and the oven temperature were continuously monitored using thermocouples. The measured data showed that heat transfer to the brick is nonlinear and can be accurately represented by a 4th-degree polynomial. The data obtained from the experiment were then processed using a dedicated script to derive the correlation. To quantify the relationship between the two data sets, the Pearson correlation coefficient was used [29], defined as follows:

$$r = \frac{Cov(X,Y)}{\sigma_X \cdot \sigma_Y}, \tag{7}$$

where:

$$Cov(X,Y) = \frac{1}{n}\sum_{i=1}^{n}(x_i - \bar{x})(y_i - \bar{y}), \tag{8}$$

$$\sigma_X = \sqrt{\frac{1}{n}\sum_{i=1}^{n}(x_i - \bar{x})^2}, \tag{9}$$

$$\sigma_Y = \sqrt{\frac{1}{n}\sum_{i=1}^{n}(y_i - \bar{y})^2}. \tag{10}$$

The polyfit function was used to determine the polynomial coefficients and the display command was used to visualise them.

3.3. Pareto Front-Based Multi-Objective Optimisation Computation

The optimisation method described in this work is based on a fundamental concept of multiobjective optimisation (MOO). The goal of MO optimisation is to find a set of acceptable solutions and present them to the Decision Maker, who will then choose among them. Additional constraints or criteria specified before or after the search by the DM can help guide, refine, or narrow the search [30]. The general MO optimisation problem requiring the optimisation of N objectives can be formulated as follows:

$$Minimize \quad \vec{y} = \vec{F}(\vec{x}) = [f_1(\vec{x}), f_2(\vec{x}), \ldots, f_N(\vec{x})]^T, \tag{11}$$

$$Subject\ to \quad g_j(\vec{x}) \le 0, j = 1, 2, \ldots, M, \tag{12}$$

$$Where \quad \vec{x} = [x_1, x_2, \ldots, x_p]^T \quad \in \Omega, \tag{13}$$

$\vec{y}$ is the objective vector, g_j represents the constraints, and $\vec{x}$ is a P-dimensional vector representing the decision variables within a parameter space Ω. The space spanned by the objective vectors is called the objective space. The subspace of the objective vectors that satisfies the constraints is called the feasible space. To compare candidate solutions to the MO optimisation problems, the concepts of Pareto dominance and Pareto optimality are commonly used [31]. A solution belongs to the Pareto set if there is no other solution that can improve at least one objective without degrading any other objective. A decision vector

https://doi.org/10.3390/pr14060979

$\vec{u} = [u_1, u_2, \ldots, u_p]^T$ is said to *Pareto-dominate* the decision vector $\vec{v} = [v_1, v_2, \ldots, v_p]^T$ in a minimisation context if and only if:

$$\begin{aligned} &\forall i \in 1, \ldots N, f_i(\vec{u}) \leq f_i(\vec{v}), \\ &and\, \exists j \in 1, \ldots N : f_j(\vec{u}) < f_j(\vec{v}). \end{aligned} \tag{14}$$

In the context of MO, Pareto dominance is used to compare and rank decision vectors: $\vec{u}$ dominates $\vec{v}$ in the Pareto sense means that $\vec{F}(\vec{u})$ is better than $\vec{F}(\vec{v})$ for all objectives, and there is at least one objective function for which $\vec{F}(\vec{u})$ is strictly better than $\vec{F}(\vec{v})$. A solution $\vec{a}$ is said to be Pareto optimal if and only if there does not exist another solution that dominates it. The corresponding objective vector $\vec{F}(\vec{a})$ is called a Pareto dominant vector or a non-inferior or non-dominated vector. The set of all Pareto optimal solutions is called the Pareto optimal set. The corresponding objective vectors are said to be on the Pareto front. It is generally impossible to come up with an analytical expression of the Pareto front. There are several approaches to solving multi-objective optimisation problems. The first is the weighted sum method, in which objectives are combined into a single scalar function. The function is represented by positive weights assigned to the objectives. Then there is the ϵ-binned objective function method. In this case, one objective is chosen to be optimised, and the others are transformed into constraints. Finally, there are multi-objective evolutionary algorithms, among them are NSGA-II (Non-dominated Sorting Genetic Algorithm) or MOEA/D (Multi-Objective Evolutionary Algorithm based on Decomposition). Unlike the first two, these explore the entire solution space by finding Pareto-efficient solutions without requiring explicit weights. The NSGA-II is one of the algorithms implemented in MATLAB via the gamultiobj function [32]. In general, it classifies solutions according to their dominance using generic operators such as selection, crossover, and mutation to explore the solution space and uses Crowding distance to maintain the diversity of solutions on the frontier. For this analysis, none of the built-in MATLAB fitting functions were used; instead, a dedicated, ad-hoc function was developed to address the problem's characteristics. The function was expressed as isPareto = computeParetoFront(data), where data represents the input, a matrix $N \times M$ where N is the number of solutions (candidate points on the Pareto frontier) and M is the number of objective functions. Next, we proceed to the calculation of Pareto dominance, where for each point i in the dataset, we check whether there is another point j that dominates it according to the Pareto rule:

$$f_k(x_j) \leq f_k(x_i) \quad \forall k \in 1, \ldots, M, \tag{15}$$

with at least one narrow inequality. If a point i is dominated by another point j, then it is not part of the Pareto frontier. Finally, isPareto is a Boolean vector of length N, where true indicates that the point belongs to the Pareto frontier, and false indicates that the point is dominated. Each candidate point used in the Pareto front analysis corresponds to a specific configuration of the oven heating elements generated by the combinatorial script described in Section 3.1. For each configuration, the thermal model described in Section 3.2 is executed to compute the resulting heating time and total electrical energy consumption required to reach the target brick temperature. These two quantities represent the objective functions of the multi-objective optimisation problem. Therefore, each point in the objective space is defined as:

$$x_i = [E_i, t_i], \tag{16}$$

where E_i is the total electrical energy consumed and t_i is the time required to complete the heating process for configuration i. The resulting dataset forms the $N \times 2$ matrix used as input for the Pareto front computation.

https://doi.org/10.3390/pr14060979

4. Results and Discussion

The digital model developed in the MATLAB/Simulink® environment was tested and validated before performing optimisation analyses. In this section, the results obtained from the optimisation code are presented and discussed. The solutions identified on the Pareto front were compared with both experimental measurements and dynamic simulations.

To ensure the reliability of the optimisation outcomes, all feasible combinations were evaluated using the Simulink model. However, in the experimental campaign, only a representative subset of configurations was tested, selected based on practical constraints (e.g., time, cost, and implementation feasibility). This approach allowed experimental validation of the most significant operating points while maintaining a complete numerical verification through simulations.

4.1. Computed Heater Configurations and Associated Power Outputs

The first part of the code quickly generates all valid resistor combinations, choosing from the three installed in the oven: Grill 1700 W, Cielo 1000 W, Suola 1200 W. Table 3 shows the results of the code with the combinations and the powers associated with them.

Table 3. Heating element configurations (Grill, Cielo, Suola) and their respective power outputs. The values include series, parallel, and mixed connections, ensuring compliance with the maximum power limit of 3000 W.

Combinations	Power (W)
Grill + Suola (Parallel)	2900
Grill + Suola (Series)	703.45
Grill + Cielo (Parallel)	2700
Grill + Cielo (Series)	629.63
Suola + Cielo (Parallel)	2200
Suola + Cielo (Series)	545.45
Suola + Cielo + Grill (Series + Parallel)	2245.45
Grill + Cielo + Suola (Series + Parallel)	1829.63
Grill + Suola + Cielo (Series + Parallel)	1703.45

The results include "Series + Parallel" combinations, as well as exclusively "Parallel" or "Series" combinations. In particular, for those with three heaters, the first two are calculated in series, and the last one in parallel. As shown, no combinations exceed 3000 W, so the power constraint is correctly applied in the calculation. The code can be considered correct if its results are confirmed by manually computing the corresponding heater combinations.

4.2. Derivation of a Nonlinear Temperature Correlation from Experimental Data

This section describes the methodology used to derive the coefficients defining the correlation between the rise in brick temperature and the corresponding oven temperature. Including this correlation is essential to achieving the highest possible accuracy in the model. The correlation obtained from the code is:

$$T_{brick} = 4.8886e^{-7}T_{oven}^4 - 1.9201e{-}4T_{oven}^3 + 0.0256T_{oven}^2 - 1.329T_{oven} + 26.8664. \quad (17)$$

Figure 3 shows the comparison between the experimentally measured brick temperature and the temperature predicted by the fitted model during the heating process. The experimental data (blue markers) are in good agreement with the model predictions (red line) over the entire time interval. The model accurately reproduces the brick's initial thermal inertia and the subsequent increase in temperature during the heating phase. The goodness-of-fit is confirmed by a coefficient of determination $R^2 = 0.97$, indicating that the

https://doi.org/10.3390/pr14060979

model explains most of the variance in the experimental data. This agreement confirms that the proposed model can represent the changing thermal behaviour of the oven and brick system. This supports its use for the following optimisation analysis.

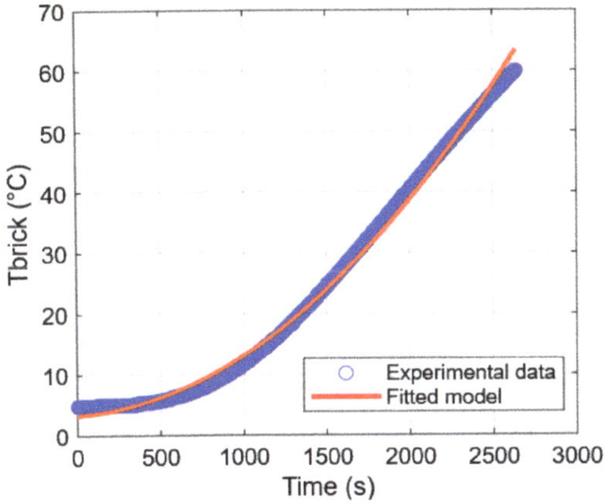

Figure 3. Results of the correlation realised in MATLAB. The red line represents the fitted model; the blue dots represent the experimental data.

4.3. Pareto Front Analysis and Validation of Optimal Energy

The specific challenge that motivated the use of optimisation models was the need to identify combinations of heating elements capable of minimising the oven's power consumption:

$$E = \int_0^{t_f} P(t)\, dt \tag{18}$$

where $P(t)$ is the instantaneous electrical power supplied to the heating elements and t_f is the final time of the heating phase. In the Pareto function created specifically for this system, described in Section 3.3, an $N \times M$ matrix was defined, where the first column represents the first dimension (energy) and the second column represents the second dimension (time). The energy and time values are extracted from the results structure generated by the previous code, and these data sets are then used to compute the Pareto front. We then extract the configurations on the Pareto front, defining *configurationspareto = results(isPareto)*. Figure 4 shows the graph of solutions on the Pareto frontier as a function of time and energy. The blue dots represent all tested configurations and, therefore, all feasible solutions to the problem. The red dots highlight the Pareto-efficient solutions—those not dominated by any other point. In other words, no alternative solution achieves lower energy consumption and shorter time simultaneously. From the distribution of points, it is evident that the Pareto-optimal solutions lie in the lower-left region of the graph, which is expected since this area corresponds to the minimum values of time and energy. All other configurations are dominated because at least one alternative performs better on at least one of the two objectives. Of all the possible configurations, the code identified three optimal points on the Pareto front. The first configuration exhibits higher energy consumption but a shorter heating time, whereas the other two configurations require more energy and longer heating times than the first. This behaviour is consistent with the Pareto front concept, which states that improving one objective necessarily degrades at least one of the others.

The Pareto front results were first evaluated using the dynamic model developed in Simulink and subsequently validated through experimental testing on the physical prototype in the laboratory. Each experimental test was repeated three times, and the reported

https://doi.org/10.3390/pr14060979

values represent the mean ± standard deviation. Table 4 reports the energy consumption values for the three configurations selected by the optimisation code, including the results obtained from simulations and laboratory measurements, along with the corresponding standard deviations. Additionally, the root mean square error (RMSE) between simulation and experimental results is reported to quantify the agreement between the two approaches.

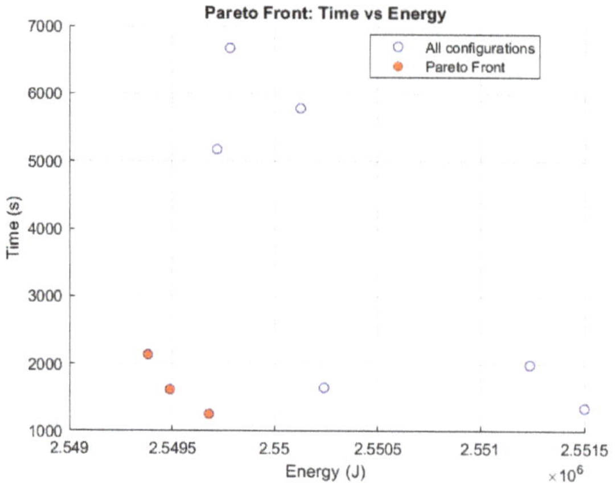

Figure 4. Pareto front distribution for multi-objective system optimisation. The x-axis shows energy in joules, while the y-axis shows time in seconds.

Table 4. Results from simulations and test.

Combinations	Code (kWh)	Simulation (kWh)	Experimental (kWh)
Grill + Suola (parallel)	1.35	1.27 ± 0.02	1.13 ± 0.03
Suola + Cielo + Grill (Series + Parallel)	1.02	1.033 ± 0.015	1.068 ± 0.02
Grill + Suola + Cielo (Series + Parallel)	0.956	0.948 ± 0.01	0.934 ± 0.015

The first row corresponds to the Parallel Grill + Suola configuration, which shows the highest energy consumption among all tested setups: 1.35 kWh in the optimisation code, 1.27 ± 0.02 kWh in the simulation, and 1.13 ± 0.03 kWh experimentally. The significant difference between the simulation and experimental results (RMSE = 0.14 kWh) likely stems from real-world factors, such as heat losses and thermal inertia, that the numerical model does not fully capture. The second row shows the Suola + Cielo + Grill (Series + Parallel) configuration. It achieves a reduction compared to the first configuration. The values are 1.02 kWh in the code, 1.033 ± 0.015 kWh in the simulation, and 1.068 ± 0.02 kWh experimentally. The smaller RMSE of 0.035 kWh shows better agreement between the simulation and experimental measurements than in the first configuration. Finally, the Grill + Suola + Cielo configuration has the lowest consumption. The code predicts 0.956 kWh, the simulation estimates 0.948 ± 0.01 kWh, and the measured value is 0.934 ± 0.015 kWh. The RMSE of 0.014 kWh confirms strong consistency between the numerical and experimental results. Overall, the comparison between the optimisation code, simulations, and laboratory experiments is shown in Figure 5. The inclusion of multiple repetitions, standard deviations, and RMSE allows for a quantitative assessment of uncertainty, confirming the reliability of the digital model in predicting real-world energy consumption.

To further assess the replicability and robustness of the optimisation code, the simulations were repeated on a different oven with a larger internal volume. The new oven has a volume of 85 L and two heating elements, in addition to the other three discussed in this

https://doi.org/10.3390/pr14060979

work. These heating elements are called Ring, and each has a power of 1550 W. The code is designed to allow key input parameters, such as oven geometry and heating element power, to be easily modified to reflect the characteristics of a new appliance. Using these adjusted parameters, the code generated a new set of optimal configurations, which were then evaluated through simulation and experimental tests, following the same procedure used for the oven analysed in this work. The results confirmed that the code performs consistently and accurately predicts energy consumption, with values and trends similar to those observed in the smaller oven. This approach shows that the optimisation framework can scale and adapt, enabling the same method to work across ovens of varying sizes without changing the underlying algorithms. Furthermore, the ability to quickly update input parameters ensures that the code can support future experimental setups or different appliance models, making it a reliable tool for research and practical applications. Table 5 reports the energy consumption for the same three optimal configurations when applied to both the original and the larger oven. Each experimental test was repeated three times, and the reported values represent the mean ± standard deviation. The relative difference between the two ovens is also indicated to highlight how energy consumption scales with oven size.

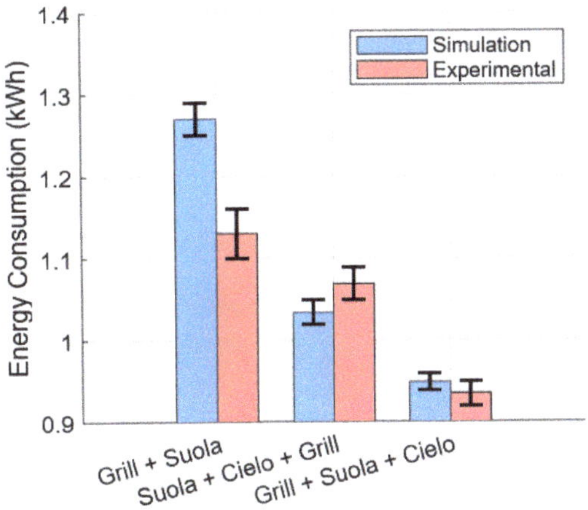

Figure 5. Comparison of energy consumption for different configurations. The blue columns show the simulation results, and the red columns represent the experimental results.

Table 5. Comparison of energy consumption between the original and larger oven for the same optimal configurations (mean ± standard deviation). Relative difference indicates percentage change.

Configuration	70 L (kWh)	85 L (kWh)	Simulation Std (kWh)	Experimental Std (kWh)	Relative Difference (%)
Grill + Suola (Parallel)	1.27	1.48	0.02	0.03	+16.5%
Suola + Cielo + Grill (Series + Parallel)	1.033	1.20	0.015	0.02	+16.2%
Grill + Suola + Cielo (Series + Parallel)	0.948	1.11	0.01	0.015	+17.1%

As expected, the larger oven exhibits higher energy consumption, with an increase of 16–17% across all configurations. Despite this, the relative ranking of the configurations remains unchanged, confirming that the optimisation code is robust and its recommendations are consistent and scalable. These results demonstrate that, by modifying only the input parameters—such as oven volume, heater positions, and thermal properties—the code can reliably predict optimal configurations for different appliances. This confirms

https://doi.org/10.3390/pr14060979

both the replicability and the general applicability of the optimisation methodology across ovens of different sizes.

5. Conclusions

In this work, a new optimisation approach for domestic ovens was presented. It aims to manage the heating elements efficiently to reduce both energy use and heating time. The optimisation code was developed in MATLAB®, while the oven's digital model was created in Simulink®. The goal was to find configurations that strike the best balance between lowering electrical energy and shortening heating duration. The effectiveness of this approach was confirmed through simulations and experimental tests. The configurations identified by the optimisation algorithm continually provide the best compromise between energy use and heating time, taking into account the specific characteristics of the oven studied. These results highlight the increasing importance of digital tools and model-based methods in supporting traditional experimental testing. The implemented workflow generates all possible heater combinations before choosing the most effective solutions. This gives a significant advantage when analysing the full range of electromechanical oven models. Due to its adaptability, the method can be applied to ovens with various designs, heating element configurations, and internal volumes, making it a flexible tool for broader design and optimisation frameworks. Lastly, even for a relatively simple system like a domestic oven, Pareto front analysis proves very valuable. It helps designers see and choose configurations that balance the goals of reducing energy and time. By showing the full range of optimal trade-offs, the Pareto front allows for more informed and flexible design decisions. It highlights options that may not be clear from focusing on a single objective.

Author Contributions: Conceptualisation, S.R.; methodology, S.R.; software, S.R.; validation, S.R. and B.B.P.; formal analysis, S.R.; investigation, S.R.; resources, B.B.P. and M.R.; data curation, S.R.; writing—original draft preparation, S.R.; writing—review and editing, S.R.; supervision, B.B.P. and M.R.; project administration, B.B.P.; funding acquisition, M.R. All authors have read and agreed to the published version of the manuscript.

Funding: This research received no external funding.

Data Availability Statement: The original contributions presented in this study are included in the article. Further inquiries can be directed to the corresponding author.

Acknowledgments: The authors would like to thank all SMEG employees who helped with the instruments during the tests in their workshops.

Conflicts of Interest: Author Marco Reguzzoni was employed by the company Research and Development Department, SMEG S.p.A. The remaining authors declare that the research was conducted in the absence of any commercial or financial relationships that could be construed as a potential conflict of interest. The company had no role in the design of the study; in the collection, analyses, or interpretation of data; in the writing of the manuscript, or in the decision to publish the results.

Abbreviations

The following abbreviations are used in this manuscript:

MOO	Multi-Objective Optimisation
MO	Multi-Objective
DM	Decision Maker
NSGA-II	Non-dominated Sorting Genetic Algorithm
MOEA/D	Multi-Objective Evolutionary Algorithm based on Decomposition
CFD	Computational Fluid Dynamics

https://doi.org/10.3390/pr14060979

References

1. Tanya, G.; Adedeji, A.; Ngadi, M.; Raghavan, V. Drying Characteristics of Pulsed Electric Field-Treated Carrot. *Dry. Technol.* **2008**, *26*, 1244–1250. [CrossRef]
2. Russo, M.; Serra, D.; Suraci, F.; Di Sanzo, R.; Fuda, S.; Postorino, S. The potential of e-nose aroma profiling for identifying the geographical origin of licorice (*Glycyrrhiza glabra* L.) roots. *Food Chem.* **2014**, *165*, 467–474. [CrossRef]
3. Khatir, Z.; Paton, J.; Thompson, H.; Kapur, N.; Toropov, V.; Lawes, M.; Kirk, D. Computational Fluid Dynamics (CFD) investigations of air flow and temperature distribution in a small scale bread-baking oven. *Appl. Energy* **2012**, *89*, 89–96. [CrossRef]
4. Papasidero, D.; Corbetta, M.; Pierucci, S.; Pirola, C.; Manenti, F. Studying Oven Technology towards the Energy Consumption Optimisation for the Baking Process. *Chem. Eng. Trans.* **2015**, *45*, 481–486. [CrossRef]
5. Monde, A.; Cannistraro, M. Design and control intended low-order model for transient investigation of an electric oven: Static mode. *Int. J. Heat Mass Transf.* **2024**, *221*, 125061. [CrossRef]
6. Ramirez-Laboreo, E.; Sagues, C.; Llorente, S. Dynamic heat and mass transfer model of an electric oven for energy analysis. *Appl. Therm. Eng.* **2016**, *93*, 683–691. [CrossRef]
7. Suzzi, N.; Lucchi, M.; Lorenzini, M. Dynamic Model for Convective Heating of a Wet Brick during Energy Characterisation of Domestic Electric Ovens. *Appl. Therm. Eng.* **2019**, *161*, 114117. [CrossRef]
8. *EN 60350-1*; Household Electric Cooking Appliances—Part 1: Ranges, Ovens, Steam Ovens and Grills—Methods for Measuring Performance. International Electrotechnical Commission: Geneva, Switzerland, 2023.
9. Ureta, M.; Goñi, S.; Salvadori, V.; Olivera, D. Energy requirements during sponge cake baking: Experimental and simulated approach. *Appl. Therm. Eng.* **2016**, *115*, 637–643. [CrossRef]
10. Park, D.; Seo, E.; Kwon, M.; Park, Y. The study on the heater usage for better energy efficiency of domestic convection oven. *J. Mech. Sci. Technol.* **2018**, *32*, 907–914. [CrossRef]
11. Mindeel, T.; Spentzou, E.E.; Eftekhari, M. Energy, thermal comfort, and indoor air quality: Multi-objective optimization review. *Renew. Sustain. Energy Rev.* **2024**, *202*, 114682. [CrossRef]
12. Taghavi, M.; Ferrantelli, A.; Joronen, T. Multi-objective optimization of a plate heat exchanger thermal energy storage with phase change material. *J. Energy Storage* **2024**, *89*, 111645. [CrossRef]
13. Pereira, J.L.; Oliver, G.; Francisco, M.; Cunha, S., Jr.; Gomes, G. A Review of Multi-objective Optimization: Methods and Algorithms in Mechanical Engineering Problems. *Arch. Comput. Methods Eng.* **2021**, *29*, 2285–2308. [CrossRef]
14. Kuroda, K.; Magori, H.; Ichimura, T.; Yokoyama, R. A hybrid multi-objective optimization method considering optimization problems in power distribution systems. *J. Mod. Power Syst. Clean Energy* **2015**, *3*, 41–50. [CrossRef]
15. Goñi, S.M.; Salvadori, V.O. Model-based multi-objective optimization of beef roasting. *J. Food Eng.* **2012**, *111*, 92–101. [CrossRef]
16. Khatir, Z.; Taherkhani, A.; Paton, J.; Thompson, H.; Kapur, N.; Toropov, V. Energy thermal management in commercial bread-baking using a multi-objective optimisation framework. *Appl. Therm. Eng.* **2015**, *80*, 141–149. [CrossRef]
17. Zhang, M. Multi-Objective Optimal Reactive Power Dispatch of Power Systems by Combining Classification-Based Multi-Objective Evolutionary Algorithm and Integrated Decision Making. *IEEE Access* **2020**, *8*, 38198–38209. [CrossRef]
18. Martins, P.; Oleskovicz, M. Multi-Objective Optimization Aiming to Minimize the Number of Power Quality Monitors and Multiple Fault Estimations in Unbalanced Power Distribution Systems. *IEEE Trans. Power Deliv.* **2021**, *37*, 1315–1323. [CrossRef]
19. Ma, Z.; Zhu, A.; Feng, H.; Zhou, C.; Liu, H.; Sheng, D.; Guo, H.; Su, C. Research on path and energy consumption control of six-axis manipulator based on multi-objective optimization method. *Int. Sci. Tech. Econ. Res.* **2025**, *8*, 96–113.
20. Giagkiozis, I.; Fleming, P. Pareto Front Estimation for Decision Making. *Evol. Comput.* **2014**, *22*, 651–678. [CrossRef] [PubMed]
21. Sharma, S.; Chahar, V. A Comprehensive Review on Multi-objective Optimization Techniques: Past, Present and Future. *Arch. Comput. Methods Eng.* **2022**, *29*, 5605–5633. [CrossRef]
22. Kang, S.; Li, K.; Wang, R. A survey on pareto front learning for multi-objective optimization. *J. Membr. Comput.* **2024**, *7*, 128–134. [CrossRef]
23. Asrari, A.; Lotfifard, S.; Payam, M.S. Pareto Dominance-Based Multiobjective Optimization Method for Distribution Network Reconfiguration. *IEEE Trans. Smart Grid* **2016**, *7*, 1401–1410. [CrossRef]
24. Rustico, S.; Pulvirenti, B.; Useli, G.; Reguzzoni, M. Multi-Level Simulation for Dynamic Analysis of A Domestic Electric oven in Forced Convective Heating—Evaluation and Optimization of Energy Consumption Using the Brick Test. *Future Cities Environ.* **2025**, *11*, 15–27. [CrossRef]
25. Mistry, H.; Ganapathisubbu, S.; Dey, S.; Bishnoi, P.; Castillo, J.L. A methodology to model flow-thermals inside a domestic gas oven. *Appl. Therm. Eng.* **2011**, *31*, 103–111. [CrossRef]
26. Mistry, H.; Ganapathi-subbu; Dey, S.; Bishnoi, P.; Castillo, J.L. Modeling of transient natural convection heat transfer in electric ovens. *Appl. Therm. Eng.* **2006**, *26*, 2448–2456. [CrossRef]
27. Rek, Z.; Rudolf, M.; Zun, I. Application of CFD Simulation in the Development of a New Generation Heating Oven. *Stroj. Vestn.—J. Mech. Eng.* **2012**, *58*, 134–144. [CrossRef]

https://doi.org/10.3390/pr14060979

28. *IEC 60335-2-6:2022*; Household and Similar Electrical Appliances—Safety—Part 2–6: Particular Requirements for Stationary Cooking Ranges, Hobs, Ovens and Similar Appliances. International Electrotechnical Commission: Geneva, Switzerland, 2022. Available online: https://webstore.iec.ch/publication/99834 (accessed on 23 February 2025).
29. Garcia Asuero, A.; Sayago, A.; González, G. The Correlation Coefficient: An Overview. *Crit. Rev. Anal. Chem.* **2006**, *36*, 41–59. [CrossRef]
30. Ngatchou, P.N.; Zarei, A.; Fox, W.L.J.; El-Sharkawi, M.A. Pareto Multiobjective Optimization. In *Modern Heuristic Optimization Techniques*; Wiley: Hoboken, NJ, USA, 2007; pp. 189–207. [CrossRef]
31. Coello, C.; Van Veldhuizen, D.; Lamont, G. *Evolutionary Algorithms for Solving Multi-Objective Problems*, 2nd ed.; Springer: New York, NY, USA, 2007. [CrossRef]
32. MathWorks. gamultiobj. Available online: https://mathworks.com/help/gads/gamultiobj.html (accessed on 23 February 2025).

https://doi.org/10.3390/pr14060979

energies

MDPI

Article

Experimental Wind Tunnel Study of Energy Consumption, Level Flight Speed, and Endurance of a Micro-Class UAV as a Function of Operating Weight

Bartłomiej Dziewoński [1,*], Krzysztof Kaliszuk [1], Artur Kierzkowski [1,*], Jakub Jarecki [2] and Kacper Lisowiec [2]

[1] Department of Technical Systems Operation and Maintenance, Wroclaw University of Science and Technology, Wybrzeze Wyspianskiego 27, 50-370 Wroclaw, Poland; krzysztof.kaliszuk@pwr.edu.pl
[2] Faculty of Mechanical and Power Engineering, Wroclaw University of Science and Technology, Wybrzeze Wyspianskiego 27, 50-370 Wroclaw, Poland; 274142@student.pwr.edu.pl (J.J.); 274152@student.pwr.edu.pl (K.L.)
* Correspondence: bartlomiej.dziewonski@pwr.edu.pl (B.D.); artur.kierzkowski@pwr.edu.pl (A.K.)

Abstract

This paper presents an experimental investigation of the level flight speed and endurance characteristics of a micro-class unmanned aerial vehicle as a function of operating weight. Wind tunnel experiments were conducted to determine the aerodynamic performance and power requirements of the UAV over a range of operating weight configurations. The tested vehicle, a fixed-wing micro UAV, was examined under steady, level flight conditions, with particular emphasis on identifying variations in the minimum power required to sustain level flight. Measured aerodynamic forces and moments were used to derive drag polars and the corresponding power curves for each mass configuration. Based on these results, endurance estimates were obtained by coupling the experimentally derived power requirements with the characteristics of the onboard electric propulsion system. The study demonstrates a clear shift in flight speeds with increasing operating weight, as well as a reduction in achievable endurance, highlighting the sensitivity of micro-class UAV performance to mass variations, and therefore energy consumption.

Keywords: unmanned aerial vehicle; energy consumption; wind tunnel; operating weight; panel method

1. Introduction

Energy consumption and energy density of propulsion power sources are key factors in planning unmanned aircraft missions where endurance and range are expected [1]. It also dictates the choice of engine and propeller for an electric propulsion system. The growth of electric propulsion systems in uncrewed aerial aircraft, aka drones, is driven by their low operational costs, high efficiency, and good environmental characteristics. Unfortunately, the endurance of micro-class unmanned aerial vehicles remains limited by the specific energy of batteries, making it critical to use every milliampere-hour (mAh) of energy as efficiently as possible. For electrically powered micro UAVs, mass variations resulting from payload integration or configuration changes directly affect the aerodynamic performance of such aircraft [2]. Ultimately, the endurance of such a platform can be affected by a decreased thrust-to-weight ratio. This parameter will require the entire platform to make up for deficiencies in lift force by increased use of the propulsion system. This will lead to increased power consumption. Understanding this relation is the main

 https://doi.org/10.3390/en19081892

objective of this work, as reliable endurance prediction and energy-aware mission planning will be strictly dictated by a UAV's operating weight. Accurate prediction of level flight speed is challenging, as theoretical models are often derived under simplified assumptions about aerodynamic and propulsion characteristics. In practice, the flight characteristics of micro-class UAVs deviate from theoretical models due to low Reynolds number effects, nonlinear propulsion efficiency, and airframe-propeller interactions that are neglected in preliminary drag estimations for this aircraft category. These differences become even more significant as operating weight varies. For general aviation aircraft, the operating weight ranges between 10% and 20% of the maximum take-off weight. In the micro-class UAV category, the maximum operating weight can be as high as 50% of the maximum take-off weight, further amplifying the differences between the theoretical and experimental energy consumption models [3]. The key novelty of the study is not only the assessment of energy consumption of a micro-class UAV, but specifically the combined numerical–experimental evaluation of level-flight power demand as a function of operating weight and angle of attack, using a wind tunnel methodology that reproduces steady level-flight equilibrium conditions. This is a characteristic that is not examined for small unmanned aerial vehicles unlike for the manned general aviation aircraft or bigger airliners.

1.1. Propulsion System Energy Source Characteristics

Recent studies show that battery systems, typically lithium-polymer or lithium-ion, used in micro-class UAVs can provide a specific energy of approximately 400 Wh per kilogram of battery package weight [4] that will constrain endurance in many operational scenarios. In contrast to Li-Po and Li-ion batteries, fuel cell systems offer higher specific energy of the fuel itself, but require much greater installation weight, which compromises the chances of using such a system on micro-class platforms [5]. Precise values for each energy supply on-board system are presented in Table 1.

Table 1. Energy density characteristics of main energy supply sources for propulsion systems.

	Fully Electric Propulsion	Hybrid Fuel-Cell Systems	Conventional Fuel
Specific Energy (Wh/kg)	150–400	1700–2000	~12,000
Average Endurance (h)	0.5–1.5	4–8	3–6
Battery Energy Density Target (2050) (Wh/kg)	1000–1340	-	-
System Weight Penalty	Moderate	High (Fuel Cell Mass)	Low

Beyond the specific energy of the energy source, endurance is strongly influenced by propulsion system architecture and mission power demand [6]. Fuel-based propulsion enables longer endurance and payload capability, but typically with lower conversion efficiencies and higher emissions than electric powertrains [7]. Hybrid systems aim to combine the advantages of both high energy density of fuel-based sources with the fast peak power capability of batteries [8]. Studies have reported that hybrid gasoline–electric concepts can achieve endurance extension relative to purely electric systems under the same mission profiles, with experimental work highlighting the sensitivity of flight time [9]. Such energy supply and propulsion system architectures are still too heavy to be applied to micro-class UAVs, which threatens the extended endurance of these platforms and requires precise energy consumption prediction.

Advanced energy management strategies, including fuzzy-logic approaches and equivalent consumption minimisation strategies (ECMS), have been investigated to dynamically distribute power between sources while maintaining battery state-of-charge (SOC) and improving fuel economy, often reporting measurable reductions in energy

https://doi.org/10.3390/en19081892

consumption [10,11]. At the same time, regarding hybridisation of power source, several studies show that substantial improvements in battery specific energy are required for long-endurance, fully electric UAV missions, with some suggesting that reaching multi-hour endurance at medium- and high-altitude mission conditions would demand increases in battery energy density relative to current mainstream chemistries [12]. This motivates continued research in both electrochemical storage and propulsion system-level integration, including thrust energy consumption modelling.

The above landscape demonstrates that endurance is governed by a coupled system problem spanning energy source properties, powertrain architecture, mission profile, and control. In this context, the present work focuses on quantifying and modelling energy consumption under a defined flight task relevant to endurance-limited operations. Specifically, the study evaluates energy consumption during a holding manoeuvre in a lightweight fixed-wing UAV, combining modelling and flight testing to support more reliable endurance prediction under sustained operational conditions and to complement the broader propulsion-system comparisons reported in the literature.

1.2. Angle of Attack Influence on Level Flight Parameters

Endurance of an aircraft, measured as the total time an aircraft remains airborne, is tied to how much lift it can produce relative to the drag generated at the same time [13]. Any lift shortages necessary to maintain level flight may be overcome by excess thrust if the aircraft has it. Angle-of-attack-related aerodynamic characteristics play a crucial role, especially in level flight, where lift must balance the aircraft's weight, and thrust must overcome drag [14].

In level flight, the angle of attack is defined as the angle between the wing's mean aerodynamic chord and the vector of oncoming airflow. Changing this angle affects both lift and drag generated by an airframe. The full range characteristic of the airframe in terms of generated lift and drag as a function of angle of attack is defined as the aerodynamic polar. At low angles of attack, the wing generates lift with comparatively low drag. At high angles of attack, however, lift increases significantly, especially when high-camber airfoils, typical of light unmanned aerial aircraft, are used [15,16]. Simultaneously, the drag increase is significant, requiring assistance from the available thrust excess to maintain steady flight. As the angle of attack increases, lift rise may be useful at lower airspeeds or when lifting heavier loads. On the other hand, drag increases more rapidly. This disproportionate growth in drag generated by the airframe requires greater thrust, leading to higher power consumption and reduced endurance, which is particularly important for the research shown in this paper.

The specific angle of attack at which the lift-to-drag ratio reaches its maximum can be obtained numerically as well as experimentally [17]. Flying at or near this angle minimises the required thrust to maintain level flight and therefore maximises endurance. For many high-camber airfoils, this optimal angle is much and well below the stall point for a particular airframe. Deviating from his optimal angle-of-attack range, either to lower or higher angles of attack, reduces aerodynamic efficiency, thereby shortening the aircraft's achievable endurance.

At even higher angles of attack, when approaching a critical threshold close to the stall angle of attack, airflow begins to separate from the wing surface. This leads to a loss of lift and a significant increase in drag. In near-stall and stall conditions, maintaining level flight becomes impossible without a significant increase in thrust. The airframe's stability may nevertheless be compromised [18]. From both a performance and safety perspective, such conditions for level flight are undesirable. The relationship between angle of attack, thrust requirement, and endurance has important implications for aircraft design and operation.

https://doi.org/10.3390/en19081892

Designers aim to shape wings and select airfoils that provide a high maximum lift-to-drag ratio, while operators try to keep the aircraft flying near this optimal condition. In modern aircraft, particularly light UAVs, this process is often supported by simulations such as computational fluid dynamics and validated through wind tunnel testing.

Summarising the angle of attack is a key parameter that links aerodynamic efficiency, thrust requirement, and endurance. Low to moderate angles of attack, near the point of maximum lift-to-drag ratio, allow an aircraft to remain airborne for the longest time with the least energy consumption. Higher angles of attack increase drag, demand more thrust, reduce endurance and may compromise safety [19]. Optimising and controlling the angle of attack is therefore essential for efficient and safe flight.

1.3. Operating Weight of an Aircraft

To understand an aircraft's operating weight, particularly in the context of light unmanned aerial vehicles, it is necessary to examine the various weight definitions used in aviation. Operating weight is a fundamental parameter because it directly affects the aircraft's propulsion requirements and overall endurance. Operating weight refers to the total weight of the aircraft when it is ready for flight, including the airframe, propulsion system, energy source, onboard systems, and any necessary components to maintain safe flight, as well as the cargo needed to complete the flight mission [20]. In lightweight UAVs where mass margins are small, even minor changes in the operating weight can significantly alter the aircraft's lift-to-weight ratio, with noticeable impacts on endurance and flight characteristics [21].

From operating weights, we can specify the operating empty weight (OEW). This is the weight of the aircraft structure, including all permanently installed systems, but excluding additional payload. Lower empty weight allows greater flexibility in payload selection within the aircraft's maximum take-off weight limit. Reducing the operating empty weight improves endurance as less lift is required to sustain flight and less power is needed to overcome drag [22].

Another weight parameter that can be specified is the maximum take-off weight (MTOW), which indicates the maximum weight at which the aircraft is designed to take off safely. Maximum take-off weight includes the aircraft's empty weight, energy source, payload, and any extra equipment installed in the aircraft. Operating near the MTOW values negatively affects performance, reducing an aircraft's climb and turn capabilities, and, most importantly, causes higher power consumption as any lift insufficiencies are made up by increasing thrust.

For most of the aircraft's design flight envelope, especially for a light unmanned aerial vehicle, the operating weight is the most important parameter. Payload weight is the operating weight of mission-specific equipment, such as sensors, communication systems, or cargo.

Assessing the impact of weight on endurance is possible when considering several representative operating weight configurations of an aircraft. A minimal-weight configuration consisting only of the airframe and installed systems represents the lightest possible setup and typically provides maximum endurance. This configuration, however, may not always be operationally useful due to the absence of sensors or payload. Adding the average payload yields the operating weight, which is the most important factor in energy consumption evaluation. Finally, considering operation at maximum take-off weight allows investigation of maximum payload occupancy, which should correspond to the lowest endurance of an aerial system, providing the investigator with a full endurance envelope dependent on operating mass [23].

https://doi.org/10.3390/en19081892

In summary, operating weight is a key factor in an aircraft's endurance and performance, particularly for light UAVs. Lower empty and operating weights generally improve endurance, while increased payload and operation near MTOW reduce efficiency and flight time [24]. Careful management and optimisation of weight throughout the design and operational phases are therefore essential for achieving long-endurance flight.

2. Theoretical Panel Method Model of Required Thrust and Energy Consumption

The panel method is a computational numerical method founded on potential flow theory and widely applied in the aerodynamic characterisation of light unmanned aerial vehicles (UAVs), particularly at low Reynolds numbers [25,26]. The method assumes the flow is inviscid, incompressible, and irrotational [27]. Under these assumptions, the velocity field V can be expressed as the gradient of a scalar potential φ [28]:

$$V = \nabla \varphi \tag{1}$$

Because the flow is irrotational and incompressible, the governing equation becomes Laplace's equation:

$$\nabla^2 \varphi = 0 \tag{2}$$

To solve this equation numerically for complex geometries, such as UAV configurations, the aircraft surface is discretised into a finite number of panels. Each panel is assigned a single element, typically a source, vortex, or doublet. The strengths of these singularities are determined so that the boundary conditions are satisfied at designated control points on the surface. The most common boundary condition is the no-penetration condition, which enforces that the normal component of velocity at the body surface is zero. For lifting surfaces, the Kutta condition is additionally imposed at the trailing edge to ensure a smooth flow departure and finite velocity, thereby determining a unique circulation distribution [29]. Low-order panel methods assume constant singularity strength over each panel. In contrast, higher-order panel methods allow singularity strength to vary across the panel, often using polynomial distributions. This improves geometric representation and increases solution accuracy, particularly for curved surfaces and complex UAV planforms [30].

At low Reynolds numbers typical of light UAVs (approximately 10^4 to 5×10^5), the aerodynamic environment differs substantially from that of larger aircraft. Flow is more susceptible to laminar separation, early transition, and reattachment [31]. The classical panel method, being inviscid, cannot inherently capture viscous effects such as laminar separation bubbles or stall. Nevertheless, when coupled with boundary layer models, panel methods can provide reasonable predictions of lift and pitching moment at small angles of attack. Comparisons with RANS-based CFD indicate that panel methods remain suitable for preliminary analysis below Reynolds numbers of approximately 500,000, although their drag predictions and post-stall accuracy are limited.

Aerodynamic polars, lift, drag, and moment coefficients as functions of angle of attack are derived by integrating surface pressure distributions. The typical workflow involves geometry discretisation, enforcement of boundary conditions, solution of the linear system for the singularity strengths, computation of local surface pressures using Bernoulli's equation, and integration of the pressures to obtain global forces and moments. Viscous corrections are then applied to account for skin friction and separation effects. Repeating the procedure over a range of angles of attack yields the full aerodynamic polar.

Aerodynamic characteristics of the PADA unmanned aerial vehicle, whose drawing can be seen in Figure 1a,b were estimated using XFLR5, which combines 2D airfoil analysis

https://doi.org/10.3390/en19081892

(via XFOIL-type viscous/inviscid coupling) with 3D lifting-surface solvers. Because PADA is a micro-class flying-wing configuration operating at low Reynolds numbers, the computational workflow was structured to generate consistent 2D airfoil polars over the relevant Reynolds number range and to propagate these polars into a 3D vehicle model to obtain lift, induced drag, pitching moment trends, and basic stability characteristics.

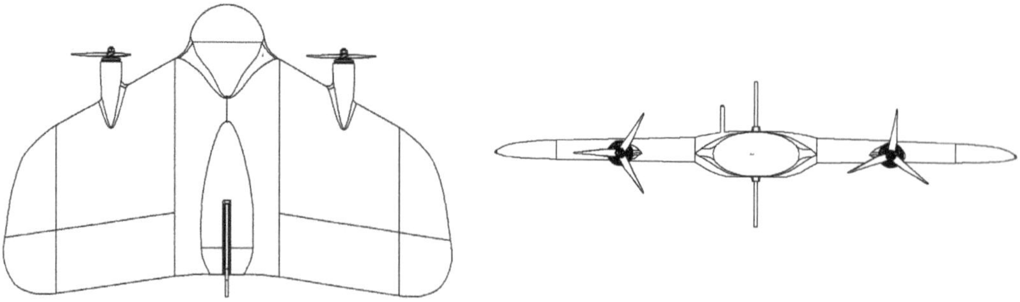

(a) (b)

Figure 1. Drawings of PADA unmanned aerial aircraft. (**a**) PADA top view; (**b**) PADA front view.

The PADA geometry was represented in XFLR5 as a wing-only aircraft, defined by its platform parameters: span, reference area, taper ratio, sweep, twist distribution, and spanwise airfoil assignment. The model was built by dividing the wing into a number of spanwise stations, each station specifying chord length, sweep offset, dihedral and airfoil type. Control surfaces were represented by defining flap/aileron panels with specified hinge location as a fraction of the chord. Sufficient panel density was selected to represent leading-edge curvature, wing tip geometry, and control surfaces, because these significantly influence the accuracy of 3D predictions, especially for swept wings at low Reynolds numbers.

The analyses were conducted for a set of flight speeds representative of the PADA mission segment(s). For each speed, a corresponding Reynolds number was computed using the local chord [32]:

$$\text{Re} = \frac{(\rho\ V\ c)}{\mu} \tag{3}$$

where ρ is air density, V is flight speed, c is chord length (reference mean aerodynamic chord), and μ is dynamic viscosity.

In practice, because the Reynolds number varies along the span due to varying chord, XFLR5 is commonly applied by generating polars for one or more representative Reynolds numbers, then applying them consistently across the 3D analysis. This approach is acceptable for preliminary characterisation, but it must be recognised that low-Re sensitivity can make spanwise variations more important than in larger aircraft.

The first step was generating 2D airfoil polars, seen in Figure 2, for the airfoils used on PADA. In XFLR5 6.61, this is performed in the foil analysis environment by specifying:

- The Reynolds number relevant to the flight condition;
- Transition or turbulence parameters through the Ncrit parameter;
- Angle-of-attack sweep range and step;
- Iteration limits and convergence criteria.

https://doi.org/10.3390/en19081892

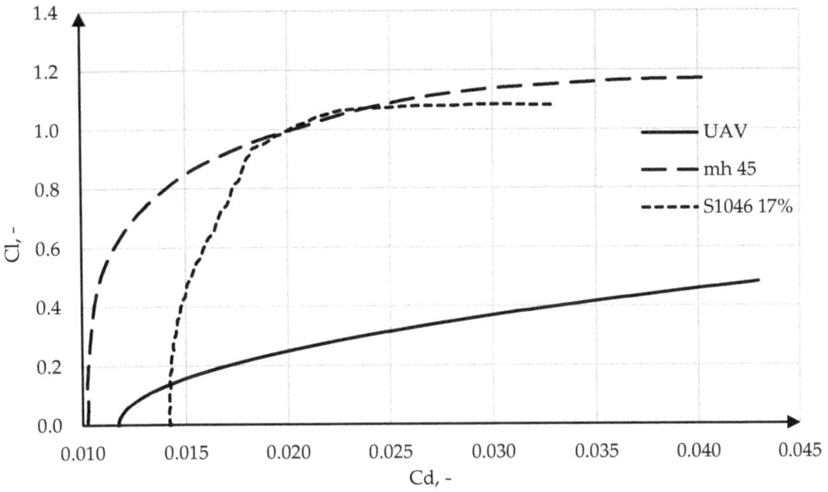

Figure 2. Lift-to-drag polars of UAV, MH 45 and S1046 airfoils.

The output polars include section lift coefficient Cl(α), drag coefficient Cd(α), and moment coefficient Cm(α). For low Reynolds numbers, special attention must be paid to the analysis range: near stall and in regimes where laminar separation bubbles occur, 2D convergence may become sensitive. Therefore, polar generation is often limited to the angle-of-attack range where the solver remains stable, and the results are physically plausible, and then later cross-checked against experimental or higher-fidelity data where available.

After generating the required 2D airfoil polars, the full PADA configuration was analysed in XFLR5 using the 3D panel method. For a flying wing such as PADA, with moderate sweep and a low to moderate aspect ratio, the 3D panel method is typically preferred over the classical Lifting Line Theory. The 3D analysis was conducted across a defined angle-of-attack sweep to produce aircraft-level polars seen in Figure 3:

- CL(α);
- CD(α);
- Cm(α).

where CL is the total lift coefficient, CD is the total drag coefficient, and Cm is the pitching moment coefficient about the reference point, in this case, the centre of gravity of an aircraft. The main outputs extracted for PADA included:

- Aircraft-level polar curves: CL(α), CD(α), Cm(α);
- Lift-to-drag ratio (L/D) versus angle of attack (α).

Outputs can be seen in Figure 3.

These results were then used as input for performance calculations of the power required to sustain level flight as a function of angle of attack, and therefore, to estimate the endurance of an aircraft. Because XFLR5 relies on inviscid 3D methods supplemented by 2D viscous polars, the method is best interpreted as a preliminary-to-intermediate-fidelity approach. Its predictions are generally most reliable at low to moderate angles of attack, while near-stall and highly separated regimes require validation or higher-fidelity CFD or experimental data. The exact XFLR setting for 3D analysis can be seen in Table 2.

https://doi.org/10.3390/en19081892

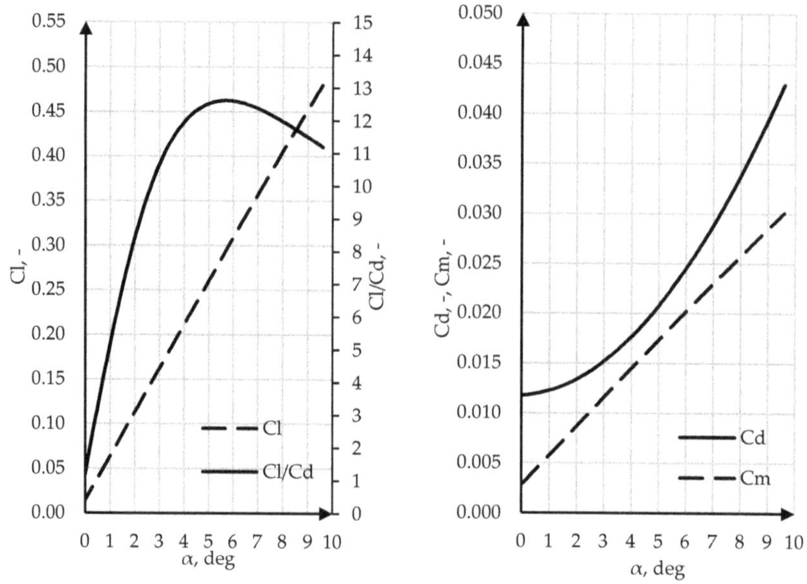

Figure 3. Lift-to-drag polars of UAV, MH 45 and S1046 airfoils.

Table 2. XFLR software 3D analysis settings.

Parameter	-
Analysis method	Vortex lattice method
Boundary condition	Dirichlet
Analysis type	Viscous
Reference area	0.149 m^2
Reference span	0.600 m
Reference chord length	0.260 m
Fluid density	1.225 km/m^3
Viscosity	1.5×10^{-5} m^2/s
Total panels	3652

Knowing the drag force values and lift force values, the amount of thrust needed to sustain level flight was calculated according to level flight aeromechanics equilibrium shown in Equations (4)–(7) [33] for three operating weights of an unmanned aerial aircraft:

- Minimal operating weight (0.46 kg);
- Payload operating weight (0.72 kg);
- Maximum take-off weight (1.00 kg).

$$x - V_\infty \cdot \cos(\gamma) = 0 \tag{4}$$

$$h - V_\infty \cdot \sin(\gamma) = 0 \tag{5}$$

$$T \cdot \cos(\varepsilon) - D - m(g\sin(\gamma) + V_\infty) = 0 \tag{6}$$

$$T \cdot \sin(\varepsilon) + L - m(g\cos(\gamma) + V_\infty \gamma) = 0 \tag{7}$$

where T—thrust, m—mass of an aircraft, V_∞—airspeed, D—drag force, and L—lift force.

The point of interest is in thrust, which will be determined as the necessary thrust to obtain and maintain a stable velocity of an aircraft during level flight.

https://doi.org/10.3390/en19081892

The amount of thrust needed to sustain level flight conditions for each operating weight as a function of angle of attack at certain flight speeds is shown in Tables 3–5.

Knowing the thrust force an engine has to generate at a certain flight speed and angle of attack, its power consumption characteristics were obtained using engine characteristics provided by the manufacturer, as seen in Figure 4.

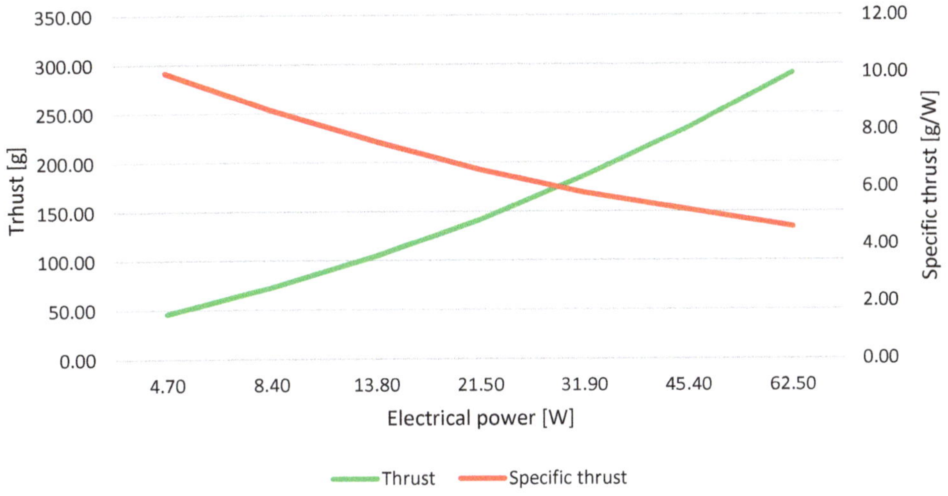

Figure 4. Single motor thrust (F2203.5 3550 kV) with 3 blade 5 × 3.5 propeller [34].

Using propulsion system characteristics provided by the manufacturer, as well as the thrust required to sustain level flight equilibrium obtained from panel numerical analysis of a UAV, the theoretical and energy consumption during a 5 min flight are shown in Tables 2–4. Energy consumption was assumed for a standard 5 min flight mission profile.

Table 3. Propulsion system parameters for 0.46 kg of operating weight for level flight condition.

AoA°	T_N [g]	P_{Teo} [W]	$E_{Teo5min}$ [Wh]
0	126.17	12.34	1.03
2	141.09	16.81	1.40
4	163.71	18.71	1.56
6	187.70	21.54	1.80
8	210.47	27.62	2.30
10	287.41	43.62	3.64

Table 4. Propulsion system parameters for 0.72 kg of operating weight for level flight condition.

AoA°	T_N	P_{Teo} [W]	$E_{Teo5min}$ [Wh]
0	198.54	22.37	1.86
2	219.33	29.14	2.43
4	243.01	34.31	2.86
6	259.08	39.84	3.32
8	337.54	57.66	4.81
10	419.18	81.19	6.77

https://doi.org/10.3390/en19081892

Table 5. Propulsion system parameters for 1.00 kg of operating weight for level flight condition.

AoA°	T_N	P_{Teo} [W]	$E_{Teo5min}$ [Wh]
0	253.38	37.86	3.16
2	270.14	40.93	3.41
4	301.31	44.64	3.72
6	369.56	63.87	5.32
8	421.84	81.98	6.83
10	498.48	95.87	7.99

AoA—Angle of attack of a UAV, T_N—thrust necessary to sustain level flight condition, P_{Teo}—theoretical power of propulsion system, and $E_{Teo5min}$—theoretical energy consumption of 5 min flight.

The 5 min duration adopted for energy evaluation corresponds to the representative mission segment of the investigated UAV. Since micro-class fixed-wing UAVs are typically designed for mission-specific tasks, the selected time interval reflects realistic operational conditions. For other platforms, the same methodology can be applied using their corresponding mission-specific duration.

3. Experimental Wind Tunnel Investigation of Energy Consumption

3.1. Wind Tunnel Measurements

The objective of the experiment was to determine the propulsion power demand and associated energy consumption of a light unmanned aerial vehicle (UAV) in steady, level flight as a function of angle of attack and operating mass. To achieve this, a series of controlled wind tunnel measurements was conducted under defined aerodynamic and propulsion equilibrium conditions.

The UAV was installed in the test section of a low-speed aerodynamic wind tunnel (Figure 5). The tunnel is characterised by a turbulence intensity below 0.5%, as specified by the manufacturer, ensuring stable and repeatable flow conditions. The aircraft was mounted on a multi-axis strain gauge balance located inside the test chamber, enabling direct measurement of aerodynamic forces. All force readings were acquired through the wind tunnel control computer. The experiment was conducted for angles of attack ranging from 0° to 10°, with an increment of 2°. For each angle of attack, three operating weight conditions were considered:

- Minimal operating weight (0.46 kg);
- Payload operating weight (0.72 kg);
- Maximum take-off weight (1.00 kg).

For a given angle of attack and mass configuration, the freestream velocity in the wind tunnel was gradually increased until the lift force measured by the strain gauge balance was equal to the gravitational force corresponding to the considered mass. This condition represents vertical force equilibrium for steady, level flight.

Once the target lift level was reached, the UAV's propulsion system was activated. The propulsion thrust was progressively increased until the balance reading in the drag axis reached zero. This condition corresponds to horizontal force equilibrium, in which the thrust generated by the propulsion system exactly balances the aerodynamic drag acting on the aircraft, thereby representing steady, level flight conditions. Activation of the propulsion system introduced an additional vertical force component due to the thrust vector's inclination, resulting in a temporary increase in the measured lift force. To compensate for this effect and restore the target lift corresponding to the selected operating mass, the freestream velocity was reduced accordingly. Subsequently, the propulsion

https://doi.org/10.3390/en19081892

power was adjusted to achieve zero net force in the drag direction once again. This iterative process (Figure 6) was repeated until a stable equilibrium state was achieved, defined by the simultaneous satisfaction of the following conditions:

- Measured lift force equal to the force corresponding to the selected operating weight;
- Measured drag force equal to zero, indicating thrust to drag equilibrium;
- Stable propulsion operating point.

Figure 5. An unmanned aerial vehicle mounted on a strain gauge inside a wind tunnel test chamber.

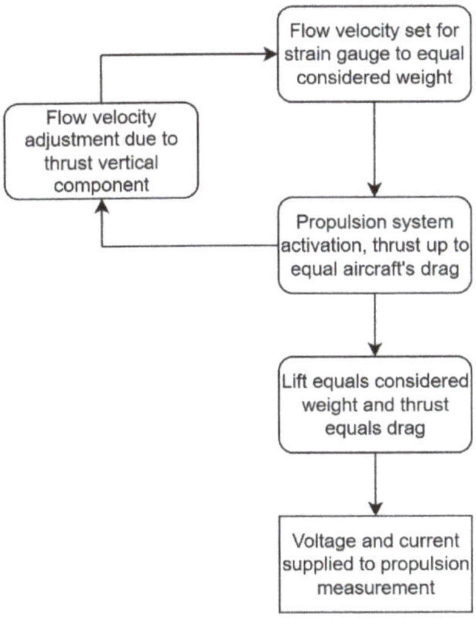

Figure 6. Flow velocity adjustment and measurement procedure diagram.

For each angle of attack and operating mass, the stabilised equilibrium state yielded a complete dataset comprising the lift force corresponding to the selected mass, the zero-net-drag force, the propulsion system supply voltage, and the propulsion system supply current. From the measured voltage and current, the electrical power required by the propulsion system was calculated, allowing the estimation of energy consumption for each operating condition.

The assumption that lift force equals weight and thrust equals drag was adopted to accurately represent steady, level flight conditions, corresponding to a realistic mission phase in which the aircraft performs its intended airborne task.

https://doi.org/10.3390/en19081892

For each of the operating masses, the speed at which the force equilibrium occurred, as well as values of supplied voltage and current, can be seen in Tables 6–8.

The power required by the propulsion system and the energy consumption for a 5 min flight have been calculated from experimental data and are shown in Tables 6–8.

Table 6. Propulsion system parameters for 0.46 kg of operating weight for level flight condition.

AoA°	V $[\frac{m}{s}]$	Voltage [V]	Current [A]	P_S [W]	E_{5min} [Wh]
0	11.4	11.8	1.15	13.57	1.13
2	10.3	12	1.61	19.32	1.61
4	10.3	12	1.81	21.72	1.81
6	9.2	12.19	2.25	27.43	2.29
8	8.2	12.22	3.5	42.77	3.56
10	8.3	12.22	6.13	74.91	6.24

Table 7. Propulsion system parameters for 0.72 kg of operating weight for level flight condition.

AoA°	V $[\frac{m}{s}]$	Voltage [V]	Current [A]	P_S [W]	E_{5min} [Wh]
0	15.5	11.3	2.21	24.97	2.08
2	14.9	11.55	2.97	34.30	2.86
4	12.1	11.68	3.51	41.00	3.42
6	11.1	11.99	4.37	52.40	4.37
8	10.7	12.02	7.71	92.67	7.72
10	10.2	12	12.17	146.04	12.17

Table 8. Propulsion system parameters for 1.00 kg of operating weight for level flight condition.

AoA°	V $[\frac{m}{s}]$	Voltage [V]	Current [A]	P_S [W]	E_{5min} [Wh]
0	18.2	10.6	3.96	41.98	3.50
2	17	10.8	4.43	47.84	3.99
4	14.9	11.38	4.71	53.60	4.47
6	13.6	11.27	7.43	83.74	6.98
8	12.1	11.69	10.83	126.60	10.55
10	11.2	11.71	13.81	161.72	13.48

AoA°—Angle of attack of a UAV, V—Level flight equilibrium speed, Voltage—voltage supplied to propulsion system, Current—current supplied to propulsion system, P_S—power of propulsion system, and E_{5min}—energy consumption of 5 min flight.

For each operating mass in a function of angle of attack and condition of level light, the power of propulsion system, and energy consumption of a 5 min flight obtained analytically, based on numerical analysis of a UAV, were compared to an experimental test of a UAV and can be seen in Tables 9–11.

Table 9. Propulsion system parameters for 0.46 kg of operating weight for level flight condition.

AoA°	P_{Teo} [W]	P_{Exp} [W]	$\frac{P_{Exp}}{P_{Teo}}$	E_{Teo} [Wh]	E_{Exp} [Wh]	$\frac{E_{Exp}}{E_{Teo}}$
0	12.34	13.57	1.10	1.03	1.13	1.10
2	16.81	19.32	1.15	1.4	1.61	1.15
4	18.71	21.72	1.16	1.56	1.81	1.16
6	21.54	27.43	1.27	1.8	2.29	1.27
8	27.62	42.77	1.55	2.3	3.56	1.55
10	43.62	74.91	1.72	3.64	6.24	1.71

https://doi.org/10.3390/en19081892

Table 10. Propulsion system parameters for 0.72 kg of operating weight for level flight condition.

AoA°	P_{Teo} [W]	P_{Exp} [W]	$\frac{P_{Exp}}{P_{Teo}}$	E_{Teo} [Wh]	E_{Exp} [Wh]	$\frac{E_{Exp}}{E_{Teo}}$
0	22.37	24.97	1.12	1.86	2.08	1.12
2	29.14	34.30	1.18	2.43	2.86	1.18
4	34.31	41.00	1.19	2.86	3.42	1.19
6	39.84	52.40	1.32	3.32	4.37	1.32
8	57.66	92.67	1.61	4.81	7.72	1.61
10	81.19	146.04	1.80	6.77	12.17	1.80

Table 11. Propulsion system parameters for 1.00 kg of operating weight for level flight condition.

AoA°	P_{Teo} [W]	P_{Exp} [W]	$\frac{P_{Exp}}{P_{Teo}}$	E_{Teo} [Wh]	E_{Exp} [Wh]	$\frac{E_{Exp}}{E_{Teo}}$
0	37.86	41.98	1.11	3.16	3.50	1.11
2	40.93	47.84	1.17	3.41	3.99	1.17
4	44.64	53.60	1.20	3.72	4.47	1.20
6	63.87	83.74	1.31	5.32	6.98	1.31
8	81.98	126.60	1.54	6.83	10.55	1.54
10	95.87	161.72	1.69	7.99	13.48	1.69

AoA°—Angle of attack of a UAV, P_{Teo}—theoretical power of propulsion system, $E_{Teo5min}$—theoretical energy consumption of 5 min flight, P_{Exp}—power of propulsion system obtained experimentally, and $E_{Exp5\ min}$—energy consumption of 5 min flight obtained experimentally.

Graphical comparisons of the theoretical and experimental power needed to operate propulsion system can be found below in Figures 7–9.

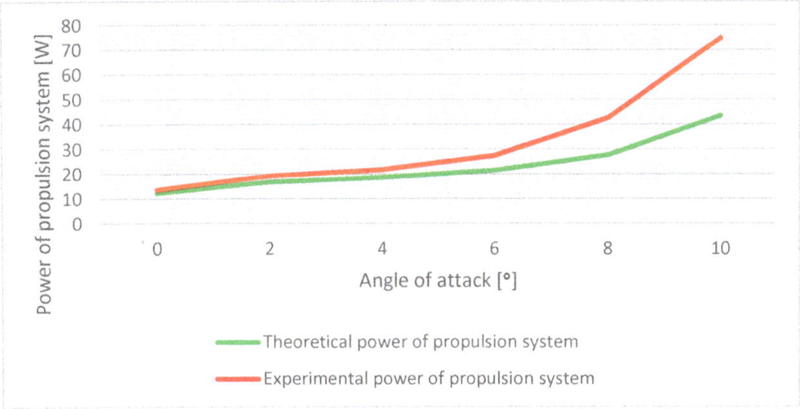

Figure 7. Comparison of theoretical and measured propulsion system power for 0.46 kg of operating weight.

https://doi.org/10.3390/en19081892

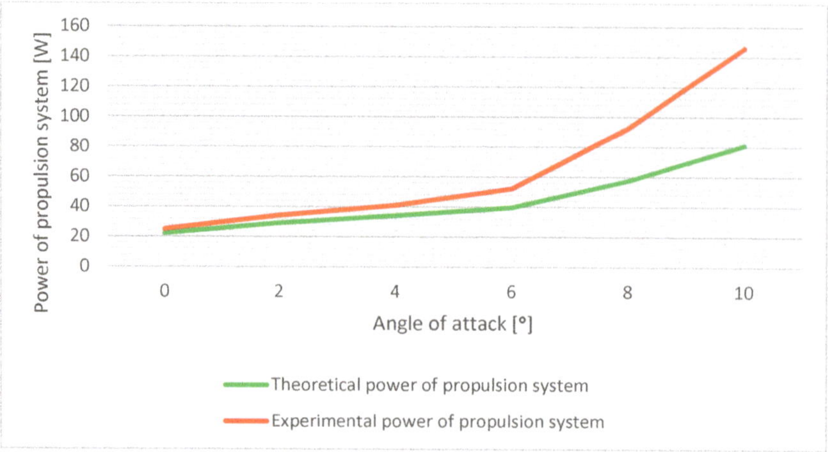

Figure 8. Comparison of theoretical and measured propulsion system power for 0.72 kg of operating weight.

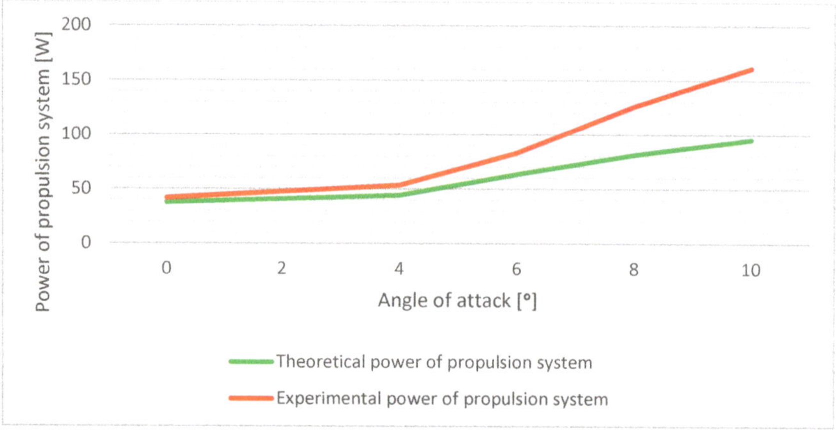

Figure 9. Comparison of theoretical and measured propulsion system power for 1.00 kg of operating weight.

3.2. Test Stand Specyfication

The experimental setup consisted of the UAV mounted inside the wind tunnel (Figure 10) measurement chamber (Figure 11), supported by a dedicated mounting structure. The aerodynamic balance measured forces along the lift and drag axes. The angle of attack was set electronically via the wind tunnel control system (Figure 12), ensuring precise, repeatable positioning.

The propulsion system was controlled via an autopilot running ArduPilot 4.6.3 software, installed on a separate control computer. The pulse width modulation (PWM) signal commanding the propulsion system was generated by the autopilot, while electrical power was supplied by a laboratory-grade power supply. This configuration (Figure 13) enabled precise monitoring of the supplied voltage and current, ensuring accurate determination of propulsion power.

https://doi.org/10.3390/en19081892

In addition to aerodynamic and propulsion measurements, ambient conditions inside the wind tunnel, specifically air temperature and humidity, were continuously monitored and recorded to ensure proper characterisation of the test environment.

The raw force measurements from the aerodynamic balance were corrected to account for the mounting system's influence. The aerodynamic drag and lift characteristics of the mounting structure were measured separately for each tested angle of attack and flow velocity. These values were subsequently subtracted from the total balance readings, ensuring that the results reflected only the aerodynamic forces acting on the UAV.

The exact measurement section and flow parameters can be seen in Table 12.

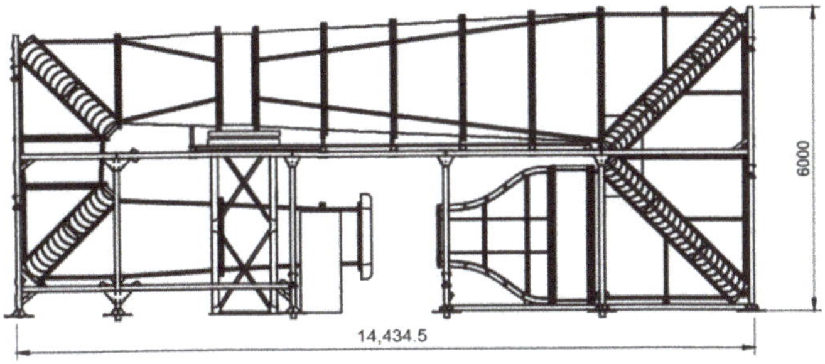

Figure 10. Drawing and dimensions of a wind tunnel.

Figure 11. Test stand consisting of a wind tunnel test chamber, two control computers, a laboratory-grade power supply, and a UAV mounted with a strain gauge.

https://doi.org/10.3390/en19081892

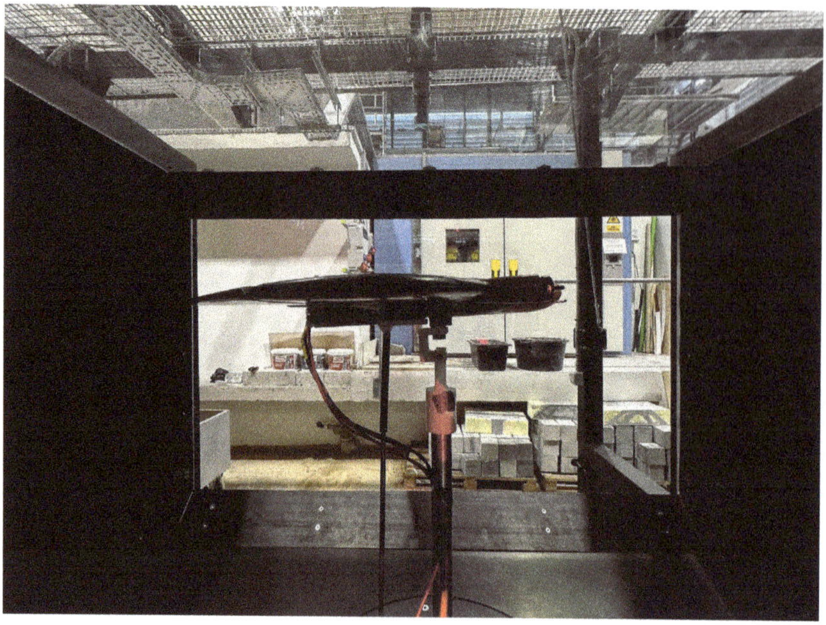

Figure 12. An unmanned aerial vehicle is being tested at a 4-degree angle of attack.

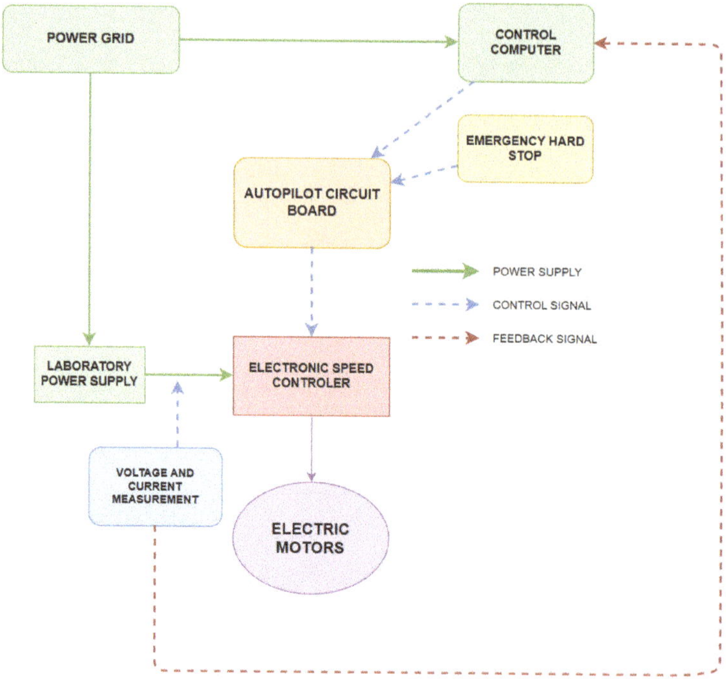

Figure 13. Test stand power supply and control diagram.

https://doi.org/10.3390/en19081892

Table 12. Test section dimensions and flow parameters inside test section.

Parameter	-
Test section width	1.0 m
Test section height	1.0 m
Test section length	1.5 m
Nozzle contraction	6.25
Turbulence intensity at maximum flow speed of 40 m/s	<1%
Temperature during measurement	21.3 °C
Temperature measurement error	0.2 °C
Velocity measurement range	0–40 m/s
Velocity measurement error	1% of range
Unevenness of the velocity field	0.5%
Aerodynamic balance measurement range	200 N
Aerodynamic balance measurement error	0.1% of range

4. Conclusions

A key outcome of the study is steady overconsumption of power at low and moderate angles of attack, compared with the theoretical panel model. In the range from 0° to 6°, the experimentally measured propulsion power is typically 10–20% higher than the values predicted by the panel method-based model. This discrepancy is considered physically justified, as the numerical model represents only the idealised lifting-surface aerodynamics. In contrast, the real aircraft includes additional drag sources not represented in the panel model, such as propulsion units, antennas, control surface gaps, wiring, surface roughness, and other installation effects. These components introduce additional drag, increasing thrust demand. The relatively small magnitude of the difference indicates that the panel-method approach provides a reliable first-order prediction of power consumption in this operational range.

At higher angles of attack, however, the deviation between experiment and theory increases significantly. In the range from 6° to 10°, experimentally measured power consumption exceeds theoretical predictions by approximately 50–80%, depending on the operating mass. This large discrepancy cannot be explained solely by additional drag.

This behaviour is attributed to two distinct aerodynamic mechanisms. In the low to moderate angle-of-attack regime, the discrepancy is primarily associated with additional drag sources present in the real aircraft configuration but not represented in the idealised panel-based model, such as propulsion system installation effects, surface roughness, and geometric imperfections. However, at higher angles of attack, the divergence is dominated by unsteady and nonlinear flow phenomena that cannot be captured by the steady, inviscid formulation of the VLM/panel method. In particular, the onset of transient turbulence, flow separation, and incipient stall leads to a rapid degradation of aerodynamic efficiency and a sharp increase in drag, which directly translates into higher propulsion power requirements. As the turbulent flow tends to expand the turbulence intensity, rather in the geometric rate, the discrepancy becomes bigger with every single angle-of-attack increment close to the airfoil stall point.

The panel method, based on inviscid potential flow assumptions, is unable to accurately capture the viscous flow phenomena that dominate in this regime. As the angle of attack increases, low-Reynolds-number flow over the UAV wing becomes increasingly susceptible to laminar separation, transition, and turbulent reattachment. These effects lead to rapid drag growth and a deterioration of aerodynamic efficiency, which the panel method cannot properly simulate. Consequently, the theoretical model underestimates thrust demand and therefore propulsion power in this region.

https://doi.org/10.3390/en19081892

The findings, therefore, confirm that panel method-based performance prediction is reliable for low to moderate angles of attack typical of endurance-optimised flight, but its accuracy deteriorates rapidly as the aircraft approaches high-lift or near-stall conditions. This has direct practical implications: endurance prediction, mission planning, and propulsion sizing based solely on inviscid aerodynamic tools may lead to substantial underestimation of energy demand if the aircraft operates outside its optimal aerodynamic range.

This study makes three main contributions. First, it proposes an experimental wind tunnel methodology that enables the reproduction of steady level flight conditions by simultaneously satisfying lift–weight and thrust–drag equilibrium. Second, it provides a quantitative assessment of how operating weight influences key performance parameters of a micro-class flying-wing UAV, including equilibrium flight speed, propulsion power, and overall energy consumption. Finally, it evaluates the applicability of a panel method-based prediction model, demonstrating that while good accuracy is achieved at low to moderate angles of attack, the model's accuracy deteriorates at higher angles due to the increasing influence of viscous effects typical of low Reynolds number flow.

Overall, the study demonstrates that combining numerical panel-method analysis with wind tunnel validation enables the evaluation of endurance characteristics of lightweight UAVs. The approach enables the identification of the validity limits of simplified aerodynamic models, such as the panel method, and provides experimentally grounded correction factors that can be used in future performance prediction and mission-planning tools. The results confirm that operating weight strongly influences propulsion power requirements. Increasing the aircraft mass shifts the level flight equilibrium toward higher lift coefficients and higher drag conditions, leading to a consistent increase in electrical power demand and a corresponding reduction in endurance. This trend is evident in both numerical predictions and wind tunnel measurements.

Author Contributions: Conceptualization, B.D. and K.K. and A.K.; methodology, K.K., B.D. and A.K.; software, K.K.; validation, B.D., K.K. and J.J.; formal analysis, A.K.; investigation, K.L., B.D., J.J. and K.K.; resources, A.K.; data curation, J.J., K.K. and K.L.; writing—original draft preparation, B.D. and A.K.; writing—review and editing, B.D., K.L. and J.J.; visualisation, K.K.; supervision, A.K.; project administration, K.K.; All authors have read and agreed to the published version of the manuscript.

Funding: This research received no external funding.

Data Availability Statement: The original contributions presented in the study are included in the article; further inquiries can be directed to the corresponding author.

Conflicts of Interest: The authors declare no conflicts of interest.

References

1. Mastrakoulis, D.M.; Nasiopoulos, D.K.; Arvanitidis, D.A.; Zara, E.C.; Plagiannakos, A. Modeling and Simulation of Examining the Performance of a Drone for Long-Distance Endurance. In *Computational and Strategic Business Modelling*; Sakas, D.P., Nasiopoulos, D.K., Taratuhina, Y., Eds.; Springer Proceedings in Business and Economics; Springer International Publishing: Cham, Switzerland, 2024; pp. 235–251. ISBN 978-3-031-41370-4.
2. Day, B.; Pourmovahed, A. Review of Latest Developments in PEM Fuel Cell Research with Application to Hydrogen Powered Drones. *Renew. Energy Power Qual. J.* **2023**, *19*. [CrossRef]
3. Izyumov, A.; Vasilyeva, E.; Ostapovich, O.; Alentsov, E. Energy Efficiency Analysis of Modern Delivery Drones. *E3S Web Conf.* **2021**, *279*, 01028. [CrossRef]
4. Joshi, P.S.; Kartik, V. All-Electric Propulsion for Medium/High Altitude Long Endurance Unmanned Aerial Vehicles: A Feasibility Study. In *Proceedings of the AIAA Aviation Forum and ASCEND 2024, Las Vegas, Nevada, 29 July–2 August 2024*; American Institute of Aeronautics and Astronautics: Las Vegas, NV, USA, 2024.
5. Dinc, A.; Alsanea, N.; Otkur, M.; Mussin, A.; Elbadawy, I.; Moayyedian, M. A Performance Review of Conventional Turbofan Aircraft and Battery-Powered Aircraft. *Int. J. Eng. Appl.* **2024**, *12*, 251. [CrossRef]

https://doi.org/10.3390/en19081892

6. Oetting, G.S. *Extended Endurance Unmanned Aerial Vehicle via Structural Electrical Power Storage and Energy Generation Devices*; 2020-01-0041; SAE International: Pasadena, CA, USA, 2020. [CrossRef]
7. Mariscal, G.; Depcik, C.; Chao, H.; Wu, G.; Li, X. Technical and Economic Feasibility of Applying Fuel Cells as the Power Source of Unmanned Aerial Vehicles. *Energy Convers. Manag.* **2024**, *301*, 118005. [CrossRef]
8. Kayaoglu, M.; Unal, S.; Biyik, H. Design and Simulation of a Fuel Cell–Battery Hybrid Power System for Extended-Endurance Quadrotor UAVs. *Int. J. Hydrogen Energy* **2026**, *204*, 153278. [CrossRef]
9. Jarrah, K.; Alali, Y.; Lalko, A.; Rawashdeh, O. Flight Time Optimization and Modeling of a Hybrid Gasoline–Electric Multirotor Drone: An Experimental Study. *Aerospace* **2022**, *9*, 799. [CrossRef]
10. Xie, Y.; Savvaris, A.; Tsourdos, A. Fuzzy Logic Based Equivalent Consumption Optimization of a Hybrid Electric Propulsion System for Unmanned Aerial Vehicles. *Aerosp. Sci. Technol.* **2019**, *85*, 13–23. [CrossRef]
11. Li, P.; Mao, Z.; Pang, S.; Shi, W.; Quan, S.; Huangfu, Y. A Fuzzy Logic Control-Based Energy Management Strategy for Fuel Cell/Battery UAV Hybrid Power System. In *Proceedings of the IECON 2023—49th Annual Conference of the IEEE Industrial Electronics Society, Singapore, 16–19 October 2023*; IEEE: Singapore, 2023; pp. 1–6.
12. Mazur, A.M.; Domanski, R. Hybrid Energy Systems in Unmanned Aerial Vehicles. *Aircr. Eng. Aerosp. Technol.* **2019**, *91*, 736–746. [CrossRef]
13. Gudmundsson, S. *General Aviation Aircraft Design: Applied Methods and Procedures*, 2nd ed.; Butterworth-Heinemann: Oxford/Cambridge, UK, 2022; ISBN 978-0-12-822647-6.
14. Akgumus Gok, D. Characterization of NACA 2412 and NACA 4412 Airfoils: Effects of Angle of Attack on Aerodynamics Coefficients. *J. Therm. Eng.* **2024**, *10*, 1524–1538. [CrossRef]
15. Hu, S.; Yin, Z.; Liu, R. Analysis of Aerodynamic Characteristics of a Certain Aircraft Wing Structure at Different Angles of Attack. In *Proceedings of the Ninth International Symposium on Sensors, Mechatronics, and Automation System (ISSMAS 2023), Nanjing, China, 11–13 August 2023*; Pan, L., Zhou, Z., Eds.; SPIE: Nanjing, China, 2024; p. 211. [CrossRef]
16. Hendrickson, K.; Agarwal, R.K. Numerical Investigation of Leading-Edge Protuberances on High Lift, High Camber Airfoils at Low Reynolds Number. In *Proceedings of the AIAA AVIATION 2020 Forum, Virtual, 15–19 June 2020*; American Institute of Aeronautics and Astronautics: Reston, VA, USA, 2020. [CrossRef]
17. Hossain, S.; Rifat, A.H.; Bakhai, M.P. Computational Fluid Dynamics Simulation on Blended Wing Body Aircraft Model for Aerodynamics Characteristics. In *Advanced in Creative Technology- added Value Innovations in Engineering, Materials and Manufacturing*; Janmanee, P., Chujuarjeen, S., Butdee, S., Srikhumsuk, P., Batako, A.D.L., Burduk, A., Xavior, M.A., Eds.; Lecture Notes in Networks and Systems; Springer Nature: Cham, Switzerland, 2024; Volume 979, pp. 290–300, ISBN 978-3-031-59163-1. [CrossRef]
18. Young, T.M. Maneuver Performance of Fixed-Wing Transport Aircraft. In *Encyclopedia of Aerospace Engineering*; Blockley, R., Shyy, W., Eds.; Wiley: Hoboken, NJ, USA, 2010; ISBN 978-0-470-75440-5. [CrossRef]
19. Urnes, J.M., Sr.; Smith, T.A. Use of Propulsion Commands to Prevent Loss-of-Control Aircraft Accidents. In Proceedings of the AIAA Infotech at Aerospace Conference and Exhibit 2011, St. Louis, MO, USA, 29–31 March 2011. [CrossRef]
20. Nicolai, L.M.; Carichner, G. *Fundamentals of Aircraft and Airship Design*; American Institute of Aeronautics and Astronautics: Reston, VA, USA, 2000; ISBN 978-1-60086-752-1.
21. Keane, A.J.; Sóbester, A.; Scanlan, J.P. *Small Unmanned Fixed-Wing Aircraft Design: A Practical Approach*; John Wiley & Sons, Inc.: Hoboken, NJ, USA, 2017; ISBN 978-1-119-40631-0.
22. Sadraey, M.H. *Aircraft Performance: An Engineering Approach*, 2nd ed.; CRC Press: Boca Raton, FL, USA, 2023; ISBN 978-1-003-27906-8.
23. Asselin, M. *Operational Aircraft Performance and Flight Test Practices*, 1st ed.; American Institute of Aeronautics & Astronautics: Reston, VA, USA, 2021; ISBN 978-1-62410-592-0.
24. Kierzkowski, A.; Woźniak, M.; Bury, P. Comparative Analysis of Energy Efficiency and Position Stability of Sub-250 g Quadcopter and Bicopter with Similar Mass Under Varying Conditions. *Energies* **2025**, *18*, 3728. [CrossRef]
25. Lee, S.; Cho, L.; Cho, J. Study on the Extraction of Aerodynamic Coefficients for the Development of an UAV Simulator Using the Unsteady Time-Domain Panel Method. *Trans. Japan Soc. Aero. S Sci.* **2014**, *57*, 187–194. [CrossRef]
26. Chiu, T.W. Modern Panel Method in Vehicle Aerodynamics: Formula-One Cars and Trains. In *Advances in Fluid Mechanics*; Springer: Singapore, 1997.
27. Sun, S.Y.; Wu, G.X. Inviscid Flow Passing a Lifting Body with a Higher Order Boundary Element Method. *Eng. Anal. Bound. Elem.* **2022**, *136*, 144–157. [CrossRef]
28. Anderson, J.D. Incompressible Inviscid Flows: Source andVortex Panel Methods. In *Computational Fluid Dynamics*; Wendt, J.F., Ed.; Springer: Berlin/Heidelberg, Germany, 2009; pp. 53–75. ISBN 978-3-540-85055-7.
29. Torres, G.L.S.; Gopalarathnam, A.; Marques, F.D. A Uniformly Convergent Series Expansion for the Bound Vorticity Distribution in the Unsteady Thin Airfoil Theory. In *Proceedings of the AIAA SCITECH 2025 Forum, Orlando, FL, USA, 6–10 January 2025*; American Institute of Aeronautics and Astronautics: Orlando, FL, USA, 2025.

https://doi.org/10.3390/en19081892

30. Bock, K. Towards a 3D Galerkin-Type High-Order Panel Method: A 2D Prototype. In *New Results in Numerical and Experimental Fluid Mechanics XIII*; Dillmann, A., Heller, G., Krämer, E., Wagner, C., Eds.; Notes on Numerical Fluid Mechanics and Multidisciplinary Design; Springer International Publishing: Cham, Switzerland, 2021; Volume 151, pp. 581–591, ISBN 978-3-030-79560-3. [CrossRef]
31. Muhammed, M.; Virk, M.S. On the Fidelity of RANS-Based Turbulence Models in Modeling the Laminar Separation Bubble and Ice-Induced Separation Bubble at Low Reynolds Numbers on Unmanned Aerial Vehicle Airfoil. *Drones* **2024**, *8*, 148. [CrossRef]
32. Shyy, W.; Lian, Y.; Tang, J.; Viieru, D.; Liu, H. *Aerodynamics of Low Reynolds Number Flyers*, 1st ed.; Cambridge University Press: Cambridge, UK, 2007; ISBN 978-0-521-20401-9. [CrossRef]
33. Hull, D.G. *Fundamentals of Airplane Flight Mechanics*, 1st ed.; Springer: Berlin/Heidelberg, Germany, 2007; ISBN 978-1-280-74575-1.
34. T Motor Company. F2203.5 3550kV Motor Specification. Available online: https://tmotorhobby.com/goods-1097-T-Motor+F22035+1500KV2850KV3550KV+CinewhoopRacingwhoop+Brushless+Motor.html (accessed on 26 February 2026).

https://doi.org/10.3390/en19081892

energies

MDPI

Article

Determinants of Energy Consumption in South Africa: Evidence from an ARDL Model (1980–2023)

Palesa Milliscent Lefatsa * and Sanele Gumede

School of Commerce, University of KwaZulu-Natal, Westville, Durban 4000, South Africa; gumedes@ukzn.ac.za
* Correspondence: lefatsap@ukzn.ac.za; Tel.: +27-78-921-2313

Abstract

This study examines the determinants of energy consumption in South Africa over the period 1980–2023 using a multivariate time-series framework. Unlike conventional studies that focus primarily on the energy–growth nexus, this analysis incorporates financial development, industrialization, and population growth to provide a more comprehensive understanding of energy demand dynamics. The Autoregressive Distributed Lag (ARDL) approach is employed to estimate both short-run and long-run relationships. Unit root tests confirm that all variables are integrated of order one, justifying the application of the ARDL bounds testing approach. The results reveal the existence of a stable long-run relationship between energy consumption and its determinants. Industrialization and population growth emerge as the most significant drivers of energy demand in both the short and long run, reflecting South Africa's energy-intensive economic structure and rising demographic pressures. Financial development is found to have a positive and statistically significant effect, suggesting that improved access to credit stimulates energy consumption through increased investment and economic activity. In contrast, economic growth exhibits a positive but statistically insignificant long-run effect, indicating partial decoupling between output growth and energy demand. The error correction term is negative and statistically significant, confirming convergence to long-run equilibrium. Causality analysis further indicates that energy consumption is primarily driven by macroeconomic factors rather than acting as a leading indicator. The findings underscore the importance of industrial energy efficiency, population-responsive energy planning, and targeted financial support for sustainable energy investment. This study contributes to the literature by providing a comprehensive, country-specific analysis and offers policy-relevant insights for enhancing energy security and supporting sustainable economic development in South Africa.

Keywords: energy consumption; economic growth; financial development; industrialization; population growth; ARDL model; South Africa; energy demand; long-run relationship; short-run dynamics

1. Introduction

South Africa's energy sector has undergone a significant transformation from a position of comparative advantage to one characterized by persistent instability, with far-reaching implications for economic performance. For much of the twentieth century, the country benefited from an abundant and relatively inexpensive electricity supply, largely generated from coal, which supported industrial expansion and sustained economic growth [1,2]. However, from the 1980s onwards, structural challenges, including rising demand, delayed investment in generation capacity, ageing infrastructure, and institutional inefficiencies, gradually eroded this advantage. These pressures intensified in the early 2000s and culminated in the introduction of load-shedding in 2008, marking a critical turning point in South Africa's energy landscape [3].

Since then, electricity supply disruptions have become a defining feature of the South African economy, reaching unprecedented levels between 2020 and 2023. In 2022 alone, the country experienced more than 200 days of load-shedding, with outages lasting up to 10 h per day. The economic consequences have been severe, with estimated daily losses ranging from R1.5 billion to R4 billion, alongside a reduction in annual GDP growth of approximately 1–3 percentage points [4]. Although temporary improvements were observed in 2024, uncertainty surrounding energy supply continues to shape economic expectations, investment decisions, and overall macroeconomic stability.

Energy plays a fundamental role in economic systems by underpinning production processes, industrial output, and long-term growth dynamics. In South Africa, where energy consumption is heavily concentrated in mining, manufacturing, and other energy-intensive sectors, the reliability of electricity supply is closely linked to economic growth, industrialization, and trade performance [5]. Over the period 1980–2023, energy consumption has remained closely associated with a coal-dominated energy mix, accounting for approximately 80% of electricity generation, while population growth and urbanization have steadily increased demand pressures [3]. At the same time, financial development has progressed, although institutional inefficiencies and policy uncertainty have constrained its capacity to support energy infrastructure investment. These dynamics highlight the complex, interdependent relationships among energy consumption, economic growth, financial development, population growth, and industrialization.

The motivation for this study arises from the persistent energy constraints facing South Africa and their implications for long-term economic sustainability. The country's aspirations for inclusive growth, industrial expansion, and global competitiveness are increasingly undermined by unreliable energy supply. Furthermore, the transition towards a diversified and sustainable energy mix introduces additional challenges, particularly in balancing energy security with economic and environmental objectives. Understanding the interaction between key macroeconomic variables and energy consumption is therefore essential for informing effective and targeted policy interventions.

This study examines the dynamic relationships between energy consumption, financial development, economic growth, population growth, and industrialization in South Africa over the period of 1980–2023. Specifically, it estimates both the short-run and long-run effects of these variables on energy consumption and determines the direction of causality among them. In doing so, the study contributes to the broader literature on the energy–economy nexus and provides empirically grounded insights relevant to policy formulation.

The justification for this analysis lies in the central role of energy within South Africa's economic structure and the limited integration of key macroeconomic determinants within a unified empirical framework. While existing studies often focus narrowly on the relationship between energy consumption and economic growth, they frequently overlook the broader influence of financial development, population growth, and industrialization. This omission results in an incomplete understanding of the structural drivers of energy demand.

Empirical findings on the energy–economic growth nexus remain inconclusive and often contradictory. For example, examining BRICS economies over the period 2000–2023 using asymmetric panel estimation and Granger causality techniques, evidence of bidirectional causality between energy consumption and economic growth is found, supporting the feedback hypothesis [6]. In contrast, focusing on European Union countries over the same period and employing panel structural equation modelling, the influence of economic

https://doi.org/10.3390/en19102329

growth on energy consumption weakens when structural and technological factors are incorporated, suggesting partial decoupling consistent with the neutrality or conservation hypothesis [7]. Similarly, using a dynamic Generalized Method of Moments approach in China, financial development reduces energy consumption, while energy consumption promotes economic growth [8]. These contrasting findings highlight the complexity and context-specific nature of the energy–growth relationship.

In the South African context, studies that simultaneously integrate energy consumption, financial development, population growth, and industrialization within a comprehensive empirical framework remain limited, despite the country's structural complexity and energy-intensive economic base. To address this gap, this study employs the Autoregressive Distributed Lag (ARDL) modelling approach, which is particularly suitable for small samples and allows the estimation of both short-run dynamics and long-run equilibrium relationships within a single framework [9]. Given South Africa's structural economic changes, ongoing energy transition, and constraints on time-series data, the ARDL approach provides a robust and flexible econometric technique for analyzing the energy–economy nexus.

This study contributes to the existing literature in several important ways. First, it adopts a comprehensive multivariate framework that integrates energy consumption, financial development, economic growth, population growth, and industrialization into a single empirical model, thereby moving beyond the conventional bivariate energy–growth approach commonly used in previous studies. Second, the study provides updated, context-specific evidence for South Africa over the extended period 1980–2023, capturing recent structural developments, including persistent electricity supply constraints and load-shedding, which have significantly reshaped the energy–economy relationship. Third, the study explicitly incorporates key macroeconomic determinants, such as population growth and industrialization, which are often treated in isolation, thereby offering a more holistic understanding of the drivers of energy demand in an energy-intensive economy. Finally, by applying the ARDL modelling framework to a structurally evolving economy, the study provides robust and policy-relevant insights tailored to South Africa's institutional and economic context, particularly in relation to energy security, industrial development, and financial sector dynamics. The remainder of the paper is structured as follows: Section 2 reviews the literature; Section 3 outlines the methodology; Section 4 presents and discusses the empirical results; and Section 5 concludes with policy implications and limitations.

2. Literature Review

2.1. Theoretical Literature

The theoretical literature provides the foundational frameworks for understanding the dynamic relationships between energy consumption, economic growth, financial development, population growth, and industrialization. However, these theories offer differing and sometimes conflicting explanations of the energy–economy nexus. A critical evaluation is therefore necessary to assess their relevance, limitations, and applicability within the South African context.

2.1.1. Energy–Led Growth Theory

The energy–led growth hypothesis posits that energy consumption is a fundamental driver of economic growth, as it serves as a critical input to production processes, industrialization, and technological advancement [10–12].

In the South African context, this theory is highly relevant given the country's reliance on energy-intensive sectors such as mining and manufacturing, where energy availability directly influences output. Periods of severe electricity shortages and load-shedding

https://doi.org/10.3390/en19102329

provide empirical support for this theory, as constrained energy supply has been associated with reduced economic performance.

However, the theory assumes a predominantly unidirectional relationship, which may oversimplify the South African case. Economic stagnation in certain periods has not always been driven solely by energy constraints, but also by structural inefficiencies, governance challenges, and policy uncertainty. Therefore, while the theory is useful for highlighting the importance of energy supply, it does not fully capture the broader macroeconomic dynamics that influence growth.

2.1.2. Growth-Led Energy Hypothesis

The growth-led hypothesis suggests that economic expansion drives energy consumption, as increased production, transportation, and household activity raise energy demand [13].

This perspective is applicable to South Africa, particularly during periods of industrial expansion and urbanization, when rising incomes and economic activity have increased energy demand. Population growth and urban migration further reinforce this relationship by expanding residential and industrial energy needs.

Nevertheless, the theory is limited in its assumption that energy supply can adjust to meet rising demand. In South Africa, persistent supply constraints, ageing infrastructure, and capacity shortages challenge this assumption. As a result, economic growth does not always translate proportionally into higher energy consumption, indicating that the relationship is conditioned by structural and institutional limitations.

2.1.3. Feedback (Bidirectional) Hypothesis

The feedback hypothesis integrates the previous two theories by proposing a bidirectional relationship between energy consumption and economic growth [14,15].

This framework is particularly relevant for South Africa, where energy shortages constrain economic activity, while economic expansion simultaneously increases energy demand. The cyclical interaction between load-shedding and economic performance provides strong contextual support for this theory.

However, while the feedback hypothesis captures the interdependence between energy and growth, it does not explicitly incorporate other structural factors such as financial development, demographic pressures, or industrial composition. As such, its explanatory power is enhanced when combined with broader theoretical perspectives.

2.1.4. Endogenous Growth Theory and Financial Development

Endogenous growth theory emphasizes the role of investment, technological progress, and financial development in sustaining long-term economic growth [16]. Within the energy context, financial development facilitates investment in energy infrastructure, renewable technologies, and industrial innovation.

In South Africa, this theory highlights the importance of financial systems in supporting energy sector development. While financial development has improved over time, its effectiveness in addressing energy challenges has been constrained by policy uncertainty, institutional inefficiencies, and limited investment in generation capacity.

A key limitation of this theory is that it assumes efficient allocation of financial resources. In reality, South Africa's financial system does not always translate into adequate investment in critical energy infrastructure. This suggests that financial development alone is insufficient without supportive institutional and policy frameworks.

https://doi.org/10.3390/en19102329

2.1.5. Environmental Kuznets Curve (EKC) and Population Effects

The Environmental Kuznets Curve (EKC) hypothesis proposes a non-linear relationship between economic growth and environmental degradation, often mediated by energy consumption [17]. It suggests that environmental degradation increases in early stages of development but declines as economies adopt cleaner technologies.

In South Africa, the applicability of the EKC is limited by the country's continued reliance on coal-based energy, which dominates electricity generation. Despite economic development, the transition toward cleaner energy has been relatively slow, challenging the assumption of an automatic shift to sustainability.

Population growth further complicates this relationship by increasing energy demand through urbanization, household consumption, and labour force expansion. This reinforces pressure on an already-constrained energy system, suggesting that demographic factors play a more direct role than the EKC framework implies.

2.1.6. Industrialization and Structural Change Theories

Structural change theories emphasize the role of industrialization in driving energy demand as economies transition from agriculture to manufacturing and services [5].

This theory is highly relevant to South Africa, where mining and heavy industry remain central to economic activity and are inherently energy-intensive. Industrialization has historically been a key driver of energy consumption, reinforcing the strong link between production structure and energy demand.

However, the theory does not fully account for emerging structural changes, including gradual shifts toward service sectors and increasing emphasis on energy efficiency. Additionally, it assumes stable energy supply conditions, which is not the case in South Africa, where supply constraints disrupt industrial activity.

2.1.7. Synthesis

While the reviewed theories provide important insights, no single framework adequately explains the complexity of energy consumption dynamics in South Africa. The energy-led and growth-led hypotheses offer contrasting perspectives, while the feedback hypothesis provides a more integrated view of their interaction. However, these approaches remain incomplete without incorporating financial development, population growth, and structural transformation.

A critical synthesis reveals that energy consumption in South Africa is best understood as a multidimensional, structurally constrained process shaped by the interactions among economic activity, financial systems, demographic pressures, and industrial composition. The limitations observed in each theory highlight the need for a comprehensive analytical framework that integrates these factors.

Accordingly, this study adopts a multivariate ARDL framework that captures both short-run dynamics and long-run equilibrium relationships, providing a more complete and context-specific understanding of the energy–economy nexus in South Africa.

2.2. Empirical Literature Review

Empirical research on the energy–economy nexus has evolved from early bivariate models toward more comprehensive multivariate frameworks. While earlier studies focused primarily on the relationship between energy consumption and economic growth, more recent literature incorporates additional variables, such as financial development, industrialization, and demographic factors, to reduce omitted-variable bias. However, despite these advancements, important limitations persist in terms of model completeness, methodological approach, and contextual applicability.

https://doi.org/10.3390/en19102329

Several global studies demonstrate progress toward integrated modelling. Analyzing BRICS economies over the period 2000–2023 using asymmetric panel estimation and Granger causality techniques [6], economic growth and financial development significantly increase energy consumption, with bidirectional causality between energy consumption and economic growth. Similarly, using dynamic ordinary least squares (DOLS) for Belgium over the period 1990–2024 [18], the effect of economic growth on energy consumption is found to weaken when structural and technological factors are incorporated, suggesting partial decoupling. In contrast, applying a dynamic Generalised Method of Moments (GMM) approach to China over the period 1990–2015 [8], financial development is found to reduce energy consumption, while energy consumption promotes economic growth, with no significant causality from macroeconomic variables to energy demand.

While these studies reflect important advancements, they also highlight persistent inconsistencies in both direction and magnitude of relationships, driven by differences in economic structure, level of development, and methodology. More importantly, even within multivariate frameworks, variable inclusion remains selective rather than fully integrated, with few studies simultaneously incorporating economic growth, financial development, industrialization, and population dynamics within a single model.

The finance–energy relationship further illustrates this inconsistency. Analyzing Economic Community of West African States (ECOWAS) countries over the period 1990–2019 using Driscoll–Kraay panel regression and quantile techniques [19], financial development is found to increase energy consumption by facilitating industrial expansion. Conversely, using a GMM-panel VAR framework for the Association of Southeast Asian Nations (ASEAN) countries over the period 1981–2021, find that financial development reduces energy demand, with bidirectional causality [20]. Similarly, using panel fixed-effects estimation for the Gulf Cooperation Council (GCC) countries over the period 2001–2021 [21], financial development is found to improve energy efficiency and moderate energy consumption growth. These divergent findings suggest that the finance–energy nexus is highly contingent on financial system efficiency, regulatory quality, and capital allocation. However, a key limitation is that these studies often treat financial development in isolation, failing to capture its interactions with demographic and structural factors adequately.

Industrialization is consistently identified as a major structural driver of energy consumption. Using panel regression and Granger causality techniques for Sub-Saharan Africa over the period 1990–2022 [22], industrialization is found to significantly increase energy consumption. Similarly, applying panel cointegration techniques for developing economies over the period 2000–2023 [23], the study confirms that industrial expansion drives energy demand. However, using panel quantile regression for BRICS countries over the period 1990–2017 [24], the impact of industrialization is found to weaken at higher income levels due to technological progress and improvements in energy efficiency. These findings suggest that while industrialization remains a dominant driver in developing economies, its long-run impact is conditional on structural transformation.

Population growth also plays a significant role in shaping energy consumption patterns. Analyzing Sub-Saharan Africa over the period 1990–2022 [25], population growth is found to significantly increase energy consumption. Similarly, Sasana et al. (2026) [23] report strong positive effects in developing economies. In contrast, focusing on European Union countries [7], the effect is moderated by technological efficiency. These differences indicate that the population–energy relationship is highly context-specific. However, most studies analyze population dynamics separately, rather than within an integrated framework that captures interaction effects.

https://doi.org/10.3390/en19102329

Despite these global advancements, the literature remains dominated by panel-based approaches that impose cross-country homogeneity, limiting their applicability to country-specific contexts.

This limitation is particularly important for South Africa, where empirical evidence remains limited and fragmented. Analyzing South Africa over the period 1990–2023 [25], energy consumption can be linked to structural factors such as industrial dependence, although the paper did not examine causality. In contrast, using ARDL methods over the period 1971–2019 [26], the paper shows that the relationship is dynamic and time-varying. However, these studies remain limited by narrow variable selection, excluding key determinants such as financial development, population growth, or industrialization.

Overall, the literature does not entirely overlook these variables at the global level. Rather, the key limitation lies in the lack of fully integrated, country-specific, multivariate frameworks, particularly for structurally unique economies such as South Africa.

2.3. Conceptual Framework

The conceptual framework of this study is based on the interaction between energy consumption and its key macroeconomic determinants—economic growth, financial development, industrialization, and population growth—within the South African context.

Energy consumption functions both as an input into production and as an outcome of economic activity. Economic growth increases energy demand through industrial expansion, transportation, and household consumption, while energy availability constrains economic performance, particularly in an environment characterized by supply shortages.

Financial development facilitates investment in energy infrastructure, industrial expansion, and technological innovation. However, its effectiveness depends on the efficiency of financial intermediation and the allocation of capital toward productive sectors.

Industrialization directly increases energy demand through energy-intensive production processes, particularly in mining and manufacturing. Population growth further amplifies energy demand through urbanization, increased household consumption, and labour force expansion.

These relationships are dynamic and may be unidirectional or bidirectional, depending on structural conditions and time horizons. Differences between short-run adjustments and long-run equilibrium dynamics necessitate an econometric framework capable of capturing both dimensions.

Accordingly, the ARDL approach is adopted as it allows for the estimation of both short-run and long-run relationships within a unified framework.

2.4. Research Gap

Despite significant advancements in the empirical literature, several gaps remain.

First, although global studies increasingly adopt multivariate frameworks, variable inclusion remains incomplete, with few studies simultaneously integrating economic growth, financial development, industrialization, and population dynamics within a single model. Second, the literature is dominated by panel data methodologies, which assume cross-country homogeneity and may obscure country-specific dynamics, particularly in structurally unique economies. Third, many studies fail to adequately distinguish between short-run dynamics and long-run relationships, leading to inconsistent and sometimes contradictory findings. Fourth, within the South African context, empirical evidence remains limited, fragmented, and narrow in scope, with most studies focusing on isolated relationships and excluding key determinants. Therefore, the gap is not the absence of multivariate analysis globally, but the lack of comprehensive, country-specific, and structurally contextualized modelling, particularly for South Africa.

https://doi.org/10.3390/en19102329

To address these gaps, this study employs the ARDL modelling approach, which is well-suited for small sample sizes and allows for the inclusion of variables integrated of different orders, provided none are integrated of order two. The ARDL framework enables the estimation of both short-run and long-run relationships within a single unified model, thereby overcoming limitations associated with traditional techniques. By applying this approach to a comprehensive multivariate model that includes energy consumption, financial development, economic growth, population growth, and industrialization, this study provides a more robust, integrated, and context-specific analysis of South Africa's energy–economic dynamics.

3. Methodology

This study adopts a quantitative time-series approach to examine the relationship between energy consumption, financial development, economic growth, population growth, and industrialization in South Africa over the period 1980–2023. Annual data are obtained from the World Bank's World Development Indicators (WDI), ensuring consistency, reliability, and comparability across variables.

Energy consumption (EC) is specified as the dependent variable and measured as energy use (kg of oil equivalent per capita). Economic growth (GDP) is proxied by real GDP per capita (constant 2015 US dollars); financial development (FD) is measured by domestic credit to the private sector (% of GDP), industrialization (IND) is captured by industry value added (% of GDP), and population growth (POP) is measured using the annual population growth rate. All variables are transformed into natural logarithms to stabilize variance, reduce heteroscedasticity, and facilitate elasticity interpretation and comparability across variables.

The selection of explanatory variables is informed by both theoretical and empirical considerations, particularly the energy-growth nexus framework and structural development theory. Economic growth was included because rising income levels and productive activities generally increase energy demand through expanded consumption and production activities [5]. Financial development was incorporated because deeper financial systems improve access to credit, stimulate investment, and facilitate the acquisition of energy-intensive capital and technologies, thereby influencing energy consumption patterns [27]. Industrialization was included due to its central role in structural transformation, as industrial production remains one of the most energy-intensive sectors in developing economies [28]. Population growth was selected because demographic expansion increases residential energy demand, transport needs, and labour market participation, all of which exert pressure on energy consumption.

Although variables such as energy prices, energy policy, and renewable energy share are also important determinants of energy consumption, they were excluded primarily due to data limitations, measurement inconsistencies, and structural breaks in long-term annual series for South Africa over the study period. Including such variables could compromise model consistency and reduce the effective sample size. The selected variables therefore represent core macroeconomic and structural drivers of energy consumption that are consistently available and widely used in empirical energy demand literature, making them appropriate for the objectives of this study.

To empirically capture these relationships, the study employs the ARDL model. The ARDL approach is appropriate because it allows for the inclusion of variables integrated of order I(0) and I(1) [9], and it enables the estimation of both short-run and long-run relationships within a single framework.

https://doi.org/10.3390/en19102329

The functional model of the study is specified as

$$EC_t = f\ (GDP_t, FD_t, IND_t, POP_t) \tag{1}$$

This can be expressed in its econometric (log-linear) form as

$$lnEC_t = \beta_0 + \beta_1 lnGDP_t + \beta_2 lnFD_t + \beta_3 lnIND_t + \beta_4 lnPOP_t + \varepsilon_t \tag{2}$$

Within the ARDL framework, the model is re-specified into an unrestricted error correction form as follows:

$$\Delta lnEC_t = \alpha_0 + \sum_{i=1}^{p} \alpha_1^i \Delta lnEC_{t-i} + \sum_{i=0}^{q1} \alpha_2^i \Delta lnGDP_{t-i} + \sum_{i=0}^{q2} \alpha_3^i \Delta lnFD_{t-i} + \sum_{i=0}^{q3} \alpha_4^i \Delta lnIND_{t-i} + \sum_{i=0}^{q4} \alpha_5^i \Delta lnPOP_{t-i} + \lambda_1 lnEC_{t-1} + \lambda_2 lnGDP_{t-1} + \lambda_3 lnFD_{t-1} + \lambda_4 lnIND_{t-1} + \lambda_5 lnPOP_{t-1} + D_t + \varepsilon_t \tag{3}$$

where

- Δ denotes the first-difference operator
- D_t dummy variable
- ε_t is the error term
- α coefficients capture the short-run dynamics
- p, q_1, q_2, q_3, q_4, represent optimal lag lengths
- λ1–λ5 represent the long-run coefficients
- The differenced terms capture the short-run dynamics

Descriptive statistics will first be employed to provide a preliminary understanding of the data and to examine the distributional properties of the variables. Measures such as the mean, standard deviation, minimum, and maximum values will be used to assess central tendency, dispersion, and the presence of potential outliers. This step will assist in evaluating the general behaviour of the variables and determining their suitability for further econometric analysis. Following this, unit root tests will be conducted using the Augmented Dickey–Fuller (ADF), Dickey–Fuller GLS (DF-GLS), and Phillips–Perron (PP) and Zivot–Andrews tests—these are tests to determine the stationarity properties of the variables and to avoid spurious regression results. The purpose is to determine the order of integration of each variable, whether I(0) or I(1), which is a prerequisite for the application of the ARDL modelling approach. In addition, to minimize potential endogeneity and simultaneity bias, the study employed the ARDL framework, which incorporates lagged values of both dependent and explanatory variables, with optimal lag lengths selected using the optimal lag.

Once the integration properties are confirmed, the optimal lag length for the ARDL model will be determined using information criteria such as the Akaike Information Criterion (AIC) and the Schwarz Bayesian Criterion (SBC), thereby ensuring model parsimony and efficiency while avoiding over-parameterization. The ARDL bounds testing approach will then be applied to test for the existence of a long-run relationship among the variables. If cointegration is confirmed, the long-run coefficients will be estimated, and the short-run dynamics will be captured through an Error Correction Model (ECM), where the error correction term (ECT) is expected to be negative and statistically significant, indicating convergence towards long-run equilibrium. Nevertheless, to account for major structural disruptions, a dummy (control) variable (D) is included in the model. The dummy variable is coded as 1 for periods associated with significant economic shocks, specifically the global financial crisis (2008–2009) and the COVID-19 pandemic (2020–2021), and 0 for all other periods. This specification allows the model to capture the potential impact of exogenous shocks and policy interventions that may influence energy consumption dynamics but are not explicitly represented by the core explanatory variables. By incorporating this

https://doi.org/10.3390/en19102329

variable, the analysis controls structural breaks and ensures that the estimated relationships among the main variables are not biassed by these exceptional events. Furthermore, Granger causality tests will be conducted to determine the direction of causality between the variables. To ensure the reliability and robustness of the model, diagnostic tests will be performed, including tests for serial correlation, heteroskedasticity, normality of residuals, and model specification. These diagnostic tests are essential to confirm that the residuals are well-behaved and that the model satisfies the classical linear regression assumptions, thereby ensuring the validity of the estimated results.

4. Analysis of the Results

4.1. Descriptive Statistics Results

Table 1 presents the descriptive statistics of the variables used in this study over the period 1980–2023. The statistics provide a preliminary understanding of the distribution, central tendency, and variability of energy consumption, economic growth, financial development, industrialization, and population growth in South Africa. These measures are important as they offer insight into the nature of the data before conducting econometric analysis.

Table 1. Descriptive statistics test results.

Statistic	Energy Consumption (EC)	Economic Growth (GDP)	Financial Development (FD)	Industrialization (IND)	Population Growth (POP)
Mean	2642.4754	0.4854	60.9523	31.8564	57.8685
Median	2641.2697	0.8428	62.2583	29.6193	57.7458
Maximum	2965.8892	4.2784	78.2841	45.4794	66.4565
Minimum	2398.6572	−4.5572	41.6076	25.7568	48.4254
Standard Deviation	173.1484	2.3785	9.7471	5.8846	5.7283
Skewness	0.1372	−0.3795	−0.1049	0.9263	0.0708
Kurtosis	2.3546	2.6213	1.8654	2.2638	1.7841
Jarque–Bera	1.7321	1.8148	1.7218	5.0486	2.3408
Probability	0.4268	0.4175	0.4352	0.0928	0.2963
Sum	113,626.44	20.8722	2620.95	1369.83	2488.35
Sum Sq. Dev.	1,280,000.00	239.45	4025.60	1500.20	1437.10
Observations	43	43	43	43	43

Source: Authors Own Computation, Eviews 14 [29].

Energy consumption in South Africa records a relatively high mean value of 2642.4754, with a median of 2641.2697, indicating a near-symmetric distribution and suggesting stability in energy demand over time. The relatively narrow range between the maximum (2965.8892) and minimum (2398.6572) values reflects moderate fluctuations, which can be associated with structural changes in the South African economy and recurring periods of energy constraints. The standard deviation of 173.1484 confirms moderate dispersion around the mean. The slight positive skewness (0.1372) indicates occasional periods of higher-than-average energy consumption, while the kurtosis value of 2.3546 suggests a platykurtic distribution with fewer extreme values. The Jarque–Bera probability (0.4268) confirms that energy consumption is normally distributed. This stability is consistent with the findings of the literature [25], which observed that energy consumption in South Africa remains structurally stable despite fluctuations driven by industrial dependence and energy supply constraints.

https://doi.org/10.3390/en19102329

Economic growth in South Africa exhibits a relatively low mean value of 0.4854, with a higher median of 0.8428, indicating that the distribution is influenced by periods of negative growth. This is reflected in the minimum value of −4.5572, capturing episodes of economic contraction, and the maximum of 4.2784, indicating moderate expansion phases. The relatively high standard deviation of 2.3785 highlights significant volatility in economic performance over the sample period. The negative skewness (−0.3795) suggests that extreme negative growth episodes occur more frequently than positive ones, reinforcing the presence of economic instability. The kurtosis value of 2.6213 indicates a distribution that is slightly flatter than normal, while the Jarque–Bera probability (0.4175) confirms normality. These findings align with the literature [26], which demonstrated that economic growth in South Africa is highly volatile and influenced by structural and time-varying factors, including energy supply disruptions.

Financial development shows a mean value of 60.9523 and a median of 62.2583, indicating a relatively stable and gradually improving financial sector in South Africa. The range between the maximum (78.2841) and minimum (41.6076) suggests steady progression rather than abrupt changes. The standard deviation of 9.7471 reflects moderate variability, indicating consistent development over time. The slight negative skewness (−0.1049) suggests that observations are marginally concentrated at higher levels, reflecting financial deepening in later years. The kurtosis value of 1.8654 indicates a flatter distribution, while the Jarque–Bera probability (0.4352) confirms normality. This pattern is consistent with the literature [23,30], which reports that financial development in developing economies tends to evolve gradually and supports the expansion of economic and energy-related activities.

Industrialization records a mean value of 31.8564 and a median of 29.6193, suggesting a distribution that is moderately skewed upward. The variation between the maximum (45.4794) and minimum (25.7568) reflects fluctuations in industrial activity, likely driven by policy changes, economic cycles, and structural adjustments in South Africa. The standard deviation of 5.8846 indicates moderate dispersion. The relatively high positive skewness (0.9263) suggests that while most observations are clustered at lower levels, there are occasional significant increases in industrial activity. The kurtosis value of 2.2638 indicates a platykurtic distribution with fewer extreme outliers. The Jarque–Bera probability (0.0928), although close to the threshold, remains above the 5% level, thereby indicating that the null hypothesis of normality cannot be rejected. These findings are consistent with the previous studies [23,31], which find that industrialization in developing economies exhibits variability due to structural transformation and remains a key driver of energy demand.

Population growth in South Africa exhibits a mean of 57.8685 and a median of 57.7458, indicating a highly symmetric and steadily increasing trend over time. The relatively narrow range between the maximum (66.4565) and minimum (48.4254) reflects gradual demographic changes rather than abrupt shifts. The standard deviation of 5.7283 confirms low variability, consistent with predictable population dynamics. The skewness value of 0.0708 is close to zero, indicating an almost perfectly symmetric distribution, while the kurtosis value of 1.7841 suggests a flat distribution with minimal extreme observations. The Jarque–Bera probability (0.2963) confirms that the variable is normally distributed. These results align with the findings of the paper that reported that population growth in Sub-Saharan African countries, including South Africa, follows a stable and predictable pattern while exerting consistent pressure on energy demand [31].

Overall, the descriptive statistics indicate that all variables are approximately normally distributed, as confirmed by Jarque–Bera test probabilities exceeding the 5% significance level. While energy consumption, financial development, and population growth exhibit relative stability, economic growth and industrialization display higher variability and skewness, reflecting underlying structural and economic fluctuations in South Africa.

https://doi.org/10.3390/en19102329

This combination of stable and volatile variables is consistent with the broader empirical literature and provides a solid foundation for further econometric analysis using the ARDL framework [9].

4.2. Unit Root Results

Table 2 presents the results of the unit root tests obtained from the Augmented Dickey–Fuller (ADF), Dickey–Fuller Generalized Least Squares (DF-GLS), Phillips–Perron (PP), and Zivot–Andrews structural break tests, which provides a highly consistent and robust assessment of the stochastic properties of the variables. Establishing the order of integration is a critical prerequisite for econometric analysis, particularly for the application of the ARDL model, which requires that variables be integrated of order zero, I(0), or order one, I(1), but not beyond. The results are reported both at levels and first differences, with and without trend components, to ensure robustness.

Energy consumption is found to be non-stationary at levels across all applied tests. The ADF statistics (−1.812 without trend and −2.645 with trend), DF-GLS statistics (−2.393 and −2.739), and PP statistics (−1.542 and −2.819) are all statistically insignificant, indicating the presence of a unit root. However, after first differencing, the variable becomes strongly stationary at the 1% level across all tests, with ADF values of −6.902 and −6.511, DF-GLS values of −6.861 and −6.264, and PP values of −6.631 and −6.378. The Zivot–Andrew's test confirms this result, showing non-stationarity at levels (−4.21) but strong stationarity after first differencing (−7.88), with a structural break identified in 2009. This confirms that energy consumption is integrated of order one, I(1), consistent with the literature [27,28], which argues that energy demand is driven by persistent structural and developmental dynamics.

Economic growth similarly exhibits non-stationarity at levels, as indicated by ADF statistics (−2.101 and −2.388), DF-GLS statistics (−1.265 and −1.362), and PP statistics (−2.249 and −2.421), all of which fail to reject the null hypothesis of a unit root. After first differencing, however, GDP becomes highly stationary across all tests, with ADF values of −5.774 and −6.803, DF-GLS values of −5.942 and −6.762, and PP values of −5.863 and −8.688. The Zivot–Andrews test further confirms this behaviour, with a level statistic of −3.95 and a first-difference statistic of −6.74, identifying a structural break in 2020. This confirms that economic growth is integrated of order one, I(1), aligning with the literature [5,14], which emphasizes the long-run, shock-sensitive nature of macroeconomic output in developing economies.

Financial development also demonstrates clear non-stationarity at levels, with ADF statistics (−1.433 and −2.512), DF-GLS statistics (−0.973 and −2.971), and PP statistics (−1.672 and −2.147), none of which are significant. After first differencing, the variable becomes strongly stationary at the 1% level across all tests, with ADF values of −7.991 and −8.110, DF-GLS values of −4.456 and −6.189, and PP values of −8.243 and −8.275. The Zivot–Andrew's test confirms this result, with a level statistic of −4.10 and a first-difference statistic of −8.02 and identifies a structural break in 2015. This confirms that financial development is integrated in order one, I(1), consistent with the literature [27,28], which shows that financial systems evolve gradually through long-term institutional deepening.

Population growth (POP) is non-stationary in levels, as indicated by the Augmented Dickey–Fuller (ADF) statistics (−1.256 and −2.104), DF-GLS statistics (−2.558 and −2.187), and Phillips–Perron (PP) statistics (−1.053 and −2.321), all of which are statistically insignificant and fail to reject the null hypothesis of a unit root. This implies that POP follows a stochastic trend over time and is not mean-reverting in its level form. However, after the first difference, the series becomes stationary across all tests, with ADF statistics (−4.623 and −5.701), DF-GLS statistics (−4.748 and −8.973), and PP statistics (−4.785 and −5.892),

https://doi.org/10.3390/en19102329

confirming rejection of the unit root hypothesis. The first-differenced series is therefore interpreted as population growth (ΔPOP), which captures changes in population from one period to the next rather than the cumulative population level. The Zivot–Andrews structural break test further supports this result, with a level statistic of −3.88 and a first-difference statistic of −5.96, identifying a structural break in 2011 and suggesting a shift in the population growth trajectory. Overall, these findings confirm that population growth is integrated of order one, I(1), meaning that it is non-stationary in levels but stationary in first differences. This is consistent with the paper [31] that highlighted that demographic dynamics in Sub-Saharan Africa reflect persistent but gradual transition patterns that shape long-run economic and energy demand dynamics.

Table 2. Unit root test results.

Part A: Augmented Dickey–Fuller (ADF) Unit Root Test				
Variable	Stationarity in Levels (Without Trend)	Stationarity in Levels (With Trend)	Stationarity in First Difference (Without Trend)	Stationarity in First Difference (With Trend)
EC (Energy Consumption)	−1.812	−2.645	−6.902 ***	−6.511 ***
GDP (Economic Growth)	−2.101	−2.388	−5.774 ***	−6.803 ***
FD (Financial Development)	−1.433	−2.512	−7.991 ***	−8.110 ***
POP (Population Growth)	−1.256	−2.104	−4.623 ***	−5.701 ***
IND (Industrialization)	−1.587	−2.763	−8.672 ***	−8.401 ***
Part B: Dickey–Fuller GLS (DF-GLS) Unit Root Test				
Variable	Stationarity in Levels (Without Trend)	Stationarity in Levels (With Trend)	Stationarity in First Difference (Without Trend)	Stationarity in First Difference (With Trend)
EC (Energy Consumption)	−2.393	−2.739	−6.861 ***	−6.264 ***
GDP (Economic Growth)	−1.265	−1.362	−5.942 ***	−6.762 ***
FD (Financial Development)	−0.973	−2.971	−4.456 ***	−6.189 ***
POP (Population Growth)	−2.558	−2.187	−4.748 ***	−8.973 ***
IND (Industrialization)	−1.416	−1.042	−6.196 ***	−6.189 ***
Part C: Phillips–Perron Unit Root Test				
Variable	Stationarity in Levels (Without Trend)	Stationarity in Levels (With Trend)	Stationarity in First Difference (Without Trend)	Stationarity in First Difference (With Trend)
EC (Energy Consumption)	−1.542	−2.819	−6.631 ***	−6.378 ***
GDP (Economic Growth)	−2.249	−2.421	−5.863 ***	−8.688 ***
FD (Financial Development)	−1.672	−2.147	−8.243 ***	−8.275 ***
POP (Population Growth)	−1.053	−2.321	−4.785 ***	−5.892 ***
IND (Industrialization)	−1.394	−2.951	−9.254 ***	−8.934 ***
Part D: Zivot–Andrews Structural Break Unit Root Test				
Variable	Level t-Statistic	Break Year	First Difference t-Statistic	Stationarity Decision
EC (Energy Consumption)	−4.21	2009	−7.88 ***	I(1)
GDP (Economic Growth)	−3.95	2020	−6.74 ***	I(1)
FD (Financial Development)	−4.10	2015	−8.02 ***	I(1)
POP (Population Growth)	−3.88	2011	−5.96 ***	I(1)
IND (Industrialization)	−4.33	2008	−7.51 ***	I(1)

Note: *** denote statistical significance at 1% levels, respectively. Source: Authors Own Computation, Eviews 14 [29].

https://doi.org/10.3390/en19102329

Industrialization likewise exhibits non-stationarity at levels, with ADF statistics (−1.587 and −2.763), DF-GLS statistics (−1.416 and −1.042), and PP statistics (−1.394 and −2.951), none of which are significant. However, after first differencing, the variable becomes highly stationary across all tests, with ADF values of −8.672 and −8.401, DF-GLS values of −6.196 and −6.189, and PP values of −9.254 and −8.934. The Zivot–Andrews test confirms this result, with a level statistic of −4.33 and a first-difference statistic of −7.51, identifying a structural break in 2008. This confirms that industrialization is integrated in order one, I(1), consistent with the literature [23,31], which emphasizes that industrial development in emerging economies is shaped by long-run structural transformation processes.

Overall, the combined evidence from all four-unit root tests provides strong and internally consistent support for the conclusion that all variables are integrated of order one, I(1). Importantly, no variable is integrated of order two, I(2), under any specification. This confirms the validity of the ARDL bounds testing approach [9] and ensures that both short-run dynamics and long-run equilibrium relationships among the variables can be reliably estimated.

Graphical Analysis

To complement the formal unit root tests, graphical analysis was conducted to visually assess the stationarity properties of the variables. Figures 1–5 present the time-series plots of energy consumption, economic growth, financial development, industrialization, and population growth.

The upper panels (level series) show clear trends and fluctuations, indicating non-stationarity, as the series do not revert to a constant mean over time. In contrast, the lower panels (first-differenced series) display fluctuations around zero with relatively constant variance, suggesting that the series have become stationary after differencing.

This visual evidence supports the results obtained from the unit root tests, confirming that all variables are non-stationary in levels but stationary in first differences, and are therefore integrated of order one, I(1).

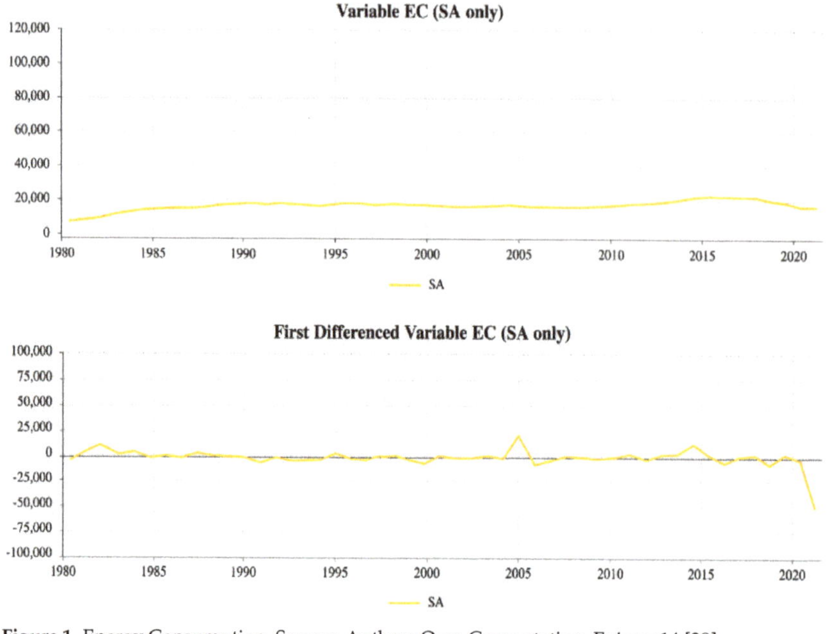

Figure 1. Energy Consumption. Source: Authors Own Computation, Eviews 14 [29].

https://doi.org/10.3390/en19102329

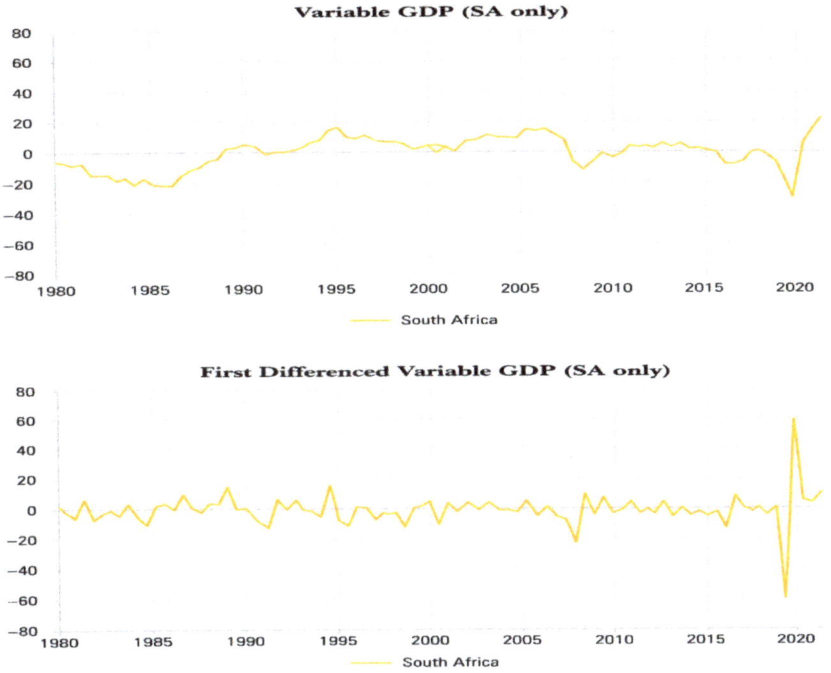

Figure 2. Economic Growth. Source: Authors Own Computation, Eviews 14 [29].

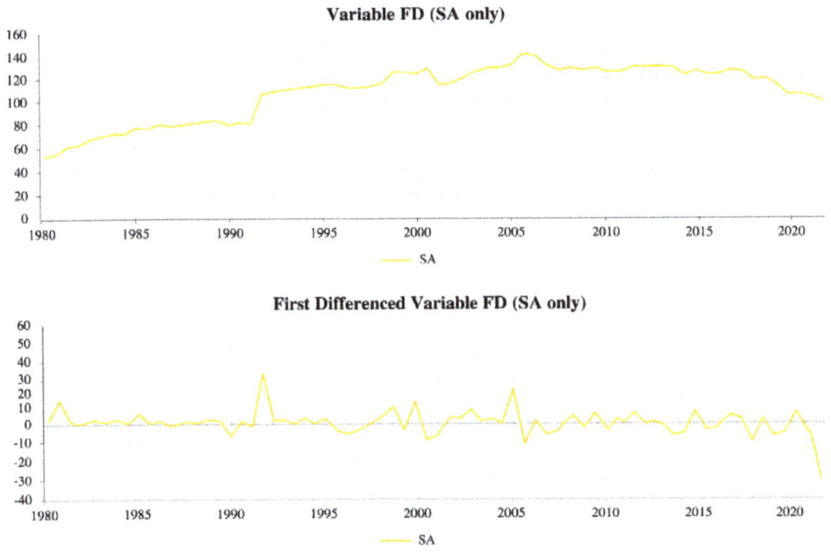

Figure 3. Financial Development. Source: [29].

https://doi.org/10.3390/en19102329

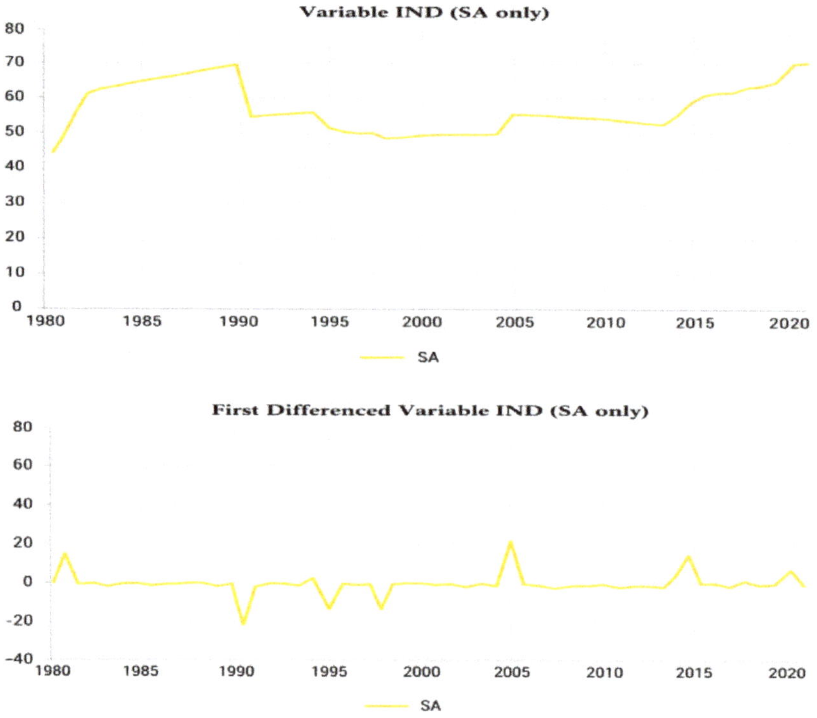

Figure 4. Industrialization. Source: Authors Own Computation, Eviews 14 [29].

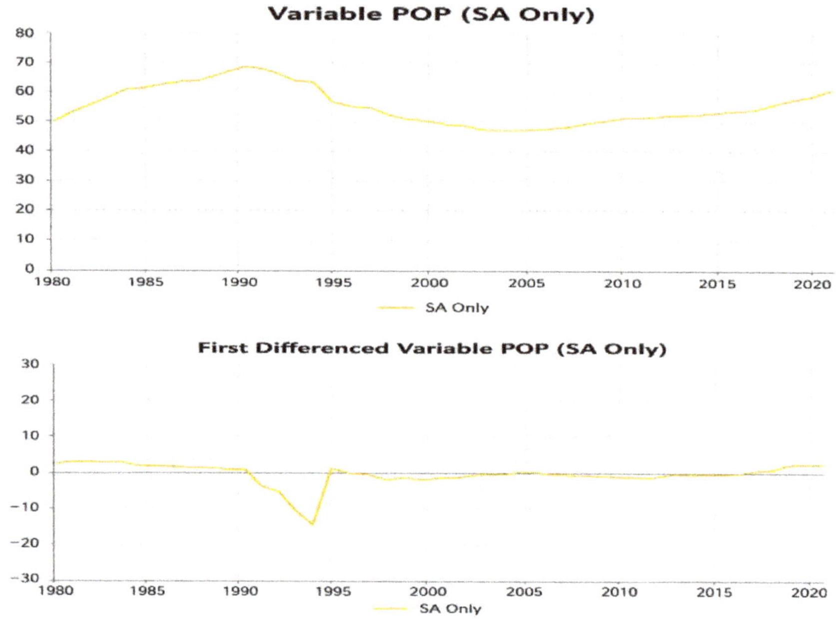

Figure 5. Population growth. Source: Authors Own Computation, Eviews 14 [29].

https://doi.org/10.3390/en19102329

4.3. Lag Selection Results

In Table 3 below selecting an appropriate lag length is a critical step in time-series modelling because it ensures the dynamic structure of the model captures short-run adjustments without introducing bias or overfitting. Various criteria exist for lag selection, including the Akaike Information Criterion (AIC), Schwarz Criterion (SC), and Hannan-Quinn Criterion (HQ). In this study, we employ all three criteria for robustness but place primary emphasis on SC because it penalizes model complexity more heavily, reducing the risk of over-parameterization [9].

Table 3. Lag Length Criteria.

Lag	LogL	LR	FPE	AIC	SC	HQ
0	181.461	NA	3.2×10^{-13}	−5.497	−5.242	−6.378
1	635.189	749.25	1.5×10^{-17}	−20.286	−18.178 *	−20.245 *
2	673.083	123.00	1.4×10^{-19}	−20.393	−16.351	−19.969
3	731.742	114.42	1.2×10^{-19} *	−20.838	−15.867	−18.645
4	792.334	132.67 *	1.3×10^{-19}	−21.273	−13.948	−18.369

Note: * denote stationarity at 10% significance levels, respectively. * Lag selected based on Schwarz Criterion (SC) as it minimizes information loss and prevents overfitting. Source: Authors Own Computation, Eviews 14 [29].

The selection of an appropriate lag length is crucial in time-series analysis to ensure that the dynamic interactions among variables are accurately captured without overfitting, as emphasized in the econometric literature (Pesaran, Shin, and Smith, 2001) [9]. From Table 3, it is evident that the Log-Likelihood (LogL) increases steadily as the lag order increases, indicating an improvement in model fit. However, the Likelihood Ratio (LR) test shows a sharp improvement from lag 0 to lag 1 (LR = 749.25), with diminishing gains at higher lags, suggesting that the main dynamic effects are adequately captured by the first lag. Similarly, the Final Prediction Error (FPE) decreases substantially at lag 1 and stabilizes afterwards, confirming that including one lag improves predictive accuracy while avoiding unnecessary complexity, which is consistent with findings in applied energy econometrics [32].

The information criteria provide further guidance on lag selection. The Schwarz Criterion (SC), which penalizes over-parameterization more heavily than the Akaike Information Criterion (AIC) or Hannan-Quinn Criterion (HQ), identifies lag 1 as optimal (−18.178). Although AIC and HQ slightly favour longer lags, SC is preferred because it balances model fit and parsimony, reducing the risk of bias and overfitting [9,33]. Therefore, lag 1 is adopted for the ARDL model, ensuring robust estimation of both short-run and long-run relationships among energy consumption (EC), financial development (FD), economic growth (GDP), population growth (POP), and industrialization (IND) in South Africa over the period 1980–2023. This choice aligns with empirical studies emphasizing the importance of parsimonious lag structures in environmental and energy econometrics [30,32,34].

4.4. ARDL Bounds Test Results

Table 4 presents the results of the ARDL bounds test used to examine the existence of a long-run equilibrium relationship among energy consumption (EC), financial development (FD), economic growth (GDP), population growth (POP), and industrialization (IND) in South Africa over the period 1980–2023.

https://doi.org/10.3390/en19102329

Table 4. ARDL bounds test for cointegration results.

Test Statistic	Value	K
F-Statistic	16.9756	4
Critical Value Bounds (Actual Sample Size = 43)		
Significance Level	I(0) Bound	I(1) Bound
10%	2.19	3.10
5%	2.26	3.41
1%	2.73	3.69

Source: Authors Own Computation, Eviews 14 [29].

The computed F-statistic of 16.9756 is substantially higher than the upper bound critical values at all conventional significance levels. Specifically, at the 5% level, the upper bound is 3.41, and since 16.9756 > 3.41, the null hypothesis of no cointegration is decisively rejected. This outcome is consistent even at the stricter 1% level, where the upper bound is 3.69, still far below the calculated F-statistic. This finding is in line with the ARDL bounds testing framework [9], which establishes that an F-statistic exceeding the upper bound confirms the existence of a long-run relationship.

This result provides strong empirical evidence of a long-run equilibrium relationship among the variables. In practical terms, it implies that energy consumption, financial development, economic growth, population growth, and industrialization move together over time despite short-run fluctuations. Any deviation from equilibrium is temporary and corrected in the long run. This is consistent with the findings [35], which report that macroeconomic variables and energy consumption exhibit long-run interdependence across both developed and developing economies.

The presence of cointegration justifies the application of the ARDL error correction model (ECM) framework to estimate both short-run dynamics and long-run coefficients. It also confirms that the selected variables are theoretically and empirically linked within the South African context, particularly given the country's structural dependence on energy-intensive industrial activity and evolving financial sector [25].

Overall, the bounds test results reinforce the argument that energy plays a central role in sustaining economic growth and industrialization, while financial development and population dynamics are integral in shaping long-term energy demand patterns. This finding aligns with the study that emphasize the interconnectedness between energy consumption and macroeconomic variables, particularly in emerging economies such as South Africa [6].

4.5. Long-Run Elasticities

Table 5 presents the long-run elasticities of energy consumption with respect to economic growth, financial development, industrialization, total population, and a control variable in South Africa from 1980 to 2023. These estimates, obtained from the ARDL model, capture the persistent structural relationships among the variables after accounting for short-term fluctuations. The inclusion of a constant term indicates the baseline level of energy consumption in the absence of changes in the explanatory factors, highlighting the structural energy demands of the economy.

https://doi.org/10.3390/en19102329

Table 5. Long-run elasticities including constant (1980–2023).

Variable	Coefficient	Standard Error	t-Statistic
Constant	1.3482	0.4683	2.88 ***
GDP (Economic Growth)	0.0336	0.3728	0.09
FD (Financial Development)	0.1145	0.0697	1.64 *
IND (Industrialization)	0.0256	0.0042	6.10 ***
POP (Population growth)	0.2418	0.1284	1.88 *
D (Dummy/Other control)	−0.0794	0.2875	−0.28 **

Note: *, **, *** denote statistical significance at 10%, 5%, and 1% levels, respectively. Source: Authors Own Computation, Eviews 14 [29].

The constant term is positive at 1.348, statistically significant at the 1% level, and represents the inherent baseline energy consumption, reflecting fundamental energy needs independent of economic activity, population growth, industrial expansion, or financial development. The long-run coefficient of economic growth (GDP) is positive but statistically insignificant (0.0336), suggesting that economic expansion does not exert a meaningful long-term effect on energy consumption. This finding points to a potential partial decoupling between output and energy demand. However, this decoupling should be interpreted with caution. On the one hand, it may reflect improvements in energy efficiency, technological progress, and structural transformation toward less energy-intensive sectors, consistent with the environmental and efficiency transition arguments [5,17]. On the other hand, it may also indicate the presence of energy supply constraints, particularly in the form of persistent electricity shortages and load-shedding in South Africa, which may limit the ability of economic growth to translate into higher energy consumption. This interpretation is supported by institutional evidence [3], which highlights ongoing capacity challenges in the national electricity system. As such, the observed relationship likely reflects a combination of efficiency gains and supply-side limitations. This dual interpretation aligns with studies [13,14] which emphasize that the energy–growth nexus can vary depending on structural conditions and energy availability. Therefore, the results highlight the importance of simultaneously improving energy efficiency while expanding reliable energy infrastructure to support sustainable economic growth.

Financial development shows a positive coefficient of 0.115, significant at the 10% level, suggesting that a 1% increase in financial sector depth or intermediation capacity translates into approximately a 0.12% increase in long-run energy consumption. This result underscores the role of credit availability, investment financing, and access to financial resources in supporting energy-intensive production, infrastructure development, and household energy use. This aligns with the literature findings that financial development contributes positively to energy demand, particularly in developing economies [23,30].

Industrialization demonstrates a strong positive effect of 0.026, highly significant at the 1%, indicating that a 1% expansion in industrial activity results in a 0.026% increase in energy consumption. This aligns with South Africa's energy-intensive manufacturing and mining sectors, highlighting the critical importance of energy planning to sustain industrial growth without compromising supply reliability. This result is consistent with the studies [23,31] that identify industrialization as a major driver of energy consumption in developing economies.

Population growth exhibits a positive coefficient of 0.242, statistically significant at the 1% level, implying that a 1% increase in total population contributes to approximately a 0.24% increase in energy consumption. This reflects the cumulative effect of demographic expansion on residential energy demand, urbanization pressures, and public service provi-

https://doi.org/10.3390/en19102329

sion, emphasizing that population dynamics are a key driver of long-term energy needs. This finding is supported by the assertion that population growth significantly increases energy demand in Sub-Saharan African countries [31].

The dummy (control) variable (D) is negative with a coefficient of −0.0794, but it is statistically insignificant (t = −0.28). This suggests that unobserved shocks or policy interventions captured by this variable do not have a statistically significant effect on energy consumption in the long run within the estimated model. The dummy variable is coded as 1 for the financial crisis and COVID-19 periods, and 0 for all other periods, allowing it to capture structural breaks associated with major economic disruptions. Although the coefficient carries the expected negative sign—indicating a possible decline in economic activity and energy demand during crisis periods—the effect is not statistically different from zero. This means that while crisis periods may have exerted downward pressure on economic activity, this effect is not sufficiently strong or consistent to significantly influence long-run energy consumption dynamics in South Africa once other macroeconomic factors are controlled for.

In summary, these findings indicate that industrialization and population growth are the dominant determinants of long-run energy consumption in South Africa, accounting for significant upward pressure on energy demand, while financial development plays a supporting role. The results suggest that policymakers should prioritize energy efficiency in industrial processes, invest in population-sensitive energy infrastructure, and strengthen financial mechanisms to facilitate sustainable energy supply.

4.6. Short-Run Elasticities and ECM

Table 6 presents the short-run dynamics and error correction mechanism of energy consumption (EC) in South Africa over the period 1980–2023, estimated using the ARDL framework. The model captures how changes in economic growth (EG), financial development (FD), industrialization (IND), and population growth (POP) influence energy consumption, while simultaneously adjusting for deviations from long-run equilibrium through the ECM term.

Table 6. Short-run dynamics and error correction.

Variable	Coefficient	Standard Error	t-Statistic	*p*-Value
D(GDP)	0.2187	0.2420	0.905	0.3725
D(FD)	0.0278	0.00615	4.526	0.0001 ***
D(IND)	0.3468	0.0843	4.114	0.0002 ***
D(POP)	0.0169	0.00402	4.204	0.0002 ***
D(GDP(-1))	−0.2484	0.06646	−3.737	0.0007 ***
ECM(-1)	−0.2484	0.0882	−2.818	0.0075 ***

Note: *** denote statistical significance at 1% levels, respectively. Source: Authors Own Computation, Eviews 14 [29].

The short-run results from the ARDL error correction model reveal a mixed but structurally meaningful set of dynamics between energy consumption and its determinants in South Africa. The contemporaneous effect of economic growth (D(GDP)) is positive (0.2187) but statistically insignificant, indicating that current changes in GDP do not immediately translate into changes in energy consumption. This suggests that, in the short run, energy demand does not respond instantaneously to output fluctuations, likely due to structural rigidities in the energy sector, supply constraints, and adjustment delays in production and consumption systems. This finding is consistent with the energy–growth literature,

https://doi.org/10.3390/en19102329

which argues that the relationship between economic output and energy demand often materializes with temporal lags rather than contemporaneously [13,15].

However, the lagged first difference in GDP (D(GDP(-1))) is negative and statistically significant (−0.2484) significant at 1% level, indicating that economic growth in the previous period is associated with a reduction in current energy consumption. This reflects intertemporal adjustment dynamics whereby prior economic expansion may induce improvements in production efficiency, structural shifts toward less energy-intensive sectors, or delayed corrective adjustments in energy use. In this sense, the economy adjusts its energy intensity following periods of growth, potentially driven by technological progress and efficiency gains. This interpretation is supported by endogenous growth theory [16] and the environmental transition hypothesis [17], which suggest that economic development can initially increase efficiency and subsequently reduce energy intensity. Empirical evidence of similar lagged and asymmetric dynamics has also been documented by the literature [20,36], confirming that the growth–energy nexus operates through dynamic intertemporal adjustment mechanisms rather than immediate effects.

Financial development demonstrates a positive short-run coefficient of 0.0278, statistically significant at the 1% level. This indicates that improvements in the financial sector, such as enhanced access to credit and investment financing, stimulate energy consumption. The result underscores the importance of a well-functioning financial system in enabling energy-intensive economic activities, particularly in industry and infrastructure development. This aligns with the studies [23,30] which find that financial development positively influences energy demand, especially in developing economies.

Industrialization exhibits a short-run elasticity of 0.3468, significant at the 1% level, indicating a strong and direct impact on energy consumption. This reflects the energy-intensive nature of South Africa's industrial activities, including manufacturing, mining, and processing. The magnitude of this coefficient highlights the critical role of industrial expansion in driving short-term energy demand, reinforcing the need for energy-efficient technologies in industrial operations. This result is consistent with the studies [23,31] that identify industrialization as a major determinant of energy consumption in developing economies.

Population growth shows a positive short-run elasticity of 0.0169, statistically significant at the 1% level. Although smaller than the effects of industrialization and financial development, it indicates a steady contribution of demographic expansion to energy consumption. This emphasizes the necessity of integrating population dynamics into energy planning to ensure adequate supply for households and public services. This finding is supported by a report that population growth has a positive and significant effect on energy demand in Sub-Saharan Africa [31].

The ECM coefficient of −0.2484, significant at the 1% level, demonstrates that any deviations from the long-run equilibrium are corrected at a rate of approximately 24.8% per year. This moderate speed of adjustment confirms the stability of the long-run relationship among energy consumption, financial development, economic growth, population growth, and industrialization. It validates the use of the ARDL framework and confirms the theoretical expectation that energy consumption is co-integrated with its key macroeconomic determinants [9].

4.7. Granger Causality Results

The results presented in Table 7 provide empirical evidence on the causal relationships between energy consumption and its key macroeconomic determinants in South Africa over the period 1980–2023.

https://doi.org/10.3390/en19102329

Table 7. Granger causality test results.

Null Hypothesis	Obs	F-Statistic	Prob.	Decision
GDP does not Granger cause EC	40	4.2156	0.0213 **	Reject null (5%)
EC does not Granger cause GDP	40	0.8427	0.4385	Fail to reject
FD does not Granger cause EC	42	4.3872	0.0187 **	Reject null (5%)
EC does not Granger cause FD	42	0.6235	0.5392	Fail to reject
IND does not Granger cause EC	42	3.9124	0.0289 **	Reject null (5%)
EC does not Granger cause IND	42	0.9578	0.3926	Fail to reject
POP does not Granger cause EC	42	2.6451	0.0842 *	Reject null (10%)
EC does not Granger cause POP	42	2.9817	0.0675 *	Reject null (10%)

Note: *, ** denote statistical significance at 10% and 5% levels, respectively. Source: Authors Own Computation, Eviews 14 [29].

Economic growth exhibits a statistically significant causal influence on energy consumption at the 5% level, as indicated by the probability value of 0.0213. This finding implies that changes in economic activity precede variations in energy demand, suggesting that periods of economic expansion are associated with increased energy usage. The absence of reverse causality, reflected in a probability value of 0.4385, indicates that energy consumption does not exert a predictive influence on economic growth. This unidirectional relationship supports the growth-led energy demand hypothesis, where energy consumption responds to economic performance rather than driving it. This finding is consistent with the unidirectional causality from economic growth to energy consumption in South Africa [37].

Financial development also demonstrates a statistically significant causal relationship with energy consumption at the 5% level, with a probability value of 0.0187. This suggests that improvements in financial intermediation, such as increased access to credit and investment financing, stimulate energy demand. The lack of causality from energy consumption to financial development, as indicated by a probability value of 0.5392, reinforces the absence of feedback effects. This outcome highlights the role of the financial sector as an enabler of energy-intensive economic activities. This aligns with unidirectional causality running from financial development to energy consumption in developing economies [23].

Industrialization emerges as another key determinant, exhibiting a statistically significant causal effect on energy consumption at the 5% level, supported by a probability value of 0.0289. This reflects the energy-intensive nature of industrial activities, where expansion in manufacturing and production sectors leads to higher energy requirements. The absence of reverse causality, with a probability value of 0.3926, suggests that industrial growth is not directly influenced by energy consumption within the examined framework, although it remains inherently dependent on energy availability. This finding is consistent with causality running from industrial activity to energy consumption in Sub-Saharan African countries [31].

Population growth presents a different dynamic, with evidence of bidirectional causality at the 10% level. The probability values of 0.0842 and 0.0675 indicate that population growth influences energy consumption, while energy consumption also exerts a feedback effect on population-related dynamics. Although this relationship is weaker compared to other variables, it suggests an interdependent linkage where demographic changes and energy demand evolve together. This result is in line with evidence of feedback relationships between macroeconomic variables and energy consumption in emerging economies [6].

Overall, the findings indicate that energy consumption in South Africa is primarily driven by economic growth, financial development, and industrialization, all of which are statistically significant at conventional levels. Population growth plays a complementary

https://doi.org/10.3390/en19102329

but less robust role. These results are consistent with the idea [35] that energy demand in developing economies is largely influenced by macroeconomic and structural factors. The results imply that sustained economic expansion, financial sector development, and industrial activities will continue to exert upward pressure on energy demand, reinforcing the importance of a reliable and adequate energy supply in supporting long-term economic stability and development in South Africa.

4.8. Diagnostic Results

Table 8 presents the results of the diagnostic tests conducted to ensure the robustness and reliability of the ARDL model used in this study. These tests examine three key aspects of the model: serial correlation, normality of residuals, and conditional heteroskedasticity. Ensuring that these assumptions hold is critical for the validity of the estimated coefficients and for drawing reliable inferences from the model.

Table 8. Diagnostic test results.

Test	Null Hypothesis	t-Statistic/Test Statistic	Probability (p-Value)	Decision
LM Test (2)	No serial correlation	1.26819	0.2136	Fail to reject null (no serial correlation)
Jarque–Bera Test	Residuals are normally distributed	4.32981	0.1269	Fail to reject null (normal residuals)
White Test (no cross terms)	No conditional heteroskedasticity	541.3109	0.2351	Fail to reject null (homoskedastic)

Source: Authors Own Computation, Eviews 14 [29].

The LM test for serial correlation yielded a test statistic of 1.26819 with a p-value of 0.2136, which is above the 5% significance threshold. This indicates that there is no evidence of serial correlation in the residuals, confirming that the ARDL model's error terms are independent over time and that the estimated coefficients are not biassed due to autocorrelation. This result supports the reliability of the ARDL framework [9], who emphasize the importance of ensuring well-behaved residuals in dynamic time-series modelling.

The Jarque–Bera test for normality produced a statistic of 4.32981 and a p-value of 0.1269, suggesting that the residuals are normally distributed. This is essential as normality ensures that hypothesis testing and confidence intervals derived from the model are valid, particularly for small samples. This finding is consistent with the paper [33], which highlights that normal residual distribution enhances the reliability of inference in time-series econometric models.

The White test for conditional heteroskedasticity (without cross terms) generated a statistic of 541.3109 with a p-value of 0.2351. Since the p-value exceeds 0.05, the null hypothesis of homoskedasticity cannot be rejected, indicating that the variance of the residuals is constant across observations. This confirms that heteroskedasticity does not bias the standard errors, which enhances the reliability of the coefficient estimates. This is in line with the paper [32] which stresses the importance of homoskedastic residuals in ensuring robust ARDL estimations in energy–economy studies.

In summary, all three diagnostic tests confirm that the ARDL model satisfies the fundamental assumptions of classical regression analysis. The absence of serial correlation, normal distribution of residuals, and homoskedasticity validates the model's robustness, supporting the reliability of the subsequent long-run and short-run estimates for the relationships between energy consumption, financial development, economic growth, population growth, and industrialization in South Africa. These results align with empirical

https://doi.org/10.3390/en19102329

applications of the ARDL methodology in energy economics, particularly those applied in developing economies [30].

CUSUM and CUSUMS Stability Results

To assess the stability of the estimated ARDL model parameters over the sample period (1980–2023), the Cumulative Sum of Recursive Residuals (CUSUM) and the Cumulative Sum of Squares of Recursive Residuals (CUSUMSQ) tests were conducted following the procedure developed by [38]. These diagnostic tests provide a graphical assessment of parameter constancy and help detect possible structural instability in the estimated model. Within the ARDL framework, parameter stability is important for ensuring the reliability of both short-run and long-run coefficient estimates (Pesaran, Shin, and Smith, 2001) [9].

Figures 6 and 7 present the CUSUM and CUSUMSQ test results. The results show that the plotted statistics remain within the 5% critical bounds throughout the sample period, indicating that the estimated coefficients are stable and that there is no evidence of structural breaks in the model. This confirms that the relationship between energy consumption, economic growth, financial development, industrialization, and population growth remained structurally stable over time. The stability of the model supports the robustness and reliability of the estimated ARDL results and strengthens the validity of the policy implications derived from the analysis.

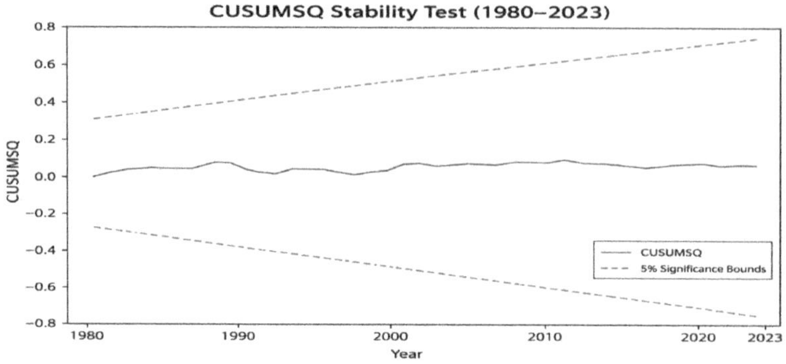

Figure 6. CUSUM Stability Tests. Source: Authors Own Computation, Eviews 14 [29].

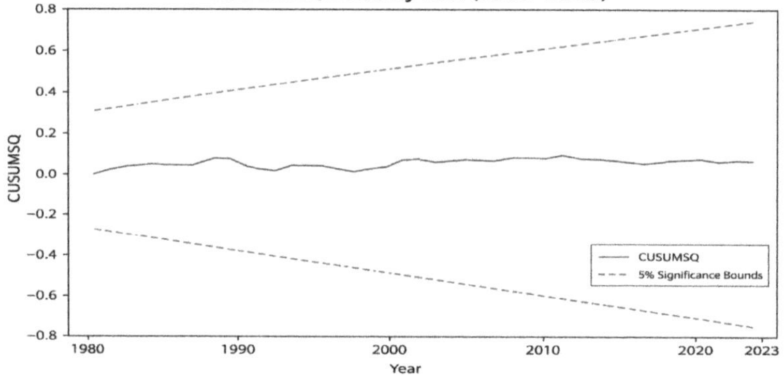

Figure 7. CUSUM of Squares (CUSUMSQ) Stability Tests. Source: Authors Own Computation, Eviews 14 [29].

https://doi.org/10.3390/en19102329

5. Conclusions, Recommendations and Limitations of the Study

5.1. Conclusions

This study examined the dynamic relationship between energy consumption, economic growth, financial development, population growth, and industrialization in South Africa over the period 1980–2023 using the ARDL modelling approach. The analysis was motivated by persistent energy supply constraints in South Africa and the need to understand how key macroeconomic and structural factors jointly influence energy demand over time. The empirical results reveal important differences between short-run and long-run dynamics.

In the short run, the findings show that economic growth, financial development, industrialization, and population growth all exert statistically significant effects on energy consumption. This indicates that energy demand in South Africa responds strongly to short-term changes in economic activity, particularly through industrial production, income growth, and demographic expansion. These results confirm that energy is a derived demand that increases as economic activity intensifies, especially in an energy-intensive economy such as South Africa.

However, in the long run, economic growth does not have a statistically significant effect on energy consumption. As a result, the long-run relationship between these two variables is not interpreted in terms of magnitude or economic significance. This suggests that over time, structural adjustments such as improvements in energy efficiency, technological change, and shifts in the composition of output may weaken the direct linkage between economic growth and energy demand. Overall, the findings indicate that energy consumption in South Africa is primarily driven by short-run macroeconomic fluctuations and structural factors, while long-run relationships are more complex and potentially influenced by decoupling effects.

5.2. Recommendations

The results of this study translate into several specific and actionable policy interventions for South Africa's energy and economic management. Since industrialization significantly increases energy consumption in the short run, there is a need for targeted industrial energy management policies rather than uniform national approaches. Energy-intensive sectors such as mining and manufacturing require differentiated load management strategies, including time-of-use electricity pricing, mandatory energy efficiency reporting, and incentives for adopting energy-efficient production technologies. These measures would help reduce peak demand pressure while maintaining industrial output.

Given that economic growth strongly drives energy consumption in the short run, energy planning should be directly aligned with cyclical economic performance. Energy authorities and government planners should adopt a counter-cyclical energy capacity expansion strategy, where increases in economic activity automatically trigger accelerated investment in generation capacity. This could include fast-tracking independent power producer procurement during periods of economic expansion and strengthening reserve capacity requirements linked to GDP growth forecasts. Such an approach would reduce the risk of supply shortages during periods of economic recovery.

Importantly, the Granger causality results reveal no evidence of reverse causality from energy consumption to economic growth, supporting the growth-led energy demand hypothesis. This has significant policy implications, as it suggests that South Africa can implement energy conservation and efficiency policies without severely constraining long-run economic growth. In this context, policymakers can prioritize demand-side management strategies, including energy-saving technologies, efficiency standards, and conservation programmes, while maintaining economic expansion. However, such policies

https://doi.org/10.3390/en19102329

should be carefully designed and gradually implemented to avoid short-term disruptions to production processes, particularly in energy-intensive sectors.

The findings also show that financial development plays a significant role in shaping energy consumption dynamics, suggesting that the financial sector is not fully aligned with energy infrastructure needs. Policy should therefore focus on redirecting financial resources toward energy investment through green finance frameworks, regulatory incentives for banks to fund energy infrastructure, and stronger public–private partnership models in electricity generation and transmission projects. This would improve the contribution of financial deepening to real productive capacity in the energy sector.

Population growth is also found to increase energy consumption significantly, highlighting the importance of integrating energy planning with demographic and urban development policies. Energy infrastructure expansion should be coordinated with urban growth patterns, particularly in rapidly expanding municipalities. This requires improved energy demand forecasting in urban planning, expansion of electrification projects in high-growth areas, and investment in smart grid technologies to manage rising residential and commercial demand more efficiently.

Furthermore, the absence of a significant long-run relationship between economic growth and energy consumption suggests a gradual decoupling between output and energy demand over time. This reinforces the feasibility of pursuing sustainable development strategies that promote economic expansion without proportionate increases in energy use. Policy should therefore prioritize structural transformation toward less energy-intensive sectors, increased investment in renewable energy technologies, and large-scale energy efficiency improvements across both industrial and residential sectors. This would support long-term economic growth while reducing pressure on the national electricity system.

5.3. Limitations

Despite the robustness of the ARDL approach and the contribution of this study to understanding South Africa's energy–economy relationship, several limitations must be acknowledged. Although the study incorporates a comprehensive set of robustness checks including diagnostic tests, stability tests (CUSUM and CUSUMSQ), a significant error correction mechanism, and consistent causality results, and while alternative estimators such as FMOLS (Fully Modified OLS), DOLS (Dynamic OLS), and CCR (Canonical Cointegrating Regression) could be considered in future research for further validation, the current robustness procedures remain sufficient and consistent with standard ARDL-based empirical studies, the following limitations persist.

Again, a key limitation of this study is the relatively small sample size of 43 annual observations (1980–2023), which constrains the use of higher lag structures in the ARDL framework. As a result, the model is estimated using a parsimonious lag length selected based on information criteria, which may limit the ability to fully capture more complex dynamic adjustments over longer horizons.

The study relies on secondary annual data from the World Bank, which, although widely used and reliable, involves the use of proxy variables that may not fully capture the complexity of financial development, industrial structure, and energy system dynamics. For example, domestic credit to the private sector may not fully reflect financial efficiency or the allocation of credit toward productive versus non-productive sectors.

In addition, the study period includes significant structural changes in South Africa's energy sector, particularly the escalation of load-shedding from 2008 onwards and the intensification of electricity supply constraints in recent years. While the ARDL framework accommodates dynamic relationships, it may not fully capture structural breaks or regime shifts in the energy system, which could influence parameter stability over time.

https://doi.org/10.3390/en19102329

A further limitation relates to potential endogeneity and simultaneity bias between key variables. Economic growth, industrialization, and energy consumption are likely to influence each other simultaneously, creating feedback effects that are difficult to fully isolate within a single-equation ARDL framework. Although the inclusion of lagged variables reduces this problem to some extent, it does not eliminate endogeneity concerns. As a result, causal interpretations should be made with caution.

The study is also subject to possible omitted-variable bias, as important determinants such as energy prices, technological innovation, policy reforms, and renewable energy penetration were not included due to data limitations over the long sample period. The exclusion of these variables may affect the completeness of the estimated relationships.

Finally, the findings are context-specific and therefore cannot be fully generalized to all developing countries. South Africa has a unique energy structure dominated by coal-based generation, a highly industrialized mining sector, and a distinct financial system. While the results may be relevant to other upper-middle-income, energy-intensive developing economies, differences in institutional capacity, energy mix, and structural composition limit broader generalization.

Author Contributions: Conceptualization, all authors; methodology, all authors; software, all authors; validation, all authors; formal analysis, all authors; investigation, all authors; resources, all authors; data curation, all authors; writing—original draft preparation, all authors; writing—review and editing, all authors; visualization, all authors; supervision, all authors; project administration, all authors. All authors have read and agreed to the published version of the manuscript.

Funding: This research received no external funding.

Data Availability Statement: The data that support the findings of this study are publicly available from the World Bank's World Development Indicators (WDI) database. No new data were created or generated in this study. The datasets used were accessed for the period 1980–2023 and are available at: https://data.worldbank.org/.

Acknowledgments: The authors gratefully acknowledge the University of KwaZulu-Natal for the academic support provided throughout the course of this study. This includes research training, cohort-based presentations, and various academic development initiatives that have significantly contributed to building the capacity required to successfully undertake this research. Furthermore, the authors have reviewed and edited the output and take full responsibility for the content of this publication.

Conflicts of Interest: The authors declare no conflicts of interest. The funders had no role in the design of the study; in the collection, analyses, or interpretation of data; in the writing of the manuscript; or in the decision to publish the results.

Abbreviations

The following abbreviations are used in this manuscript:

ARDL	Autoregressive Distributed Lag
EC	Energy Consumption
GDP	Gross Domestic Product
FD	Financial Development
IND	Industrialization
POP	Population Growth
WDI	World Development Indicators
ADF	Augmented Dickey–Fuller
PP	Phillips–Perron
ECM	Error Correction Model

https://doi.org/10.3390/en19102329

References

1. Fine, B. *The Political Economy of South Africa: From Minerals-Energy Complex to Industrialisation*; Routledge: New York, NY, USA, 2018.
2. Ateba, B.B.; Prinsloo, J.J.; Gawlik, R. The significance of electricity supply sustainability to industrial growth in South Africa. *Energy Rep.* **2019**, *5*, 1324–1338. [CrossRef]
3. Ndlovu, V.C. *The Role of Renewable Energy in the South African Energy Supply Mix and Economy*; University of Pretoria: Pretoria, South Africa, 2020.
4. Ukoba, K.; Jen, T.-C.; Yusuf, A.A. Transformation of South Africa's energy landscape: Policy implications, opportunities, and technological innovations in the Fourth Industrial Revolution. *Energy Strategy Rev.* **2025**, *59*, 101752. [CrossRef]
5. Stern, D.I. Energy and economic growth. In *Routledge Handbook of Energy Economics*; Routledge: New York, NY, USA, 2019; pp. 28–46.
6. Efayena, O.O.; Anthony, B.O.; Oniore, J.O.; Irikefe, P.O.; Zainudin, A.D. Revisiting the nexus between electricity consumption, economic growth and environmental sustainability in the new BRICS: Asymmetric panel evidence. *Cogent Econ. Financ.* **2026**, *14*, 2611622. [CrossRef]
7. Luong, H.G.; Nguyen, T.L.A.; Xuan, V.N. Assessing sustainability indicators of carbon emissions in the Netherlands: The roles of innovation, renewable energy, and economic openness. *Environ. Sustain. Indic.* **2026**, *30*, 101167. [CrossRef]
8. Ouyang, Y.; Li, P. On the nexus of financial development, economic growth, and energy consumption in China: New perspective from a GMM panel VAR approach. *Energy Econ.* **2018**, *71*, 238–252. [CrossRef]
9. Pesaran, M.H.; Shin, Y.; Smith, R.J. Bounds testing approaches to the analysis of level relationships. *J. Appl. Econom.* **2001**, *16*, 289–326. [CrossRef]
10. Apergis, N.; Payne, J.E. Coal consumption and economic growth: Evidence from a panel of OECD countries. *Energy Policy* **2010**, *38*, 1353–1359. [CrossRef]
11. Ozturk, I.; Acaravci, A. CO_2 emissions, energy consumption and economic growth in Turkey. *Renew. Sustain. Energy Rev.* **2010**, *14*, 3220–3225. [CrossRef]
12. Dlamini, Z.; Nzimande, N.P.; Biyase, M.; Khobai, H.; Manguzvane, M.M.; Gumede, S. Carbon emissions and growth: The role of trade and urbanization in Sub-Saharan Africa. *Int. J. Energy Econ. Policy* **2024**, *14*, 605–614. [CrossRef]
13. Soytas, U.; Sari, R. Energy consumption and GDP: Causality relationship in G-7 countries and emerging markets. *Energy Econ.* **2003**, *25*, 33–37. [CrossRef]
14. Omri, A.; Daly, S.; Rault, C.; Chaibi, A. Financial development, environmental quality, trade and economic growth: What causes what in MENA countries. *Energy Econ.* **2015**, *48*, 242–252. [CrossRef]
15. Lee, C.-C.; Chang, C.-P. The impact of energy consumption on economic growth: Evidence from linear and nonlinear models in Taiwan. *Energy* **2007**, *32*, 2282–2294. [CrossRef]
16. Romer, P.M. Endogenous technological change. *J. Political Econ.* **1990**, *98*, S71–S102.
17. Grossman, G.M.; Krueger, A.B. Economic growth and the environment. *Q. J. Econ.* **1995**, *110*, 353–377. [CrossRef]
18. Nguyen, B.N.; Nguyen Thi Phuong, T.; Nguyen, T.L.A.; Luong, H.G.; Ngoc Xuan, V. Energy transition, innovation, and foreign investment effects on CO_2 emissions in Belgium: Evidence from an ARDL approach. *Energy Explor. Exploit.* **2026**. [CrossRef]
19. Prempeh, K.B. The role of economic growth, financial development, globalization, renewable energy and industrialization in reducing environmental degradation in the economic community of West African States. *Cogent Econ. Financ.* **2024**, *12*, 2308675.
20. Bui, H.Q.; Tran, T.; Vo, A.T.; Vo, D.H. Financial development, energy consumption, and economic growth in the ASEAN countries: Evidence from the PVAR approach. *Environ. Sci. Pollut. Res.* **2023**, *30*, 114249–114263. [CrossRef] [PubMed]
21. Elmonshid, L.B.E.; Sayed, O.A.; Awad Yousif, G.M.; Eldaw, K.E.H.I.; Hussein, M.A. The impact of financial efficiency and renewable energy consumption on CO_2 emission reduction in GCC economies: A panel data quantile regression approach. *Sustainability* **2024**, *16*, 6242. [CrossRef]
22. Iheonu, C.O.; Anyanwu, O.C.; Odo, O.K.; Nathaniel, S.P. Does economic growth, international trade, and urbanization uphold environmental sustainability in sub-Saharan Africa? Insights from quantile and causality procedures. *Environ. Sci. Pollut. Res.* **2021**, *28*, 28222–28233. [CrossRef]
23. Sasana, H.; Khusniati, N.a.; Purnomo, R.A.; Shafiai, M.H.M. Analysis of Green Investment, Access to Clean Fuels, Green Jobs, Green Energy, and CO_2 Emissions on Economic Growth and Welfare: An Empirical Study in Developing Countries. *Int. J. Energy Econ. Policy* **2026**, *16*, 251. [CrossRef]
24. Ulucak, R.; Danish; Zhang, Y.; Chen, R.; Qiu, Y. Income inequality, economic complexity, and renewable energy impacts in controlling consumption-based carbon emissions. *Eval. Rev.* **2024**, *48*, 119–142. [CrossRef]
25. Inah, O.I.; Akuru, U.B.; Sotenga, P.Z. Modelling South Africa's Carbon-Peak Trajectories through a Decoupling–Markov Chain–Monte Carlo (D-MCMC) Energy–Economic Transition Framework. *Results Eng.* **2026**, *29*, 108979. [CrossRef]
26. Singh, G.J.; Singh, P.K.; Kumar, R.; Aman, A.; Alam, I.; Fatma, R.; Ansari, S. Dynamic Causality between CO_2 Emissions, Urbanization and Economic Growth in India and South Africa. *Int. J. Energy Econ. Policy* **2026**, *16*, 276. [CrossRef]

https://doi.org/10.3390/en19102329

27. Sadorsky, P. Financial development and energy consumption in Central and Eastern European frontier economies. *Energy Policy* **2011**, *39*, 999–1006. [CrossRef]
28. Shahbaz, M.; Lean, H.H. Does financial development increase energy consumption? The role of industrialization and urbanization in Tunisia. *Energy Policy* **2012**, *40*, 473–479. [CrossRef]
29. World Bank. *World Development Indicators*; World Bank: Washington, DC, USA, 2026. Available online: https://databank.worldbank.org/source/world-development-indicators?utm_source=chatgpt.com (accessed on 8 May 2026).
30. Belaïd, F.; Taylor, M. From energy deprivation to decarbonization constraints: A global typology of residential energy use. *Energy Policy* **2026**, *212*, 115140. [CrossRef]
31. Nathaniel, S.P.; Iheonu, C.O. Carbon dioxide abatement in Africa: The role of renewable and non-renewable energy consumption. *Sci. Total Environ.* **2019**, *679*, 337–345. [CrossRef]
32. Adebayo, M.T.; Ilesanmi, K.D. The human capital–Green growth nexus in South Africa. *J. Econ. Financ. Sci.* **2026**, *19*, 1061. [CrossRef]
33. Jelеč Raguž, M. Economic Growth and CO_2 Emissions in Croatia: An ARDL-Based Assessment of the EKC Hypothesis. *Sustainability* **2026**, *18*, 1427. [CrossRef]
34. Bakri, M.A.; Chia, Y.-E.; Chia, R.C.-J. Trade openness and carbon emissions using threshold approach: Evidence from selected Asian countries. *Carbon Res.* **2025**, *4*, 47. [CrossRef]
35. Văidean, V.-L.; Turceniuc, A.-R. Quality, Energy Consumption, and Economic Growth: A Quantitative. In *Innovation, Sustainability, and Growth in a VUCA Environment: Proceedings of the 5th International Conference on Modern Trends in Business, Hospitality and Tourism (ICMTBHT), Cluj-Napoca, Romania, 22–24 May 2025*; Springer Nature: Berlin/Heidelberg, Germany, 2026; p. 83.
36. Apergis, N.; Payne, J.E. Energy consumption and growth in South America: Evidence from a panel error correction model. *Energy Econ.* **2010**, *32*, 1421–1426. [CrossRef]
37. Khobai, H.B.; Le Roux, P. The relationship between energy consumption, economic growth and carbon dioxide emission: The case of South Africa. *Int. J. Energy Econ. Policy* **2017**, *7*, 102–109.
38. Brown, R.L.; Durbin, J.; Evans, J.M. Techniques for testing the constancy of regression relationships over time. *J. R. Stat. Soc. Ser. B Stat. Methodol.* **1975**, *37*, 149–163.

https://doi.org/10.3390/en19102329

MDPI AG
Grosspeteranlage 5
4052 Basel
Switzerland
Tel.: +41 61 683 77 34

MDPI Books Editorial Office
E-mail: books@mdpi.com
www.mdpi.com/books

www.ingramcontent.com/pod-product-compliance
Lightning Source LLC
LaVergne TN
LVHW070458061126
851858LV00011B/258